I0038064

Yonghao Du, Lining Xing, Lei Li
Satellite Scheduling Engine

Also of interest

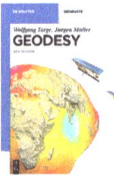

Geodesy
Torge, Müller, Pail, 2023
ISBN 978-3-11-072329-8, e-ISBN (PDF) 978-3-11-072330-4

Modern Signal Processing
X.-D. Zhang, Tsinghua University Press, X. Wang, D. Chang, L. Zhang, 2022
ISBN 978-3-11-047555-5, e-ISBN (PDF) 978-3-11-047556-2

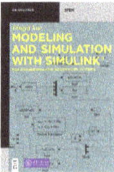

Modeling and Simulation with Simulink®.
For Engineering and Information Systems
Xue, Tsinghua University Press, 2022
ISBN 978-3-11-073904-6, e-ISBN (PDF) 978-3-11-073495-9

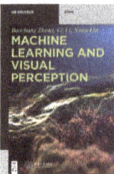

Machine Learning and Visual Perception
B. Zhang, C. Li, N. Lin, Tsinghua University Press, 2020
ISBN 978-3-11-059553-6, e-ISBN (PDF) 978-3-11-059556-7

Inertial Navigation Systems with Geodetic Applications
Jekeli, 2023
ISBN 978-3-11-078421-3, e-ISBN (PDF) 978-3-11-078432-9

Yonghao Du, Lining Xing, Lei Li

Satellite Scheduling Engine

General-Purpose Modeling and Intelligent
Optimization Methods

DE GRUYTER

清華大學出版社
TSINGHUA UNIVERSITY PRESS

Authors
Dr. Yonghao Du
College of Systems Engineering
National University of Defense Technology
P.R. China

Prof. Lining Xing
School of Electronic Engineering
Xidian University
P.R. China

Lei Li
College of Systems Engineering
National University of Defense Technology
P.R. China

ISBN 978-3-11-153599-9
e-ISBN (PDF) 978-3-11-153719-1
e-ISBN (EPUB) 978-3-11-153732-0

Library of Congress Control Number: 2024951218

Bibliographic information published by the Deutsche Nationalbibliothek
The Deutsche Nationalbibliothek lists this publication in the Deutsche Nationalbibliografie;
detailed bibliographic data are available on the Internet at http://dnb.dnb.de.

© 2025 Tsinghua University Press Limited and Walter de Gruyter GmbH, Berlin/Boston,
Genthiner Straße 13, 10785 Berlin
Cover image: (c) ismagilov/iStock/Getty Images Plus
Typesetting: Integra Software Services Pvt. Ltd.

www.degruyter.com
Questions about General Product Safety Regulation:
productsafety@degruyterbrill.com

About the book

Facing the development trend of large-scale, networked, and intelligent satellite systems, to address problems occurred during development and application of "chimney-type" and "one satellite, one system" satellite task scheduling system, this book presents a general-purpose modeling method for satellite task scheduling, as well as an adaptive parallel memetic algorithm and a distributed dynamic rolling optimization algorithm for satellite conventional and emergency task scheduling, respectively. A satellite task scheduling engine is also designed and implemented in this book with examples demonstrating its effectiveness.

The satellite task scheduling engine designed in this book can address the general-purpose modeling and optimization problems of four typical satellite tasks, contributing to the development of general-purpose task scheduling system and basic software for domestic satellites.

The authors hope that this book can help researchers, engineers, and graduate students of related majors engaged in the fields of management science and engineering, satellite operation control, and satellite tracking, telemetry, and command.

https://doi.org/10.1515/9783111537191-202

Preface

Since the twenty-first century, national aerospace industry has developed rapidly, leading to an explosive growth in satellites. To coordinate the in-orbit satellites and their management resources for better social, economic, and military applications, the satellite task scheduling is required. As the satellite management and application manners evolve over the years, cross-agency and cross-model have become the new normal, and integration and quick-response have become the new requirements. Under such circumstances, the "one satellite, one system" shortcoming in system development and application occurs, while those systems that differ in agencies, modes, and types cannot accommodate each other. This underlines the need to strengthen the generality of satellite scheduling models and algorithm. To address the "one satellite, one system" shortcoming, realize cooperative, flexible, and efficient applications and break up the monopoly of the US STK/Scheduler, a satellite task scheduling engine that contains general-purpose modeling and optimization methods is studied in this book. The main contents include

(1) A top-level framework of the satellite task scheduling engine is designed. Under the application background of the satellite task scheduling engine, four major satellite scheduling problems are defined, including the remote-sensing satellite scheduling, relay satellite scheduling, navigation satellite scheduling, and satellite range scheduling. The scope of this book and its reality-based and application-oriented principle are also determined. Then, the definition of the engine and its functional requirements are given. On this basis, a top-level engine framework that decouples the model, conventional algorithm, and emergency algorithm is designed. With this framework, a new, general-purpose, and modularized modeling and optimization manner is presented for different satellite task scheduling problems, providing a guidance for the follow-up studies of models and algorithms in this book.

(2) A general-purpose modeling method for satellite task scheduling is proposed. To address the model generality and "one satellite, one system" issues, the satellite task scheduling problems are systematically and hierarchically generalized to a task set, a resource set, a score set, and a decision matrix. Then, the decision-relation between the task set and the resources set, definitely between the satellite event and its executable opportunity, are creatively explained. The general-purpose 0–1 mixed integer decision model is then built on this relation and examples of four kinds of satellite scheduling problems are given to play the key coupling point in the loosely coupled engine framework. Based on this model and several examples, the constraints in satellite task scheduling problems are summarized and modeled via a general-purposed template, which is composed of an object, a threshold, and a relationship. The constraints can construct a network, and the constraint value is calculated incrementally and efficiently. Finally, the objective functions are formulated to complete the whole model. In summary, this general-purpose modeling method decouples the decision, constraints, and profits, which can properly model the aforementioned four-

https://doi.org/10.1515/9783111537191-203

satellite task scheduling problems in a general-purpose manner. This manner comes up with a new idea of general-purpose and refined modeling of satellite task scheduling problems, supporting the required general-purpose models for the satellite task scheduling engine in this book.

(3) **An adaptive parallel memetic algorithm (APMA) for the conventional satellite task scheduling is proposed.** To address daily and weekly conventional satellite task scheduling requirements, APMA is proposed considering several algorithm factors such as the initial solution, local optimization, global optimization, adaptivity, generality, and complexity. APMA integrates four complementary strategies, namely the heuristic-based initialization strategy, the parallel-search-based local optimization strategy, the competition-based algorithm and operator selection strategy, and the evolution-based global optimization strategy. Some benchmark experiments, including the orienteering problem and the simplified remote-sensing satellite scheduling, show APMA's generality and outperformance. As a result, APMA provides a general-purpose and efficient way to address the conventional satellite task scheduling problems, supporting the required algorithm for the satellite task scheduling engine in this book.

(4) **A distributed dynamic rolling optimization algorithm (DDRO) for the emergency satellite task scheduling is designed.** To address the urgent needs for emergency satellite task scheduling and inadequate time required by conventional scheduling algorithms in real-world situations, DDRO is designed considering the emergency scheduling requirements. DDRO also integrates four algorithm strategies, namely the dynamic contract-net-based task negotiation and assignment strategy, the rolling-based single platform re-scheduling strategy, the schedulability-based task quick-insertion strategy, and the constraint-net-based real-time deconflicting strategy. These strategies allow DDRO to perform real-time and dynamic optimization on conventional satellite task scheduling results, which offers a general-purpose and flexible way for emergency satellite task scheduling problems, supporting another required algorithm for the satellite task scheduling engine.

(5) **The satellite task scheduling engine is applied to solve real-world problems.** Based on the aforementioned contents, the satellite task scheduling engine is applied to solve the SuperView-1 remote-sensing satellite scheduling, TianLian-1 relay satellite scheduling, Beidou-3 navigation satellite scheduling, and the satellite range scheduling from the US Air Force Institute of Technology (AFIT). In these experiments, the proposed general-purpose modeling method, APMA, and DDRO are well examined. In a word, the satellite task scheduling engine is proven feasible and promising in the book, offering great potential in more real-world applications.

The authors completed this book when they were studying and working at the School of Systems Engineering, National University of Defense Technology. This book could not have been completed without the professional guidance of the authors' supervisor Prof. Xing Lining, and the great support of the scholars including Chen Yingwu, He Renjie, Yao Feng, Liu Xiaolu, Chen Yingguo, Zhang Zhongshan, Chen Yu'ning, Lv Jimin, Wang Tao, Shen Dayong, and He Lei. The authors consulted many references throughout the writing of this book and owed a great deal to contributions and inspirations of the scholars. The authors would like to extend their heartfelt gratitude to their supervisor and the scholars who have given their guidance, assistance, and inspiration. If you have any questions, suggestions, or comments, please feel free to contact the authors.

The Authors
Changsha, China
June 2022

Contents

Chapter 1
Introduction

1.1 Background and significance of the study

1.1.1 Background of the study

Since the beginning of the twenty-first century, the aerospace industry has entered a period of rapid development, and satellite technology has undergone profound changes. With the construction of major national special projects such as the high-resolution Earth observation system and the third-generation Beidou Navigation Satellite System (BDS), alongside the vigorous development of the private and commercial aerospace industry, the satellite system has experienced explosive growth, providing timely and accurate data and information for society, economy, national defense, and other important sectors. Public information shows that by the end of 2024, China had 628 satellites in orbit [1], representing over 150% growth in the past years. However, restricted by the geographical location of satellite range scheduling stations, the common problems of the satellites, such as short visible time and shortage of satellite range scheduling resources, are becoming increasingly prominent, bringing in increasing difficulties in satellite control. Furthermore, the satellite emergency response to sudden natural disasters, military struggles, and other situations has become the new normal, putting forward new and higher requirements for satellite control. Under such circumstances, satellite task scheduling technology based on modern operations study and computational science plays a crucial role to cope with the rapid growth in the scale of satellites, the limited control resources, and the new requirements for emergency response.

Satellite task scheduling technology refers to a kind of optimization technology that is driven by satellite tasks and control needs, through the modeling of tasks and resources, to avoid constraints and conflicts in the process of satellite task execution, and to maximize the benefits of satellite tasks and control, as shown in Figure 1.1. In recent years, under the impetus of many scholars and space practitioners, satellite task scheduling technology has achieved rich study results and has been widely practiced and developed in the satellite task scheduling system represented by Beidou, Gaofen, SuperView, and other domestic series of satellites. In this process, the study and application of satellite task scheduling technology have also exposed some problems:

(1) The satellite task scheduling problems are oversimplified, and related models and algorithms are difficult to meet the needs of complex and diversified applications. There is a common phenomenon of significant simplification in decision-making and constraints in studies related to satellite task scheduling, which affects the feasibility of modeling methods and algorithm results to certain degrees,

https://doi.org/10.1515/9783111537191-001

making it difficult to transform and apply some study results in engineering. In particular, at present, the complexity and diversity of satellite control have increased dramatically. For example, SuperView-1, Gaofen Multimode, and other highly maneuverable and agile remote-sensing satellites with multiple working modes have more than 100 constraints in the task scheduling process, covering multiple practical activities such as imaging, downlink, and on-board memory management. In this context, the oversimplified problem modeling methods and solving algorithms in relevant studies are not sufficient to meet the practical application needs of satellite task scheduling models and algorithms.

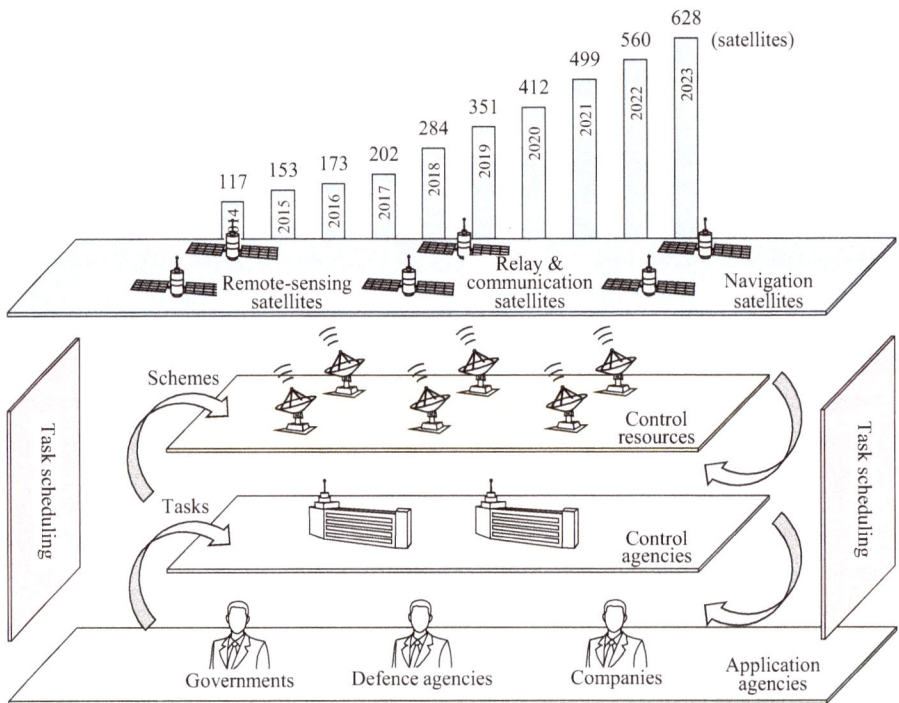

Figure 1.1: Schematic diagram of satellite control and task scheduling technology.

(2) The models and algorithms for satellite task scheduling are not sufficiently generalized, resulting in the current problem of "one satellite, one system" in the development and use of satellite control systems. Relevant studies usually focus only on the task scheduling problems of one or a few similar satellites, and the corresponding task scheduling models, algorithms, and system architectures are often difficult to modify and transplant. There are problems such as narrow applicability, poor generalizability, and insufficient expandability and integration capabilities, resulting in the status quo of "one satellite, one system" for satellite control. Consequently, it is difficult for satellite task scheduling systems managed by various agencies for different models and types of satellites to be compatible with each other. The development cost of new satellite task scheduling systems is high and the cycle is long, while the compatibility and integration of the new systems with the old systems are not satisfactory, forming a vicious cycle. It is not conducive to the "incremental" accumulation of satellite task scheduling studies and application results, and hinders the development of integrated satellite control.

(3) **The dual demands of daily task control and emergency response of satellites have led to conflicts between "conventional tasks and emergency tasks," and many algorithms have been ineffective in solving these conflicts.** Currently, with the increasing integration and complexity of the conventional control of satellites, the scale and complexity of satellite task scheduling problems have also proliferated accordingly, resulting in a consequent increase in the time for the algorithm to run (especially for calculating high-quality task scheduling solutions). But, on the other hand, the frequent demand for satellite emergency task scheduling puts forward a higher requirement for the timeliness of the algorithms. It can be seen that there is contradiction between the demand for increasing algorithm running time for satellite conventional task scheduling and the demand for algorithm timeliness for emergency task scheduling, which is called the "conventional-emergency" contradiction. In addition, due to the influence of oversimplified modeling, the satellite task scheduling algorithms in related studies are not applicable enough in practice. As a result, it is difficult for the relevant algorithms to simultaneously meet the dual demands of satellite conventional and emergency task scheduling, and their solving and application results are not ideal.

(4) **Some countries have implemented software and technology blockades, and there is no alternative and competitive domestic software or core technology for satellite task scheduling.** For a long time, as important scientific study tools for the satellite control agencies and related colleges and universities, the U.S. space task simulation and design software Satellite/Systems Tool Kit (STK) and its task scheduling plug-in STK/Scheduler provide general-purpose and efficient software services for satellite orbit projection and task scheduling. However, since 2005 (STK 7.0), the United States has banned the sale and embargo of STK to China, and the embargo has not been lifted so far. Although at the present stage, domestic task scheduling systems of various models and types of satellites can meet the basic needs of conventional control, most of the systems lack the technological kernel for general-purpose software products and user-oriented design due to the impact of insufficient generality of system models and algorithms, and domestic satellite task scheduling software or technology that can replace STK/Scheduler has not yet appeared.

1.1.2 Purpose and significance of the study

Aiming to solve the realistic problems exposed in the process of satellite task scheduling studies and applications, this book conducts a study on satellite task general-purpose scheduling engine, focusing on the general-purpose modeling methods of satellite task scheduling, as well as algorithms for conventional and emergency task scheduling. The purpose and significance of the study include:

(1) **To open up a new way for the study of modeling complex satellite task scheduling problems and promote the engineering application of the results.** In view of the oversimplified decision-making and constraints that exist in satellite task scheduling study, this book tries to systematically and completely describe the satellite task scheduling problems for constructing refined task scheduling models to truly reflect the actual optimization needs of the satellite control agencies. In combination with engineering applications, this book restores the complexity and diversity of satellite task scheduling problems to open up a new way for the study of modeling complex satellite task scheduling problems and promote the transformation and engineering applications of related study results based on task scheduling models.

(2) **To significantly improve the generality of satellite task scheduling models and algorithms to overcome the shortcoming of the current "one satellite, one system" for satellite control.** In response to the problem of insufficient generality of satellite task scheduling models and algorithms, this book decouples the models and algorithms, proposes a general-purpose satellite task scheduling framework, models, and algorithms, and flexibly applies them to different satellite task scheduling problems. Consequently, this book fundamentally breaks through the satellite control status quo of "one satellite, one system," adapts to the new normal and new requirements of integrated and flexible control of satellites, and promotes the "incremental" accumulation and flexible expansion of satellite task scheduling studies and application results. Meanwhile, necessary models and algorithms are provided in this book for the development of integrated control of satellites.

(3) **To enrich solving methods for satellite task scheduling problems to mitigate the "conventional-emergency" contradiction.** To meet the dual demands of satellite conventional and emergency task scheduling, this book builds a reasonable and effective framework for modeling and solving satellite task scheduling problems. Under this framework, conventional and emergency task scheduling algorithms are proposed and fully tested in a series of satellite task scheduling experiments to provide solutions to the contradiction between "conventional and emergency tasks" and offer necessary algorithms to play the role of the satellite system in the social, economic, and military fields.

(4) **To break through the software and technology blockade of some countries and provide necessary technology for domestic satellite task scheduling software.** Finally, in response to the software and technology blockade of some countries, this book provides a set of general-purpose satellite task scheduling modeling

and solving methods with independent intellectual property rights suitable for the national conditions, so as to get rid of the dilemma that the core technology is controlled by others and provide a technological drive for the development of domestic satellite task scheduling software to accelerate the upgrading of the country's comprehensive strength in aerospace and realization of the overall goal of becoming an aerospace power.

1.2 Current status of studies on satellite task scheduling models

In the process of solving the satellite task scheduling problem, problem modeling is often the first step, which is the key to finalize the way in which the problems are described, to guide the design of the algorithms, and to affect the complexity of the problems. At the present stage, different types of satellite tasks are often scheduled by different functional agencies. According to the different functions of the agencies, satellite tasks can be divided into two categories: operation control tasks, and tracking, telemetry, and command (TT&C) tasks, as shown in Figure 1.2. Among them, satellite operation control tasks are aimed at accomplishing satellite missions and task data downlink, with remote-sensing satellites, relay satellites, and navigation satellites having the most numerous tasks and the most frequent scheduling. TT&C tasks, on the other hand, are aimed at guaranteeing the normal operation and long-term control of satellites, and include a series of tasks such as satellite telemetry, tracking, and command.

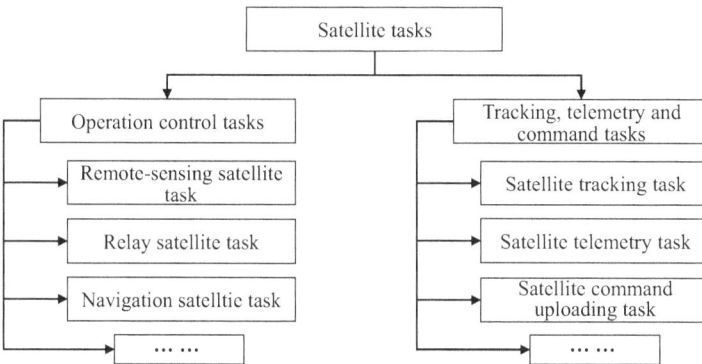

Figure 1.2: Classification of satellite tasks.

Then, this section provides a detailed overview of commonly used models for scheduling problems of satellite operation control tasks and TT&C tasks, analyzes the modeling features, decision variable forms, and model advantages and disadvantages of different satellite task scheduling problems, and clarifies the necessity of improving model generality and solving efficiency in the modeling process, so as to provide a reference for the study on general-purpose modeling of satellite task scheduling in this book.

1.2.1 Remote-sensing satellite task scheduling models

Remote-sensing satellite tasks are a type of satellite operation control task that collects information and data and transmits them to ground-based equipment through on-board remote-sensing payloads for targets in the environments of the Earth's surface, ocean, atmosphere, and space. As the largest class of satellites in orbit, remote-sensing satellites are playing an irreplaceable role in agriculture, economy, military, and other fields. Therefore, at the present stage, remote-sensing satellite tasks are also the most diverse, in-demand, and widely used type of satellite operation control tasks.

Remote-sensing satellite task scheduling is a class of combinational optimization problems with nondeterministic polynomial-time hard (NP-hard) properties [2]. The problems generally have the dual characteristics of sequence optimization and resource optimization, which involve both decision-making on the sequence of task execution and rational allocation of satellites and their payload resources for the tasks. Therefore, according to the different decision objects in the scheduling models, this section divides the remote-sensing satellite task scheduling models into two categories: resource-oriented task sequencing model (referred to as task sequencing model) and task-oriented resource allocation model (referred to as resource allocation model), and then further elaborates on the remote-sensing satellite task scheduling models.

1.2.1.1 Task sequencing models

At the end of the twentieth century, facing the brand-new problem of remote-sensing satellite task scheduling, people tried to find solutions from classical operations study models, such as vehicle routing problem (VRP) model, graph theory model, and job-shop scheduling problem (JSP) model. In these models, the decision variables mainly represent the task performing sequence of resources, reflecting the sequential logic and capability constraints in the process of consecutive execution of tasks, which plays an important role in solving the early remote-sensing satellite task scheduling problems.

(1) VRP model

The VRP model is one of the first models used to solve the remote-sensing satellite task scheduling problems. In the model, the satellites are considered as vehicles, and the task targets are considered as the cities that would be visited by the vehicles, as shown in Figure 1.3(a). Cordeau and Laporte [3] transformed the visibility of the remote-sensing satellites to the targets into the time window constraints of the tasks, and used a VRP solving algorithm [4] to solve task scheduling problems of a single-orbit remote-sensing satellite. Subsequently, Bianchessi et al. [5] applied the algorithm to multi-orbit satellites and proposed a three-stage remote-sensing satellite task scheduling architecture of "demand analysis, task distribution, and single-satellite scheduling," which is still adopted by many satellite scheduling systems. Li and Tan [6], Guo et al. [7], and Cai [8] pointed out that imaging and data downlink actions in remote-sensing satellite tasks can be regarded as loading and unloading actions in the VRP model to solve the task scheduling problems under the constraints of memory capacity. Although the above studies have made many simplifications in terms of satellite tasks, working modes, and memory constraints, the modeling idea of task sequencing in the VRP model opens up a new way for the study of satellite task scheduling.

(2) Graph theory model
The graph theory model shown in Figure 1.3(b) intuitively describes the sequence and conflict relationships between satellite tasks by means of points and lines and has also been widely applied in remote-sensing satellite task scheduling. For example, Gabrel and Vanderpooten [9], Bianchessi and Righini [10], and Chen et al. [11], respectively, abstracted satellite imaging scheduling problems into a directed acyclic graph model for optical satellites, SAR satellites, and electronic satellites. Chen and Wu [12] described the sequence and execution resources of satellite downlink, respectively, by constructing task scheduling location maps and task scheduling relationship maps. For multi-satellite joint scheduling problems, Xu et al. [13] established a graph theory model based on a minimum spanning tree, and Zhang et al. [14] and Wang [15] constructed a graph theory model with layered optimization. However, the graph theory models have a very obvious drawback: due to the limited constraint expression ability of the model, task scheduling problems are often oversimplified, and complex task constraints such as area target imaging, integrated scheduling of imaging and downlink, and memory erasure are difficult to represent in a traditional graph theory model.

(3) JSP model
The constraints on workpiece processing sequence in models such as JSP model and FSP (flow-shop scheduling problem) model are applicable to the practical requirements such as satellite area target imaging and 3D imaging. Therefore, the two models are commonly used in modeling remote-sensing satellite task scheduling problems. As early as 1994, Hall and Magazine [16] provided a JSP model for remote-sensing satellite task scheduling. Cordeau and Laporte [3], Li and Tan [6], Gu and Chen [17], and He and Gu [18] pointed out that the task execution process of remote-sensing satellites can be viewed as the process of assigning tasks (imaging tasks and downlink tasks) to different types of machines (satellites and ground stations), as shown in Figure 1.3(c).

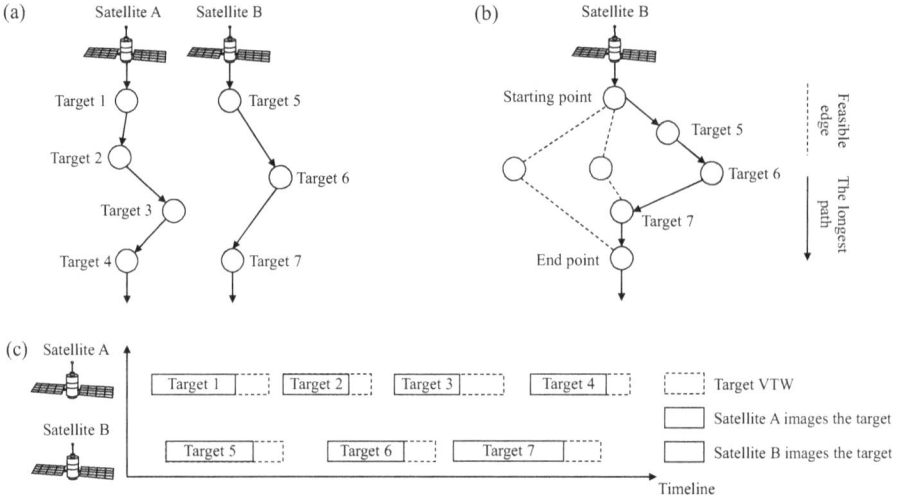

Figure 1.3: Example of remote-sensing satellite task sequencing model: (a) dual-satellite VRP model; (b) single-satellite graph theory scheduling model; and (c) dual-satellite JSP model based on the principle of earliest start.

In addition, Xiao et al. [19] designed a two-stage task scheduling framework to model the integrated scheduling problems of satellite imaging and downlink through the FSP model.

The above task sequencing models are convenient for checking the constraints of transition time between two consecutive tasks, which is close to the actual situation of satellite control. Additionally, almost all of the early remote-sensing satellites used a "sequential playback" working mode, that is, the sequence of downlink is consistent with the sequence of imaging. Therefore, many other remote-sensing satellite scheduling literatures also follow such a modeling idea. For example, Lemaître et al. [20] studied the French agile constellations Pleiades by representing the sequence of satellite tasks through decision variables, and pointed out that the constraints of satellite battery level and memory capacity can also be satisfied by adjusting the sequence of satellite tasks. Xu et al. [21] represented the positions of the tasks in the satellite execution sequence through decision variables. Lin et al. [22], Cheng et al. [23], Ren et al. [24], Sun et al. [25], and Cui et al. [26] all represented the routing or sequence between remote-sensing satellite task targets through decision variables.

However, there is an obvious decoding process in the task sequencing model, which usually follows the typical VRP model, graph theory model, and JSP model to execute tasks as early as possible, as shown in Figure 1.3(c). Nevertheless, at the current stage, the decoding process may lose high-quality solutions under the complex constraint logic of satellites (e.g., time-dependent constraints and nonlinear profits), that is, the "optimal sequence" may not be necessarily equal to the "optimal result."

Taking the constraint of a maximum of two consecutive maneuvers of a satellite as an example, as shown in Figure 1.4, in Scheme A, the task is executed based on the principle of the earliest start, while in Scheme B, target 2 is not executed at the earliest start time, that is, to say, it may happen that target 4 cannot be accomplished in Scheme A while it can be accomplished in Scheme B. As a result, the more complex the problem constraints are, the more likely it is that the task sequencing model loses high-quality solutions, and the higher the time cost of the decoding process is.

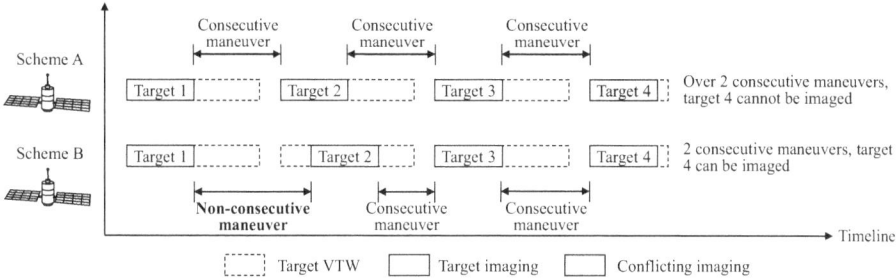

Figure 1.4: Example of shortcomings of remote-sensing satellite task sequencing model.

Moreover, with the development of satellite on-board memory technology, many remote-sensing satellites launched in recent years (e.g., "SuperView-1") no longer have the constraint of "sequential playback," that is, the satellite imaging sequence and downlink sequence can be inconsistent. When there are few downlink windows or special requirements for certain tasks, it is likely that optimizing satellite imaging sequences and downlink sequences separately may be more conducive to maximizing task profits. In this regard, scholars have proposed a task-oriented resource allocation model, in which not only the modeling idea of task sequences is retained but also the specific time of task execution is directly determined. With such advantages, the model has become an important model in the studies of task scheduling of various remote-sensing satellites.

1.2.1.2 Resource allocation models

In contrast to the task sequencing model, resource allocation models no longer determine the task order but determine the resources to be executed based on task requirements orientation, as shown in Figure 1.5(a). Due to the fact that the visibility between the satellites and the task targets is a prerequisite for task execution, their visible time windows (VTWs) have always been regarded as key resources in satellite scheduling. Therefore, the resource allocation model can also be called the VTW allocation model. For example, Bensana et al. [27], Gabrel [28], Jin et al. [29], Wang et al. [30], Liu et al. [31], Jiang et al. [32], and Wu et al. [33] expressed the VTW selected for remote-sensing satellites to execute their tasks with 0–1 decision variables for establishing task scheduling models for the remote-sensing satellites, including SPOT of France, SATellite-2 of South Korea, HJ series of China, and CBERS series of China-Pakistan. In the above literature, since the VTWs represented by 0–1 variables contain both satellite payload information and task execution time to reflect the execution sequences of satellite tasks, it is not necessary for the models to have the decoding process in VRP and JSP models, making the model representation in the literature more concise.

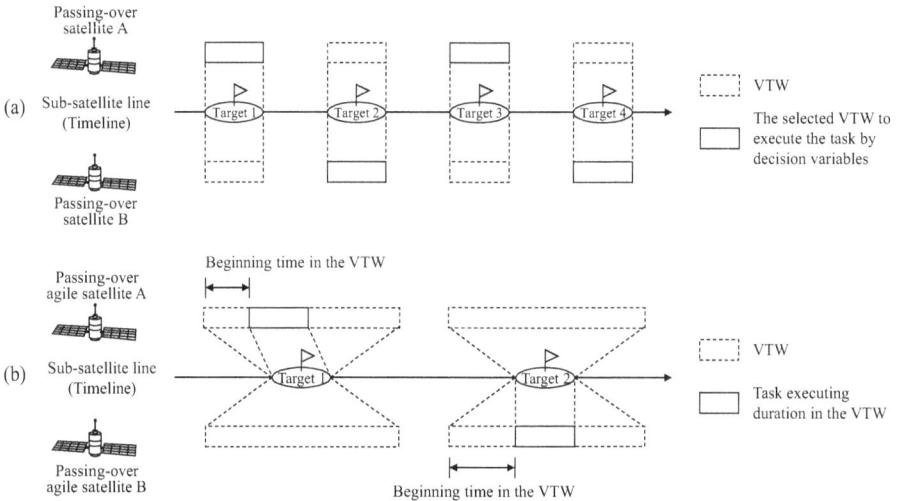

Figure 1.5: Examples of VTW allocation model for remote-sensing satellites: (a) VTW allocation model for non-agile satellite and (b) VTW allocation model for agile satellite.

However, with the increasing development of agile remote-sensing satellite technology, the traditional VTW allocation model has revealed certain limitations. Remote-sensing satellites launched in recent years are usually equipped with the capability of imaging with a pitch angle, that is, the satellites are capable of performing long-range tilt-angle imaging before or after passing over the targets, bringing longer VTWs and more imaging opportunities for the satellites [34], as shown in Figure 1.5(b). Therefore, in the agile remote-sensing satellite task scheduling models, it is necessary to decide not only each VTW for task execution but also the specific task beginning time for each VTW. To address this problem, scholars have now improved the model mainly through the following three methods.

(1) Rule-based VTW allocation model
Similar to the decoding idea in the task sequencing model in Section 1.2.1.1, Lemaître et al. [20], He et al. [35, 36], Mok et al. [37], Chu et al. [38], and Song et al. [39] scheduled each satellite task at the earliest position within the VTW that satisfies the constraints based on the earliest start principle. Zhang and Li [40] and Liu et al. [41] scheduled

the tasks based on the principle of prioritizing the imaging quality and clarified the constraints between time-dependent attitude transition and step-type imaging quality in the model. Xu et al. [21] designed specialized satellite window update strategies based on possible scenarios that may arise during the task scheduling process. Kim and Chang [42] only considered the side-view imaging mode to solve the imaging scheduling problems of SAR constellations, that is, the imaging tasks were scheduled at the midpoint of each VTW. Wu et al. [43] scheduled the imaging tasks with principles of merged tasks first and earliest start and concluded that the constraints associated with the task merging led to the nonlinear features of their model. However, although these decoding processes can help the VTW allocation model to be applicable to agile remote-sensing satellite scheduling problems, the rule-based decoding approaches to some extent limit the algorithms to explore larger solution spaces, and the time cost may be higher in the case of complex constraints.

(2) VTW allocation model with multiple decisions
Lemaître et al. [20], She et al. [44], Chen et al. [45], and Frank et al. [46] not only expressed VTWs of tasks with 0–1 variables but also used an integer variable to express the beginning time of the VTW for each task. Sarkheyli et al. [47] represented the VTWs and beginning time of agile remote-sensing satellite tasks through dual decision variables and considered the operational constraint of erasing the entire on-board memory, which has been rarely addressed in other studies. Niu et al. [48] and Chen et al. [49] also used the same decision variables and computed the constraint of nonlinear attitude transition time. In this model, a 0–1 variable determines the satellite payload and VTW for executing a task, while an integer variable determines the beginning time of the task. Therefore, this model has better model completeness since there is no rule-based decoding process involved in the above-mentioned model.

(3) Discretized VTW allocation model
Wang et al. [30], Valicka et al. [50], Nag et al. [51], Zhu et al. [52], He et al. [53], Li et al. [54], and Du et al. [55] discretized VTWs for imaging or downlink tasks of remote-sensing satellites while directly determining the discrete VTWs (i.e., start and end moments) of the tasks through 0–1 variables, as shown in Figure 1.6. This method is essentially the same as the VTW allocation model with multiple decisions but reduces the number of decision variables, making the model more concise.

Compared with the rule-based VTW allocation model, the latter two models are free from the influence of heuristic rules in the expression of solutions, which can guarantee the diversity of solutions under complex constraints and are closer to the actual situation of agile remote-sensing satellite control. However, these two models greatly expand the solution space of the problems, which may produce more infeasible solutions, and the difficulty of solving increases subsequently. Therefore, these two models usually require the support of high-performance optimization strategies and algorithms. Meanwhile, the degree of VTW discretization is controllable (the minimum accuracy of remote-sensing satellite tasks at present is at the second level), so a reasonable VTW discretization precision also helps to improve the effectiveness of the models. Additionally, some scholars have abstracted the remote-sensing satellite task scheduling problems as the knapsack problem[3, 56–58], multi-agent model [59–63], machine learning model [64–67], and so on.

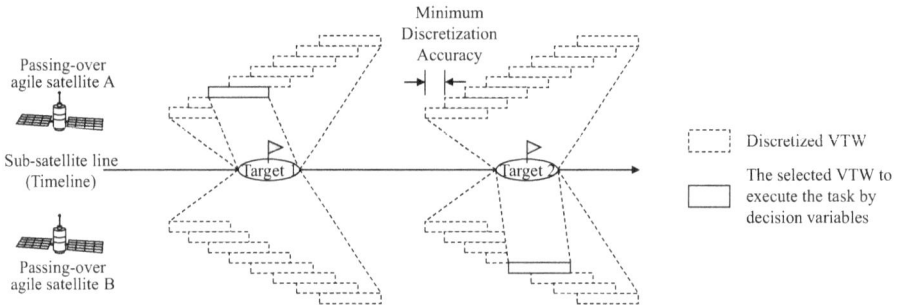

Figure 1.6: Example of discretized VTW allocation model for agile remote-sensing satellite tasks.

1.2.2 Relay satellite task scheduling models

As the number of remote-sensing satellites and the volume of the Earth observation data continue to rise, it is difficult for ground-based downlink stations to meet the large-scale and high-frequency downlink requirements of remote-sensing constellations at this stage. Therefore, relay satellites, which provide data relay services for satellites in orbit and ground-based stations, play a very important role. Because their tasks in orbit are similar to that of ground-based downlink stations, relay satellites are also called "space-based downlink stations." The task scheduling problems of relay satellites are very similar to the downlink scheduling problems of task scheduling of remote-sensing satellites in many aspects: 1) Both of them perform the downlink task to the targets on the basis of VTWs, as shown in Figure 1.7; 2) they have similar constraints in the working mode, power, memory, and so on; and 3) the single-access link and multiple-access link of the relay link support different link numbers and frequency bands, which have similar characteristics with different imaging payloads of remote-sensing satellites. Therefore, the task scheduling models of relay satellites can be categorized with reference to the task scheduling models of remote-sensing satellites.

(1) In terms of the task sequencing model
As early as 1986, Reddy and Brown [68] and Reddy [69] described the prioritization relationship between relay satellite tasks in terms of points and arcs, respectively. Rojanasoonthon et al. [70], Rojanasoonthon and Bard [71], Zhuang et al. [72], Zhuang [73], He et al. [74], and Liu et al. [75] pointed out that relay satellite task scheduling problems can be regarded as a class of VRP problems or parallel machine scheduling

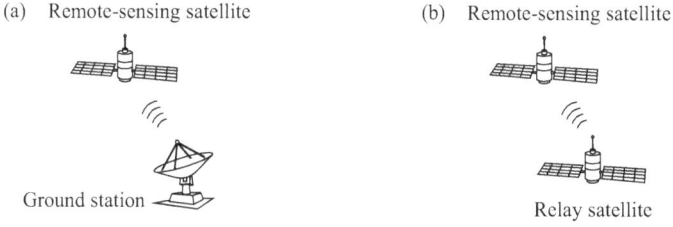

Figure 1.7: Example of downlink task of remote-sensing satellite compared to that of relay satellite: (a) downlink task of remote-sensing satellite and (b) downlink task of relay satellite.

problems and reflected the execution sequence of satellite tasks through decision variables. Guo et al. [76] defined the comprehensive priority of satellite tasks based on their urgency, difficulty, and VTW attributes and assigned the VTW and beginning time to each task based on the priority order. It can be seen that this type of model has the same modeling characteristics of the task sequencing model of remote-sensing satellites and has consistency in terms of the expression of decision variables and the principle of decoding.

(2) In terms of the VTW allocation model

In the process of task execution by relay satellites, there are often cases where the length of the VTW exceeds the duration of the corresponding task, that is, the "sliding window" mentioned in related studies in which heuristic rules are typically used to determine the beginning time of a task within its VTW. For example, Gu [77] and Zhao et al. [78–80] expressed the VTWs for executing relay tasks through 0–1 variables and calculated the beginning time of the tasks in sequence based on the priority order and the earliest start principle. Fang et al. [81–83] introduced expert system rules based on heuristic rules. He et al. [84] determined the beginning time with the mini-mum-loss opportunity of the corresponding tasks. He et al. [85] designed the strategies of the earliest, latest, and random start.

In other literature, the task scheduling problems of relay satellites are not con-sidered separately but are included in the scope of task scheduling of other types of satellites. For example, in the study of integrated task scheduling of imaging and downlink of remote-sensing satellites, Li et al. [54] considered the relay satellites as ground-based downlink stations by calculating the VTWs between the remote-sensing satellites and the relay satellites, which is also a common method in the re-mote-sensing satellite scheduling system currently. In addition, the relay satellites sometimes undertake the range scheduling tasks of some satellites, so a consider-able number of relay satellite tasks are included in the scope of satellite range scheduling tasks [86]. It can be seen that relay satellites are often regarded as a kind of common resource to guarantee the scheduling needs of other satellite tasks. Therefore, considering the missions and the control features of relay satellites, as well as the similarity between their scheduling model and other satellite scheduling problems, the relevant studies should focus on the integration and synergy with other types of satellite task scheduling problems.

1.2.3 Navigation satellite task scheduling models

At present, the major navigation systems in the world include the Global Positioning System (GPS) of the United States, the Galileo Navigation Satellite System of Europe, the global navigation satellite system (GLONASS) of Russia, and the BDS of China. The control task of navigation satellites is mainly to establish communication links between satellites and the ground or other satellites so as to guarantee the positioning accuracy and timing accuracy of the navigation system, as well as the real-time downlink of navigation messages, ephemeris information and other data. Therefore, the scheduling problems of navigation satellite tasks can be regarded as scheduling problems of satellite-ground links and inter-satellite links. Below is a brief introduction to the task scheduling models of the above-mentioned navigation satellite systems in related studies:

(1) GPS and GLONASS
As the earliest navigation and positioning system built in the world, GPS completed its worldwide deployment of ground stations in 2005, allowing for the establishment of satellite-ground links at any time. Therefore, it is usually not necessary to consider VTW constraints for it [87]. GLONASS has also deployed a large number of ground stations in Russia and neighboring countries [88], and the visibility of the satellites is also high. As a result, there is not much related public literature due to the abundance of satellite-ground communication links in GPS and GLONASS systems and the low degree of need for task and resource scheduling.

(2) Galileo Navigation Satellite System
As early as 1994, Gershman et al. [89] and Iv [90] discussed the task scheduling problems of Galileo navigation systems. Toribio [91] presented the main equipment information and organizational structure in the task scheduling process of the Galileo navigation system, and provided the concepts of short-term planning, long-term planning, and emergency planning. Hall et al. [92] introduced the scheduling system of the Galileo navigation system, that is, the Spacecraft Constellation Planning Facility, and provided the main process of task scheduling. They pointed out that the system schedules about 1,500 tasks per week, with a weekly planned scheduling time of 10 min. However, none of the above studies have provided specific models and algorithms for navigation satellite task scheduling. Marinelli et al. [93] used 0–1 variables of discrete VTWs to represent the start and end moments of navigation tasks and scheduled the weekly plan of the satellite-ground link for 30 Galileo navigation satellites.

(3) BDS

To solve the satellite-ground link scheduling problems of BDS, Long et al. [94] expressed the ground-based stations and VTWs executing the tasks based on 0–1 variables and expressed the start and end moments of the tasks through two other integer variables. They handled practical constraints such as the number of links built by the ground stations and antenna transition time and optimized the objectives such as time to build links and task completion. Tang et al. [95] expressed the beginning time of navigation tasks through discrete VTWs and optimized objectives such as task completion and payload balancing. Considering the dual time constraints of task waiting time and equipment availability, Yan et al. [96] abstracted the navigation satellite task scheduling problems into JSP problems with dual VTW constraints. Zhang et al. [97] also considered inter-satellite links on top of the satellite-ground link scheduling and proposed a two-stage task scheduling scheme based on the task sequencing model. For China's BDS, Yang et al. [98], Sun et al. [99, 100], and Liu et al. [101] constructed an inter-satellite link scheduling model with the scheduling cycle of a superframe (1 min) and the time unit of a timeslot (3 s) and designed algorithms such as differential evolution, genetic, and simulated annealing (SA) with the optimization objective of reducing the average delay of the system, completing the weekly scheduling plan for the inter-satellite link of 30 navigation satellites.

Although there is limited public literature on the scheduling of navigation satellite tasks, the scheduling of satellite-ground links and inter-satellite links of navigation satellites has significant similarities with that of ground-based satellite TT&C tasks and space-based satellite TT&C tasks, respectively. Some studies have also included the scheduling of navigation satellite tasks within the scope of satellite TT&C task scheduling. Therefore, some models for satellite range scheduling are introduced below.

1.2.4 Satellite TT&C task scheduling models

Satellite TT&C tasks refer to a series of tasks such as satellite TT&C, which are important prerequisites for guaranteeing the normal operation and long-term control of satellites. Satellite TT&C tasks are dominated by ground-based TT&C tasks but also include some space-based (based on relay satellites) TT&C tasks. Figure 1.8 shows that, similar to the satellite operation control task scheduling, the VTWs of resources and targets are also a key resource for satellite range scheduling. The length of VTWs for low-orbit target TT&C tasks is usually equal to the task duration, while the length of VTWs for high-orbit target TT&C tasks is usually greater than the task duration. Therefore, the studies on TT&C task scheduling (also known as satellite range scheduling) largely retain the problem description methods and modeling ideas similar to those of operation control task scheduling, and priority sequencing model, graph theory model, and VTW allocation model are mainly used in the studies.

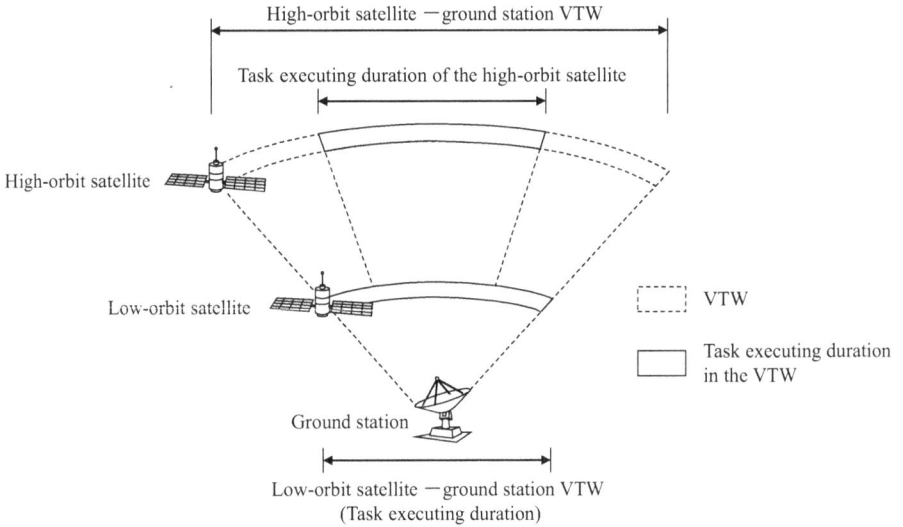

Figure 1.8: Example of satellite TT&C task scheduling.

(1) Priority sequencing models

The priority sequencing models belong to a type of resource-oriented task sequencing model, where the priority of a task refers to the priority of the resources allocated to the task. This kind of model is similar to task ordering models such as VRP, but instead of specifying task resources through decision variables, resources are directly allocated in sequence based on task sequencing results. For example, Parish [102], Barbulescu et al. [103–105], Zheng et al. [106], and Zhang et al. [107] expressed a sequence of tasks to be scheduled by 0–1 variables and assigned VTW and beginning time to each task based on the earliest start principle, converting the satellite range scheduling problem into an optimal sequencing problem for the task set. Barbulescu et al. [108] also pointed out that the satellite range scheduling problem can be viewed as a single-machine scheduling problem with preparation time, proving the NP-complete nature of the problem. Li et al. [109] treated each TT&C VTW as a task to be executed and expressed the sequence of tasks to be scheduled through 0–1 variables. The model is favored by many people because of its simple coding form, but it may not perform well in complex satellite range scheduling problems because it only determines the sequence of task resource allocation and relies solely on heuristic rules to allocate resources and time for the tasks one by one.

(2) Low-orbit satellite graph theory models

As another task sequencing model, graph theory models are also widely used in satellite range scheduling studies. As early as 1985, Arbabi et al. [110] provided a graph theory model for satellite TT&C tasks. Zufferey et al. [111] and Blöchliger and Zufferey [112] regarded satellite TT&C tasks as graphs, described TT&C resources as different colors, and transformed task scheduling problems into graph coloring problems, where the task beginning times are also determined based on the earliest start principle. Zhang et al. [113], Xu [114], Zhang and Feng [115], Zhang et al. [116–118], Chen and Wu [119], and Chen [120] expressed satellite VTWs as nodes and described the conflict relationship between the tasks in different VTWs through edges. They addressed the constraints such as uniqueness and non-overlap of task execution, transforming the satellite range scheduling problems into a maximum independent set problem. Wang et al. [121] and Vazquez and Erwin [122], based on the visibility relationship diagram between satellites and ground stations, expressed the beginning time of TT&C tasks as nodes and expressed the transition process of continuous execution of TT&C tasks at the same ground station as edges. Vázquez and Erwin [123, 124] then provided models for distributed scheduling, robust scheduling, and real-time scheduling. In addition, the Petri net model has also been used in studies on satellite range scheduling [125–128]. However, most of the above studies only considered low-orbit satellites. As the durations of low-orbit satellite TT&C tasks are equal to the length of the corresponding VTWs, the constraints of resource conflicts, antenna elevation angles, and transition time between the tasks are relatively certain and easy to describe. However, in high-orbit satellite TT&C tasks, the task VTWs are usually greater than the required TT&C time of the tasks. Hence, for the scheduling problems of TT&C tasks involving high-orbit satellites, a more suitable model is needed to accurately describe the problems.

(3) High-orbit satellite VTW allocation models
To address the joint range scheduling problems of high-orbit and low-orbit satellites, Gooley [129] and Gooley et al. [130] established a two-stage VTW allocation model, which expressed the TT&C task VTWs through 0–1 variables, prioritized the range scheduling of low-orbit satellites, and then expressed the beginning time of high-orbit TT&C tasks within the corresponding VTWs through integer variables. He and Tan [131], Liu et al. [132], and Luo et al. [133] also used the same decision variables to describe the joint TT&C task scheduling (also known as range scheduling) problems of high-orbit and low-orbit satellites and established an integrated range scheduling model. Xhafa et al. [134–137] and Valicka et al. [138] directly expressed the beginning time of satellite TT&C tasks through integer variables. Marinelli et al. [93] established a range scheduling model for the Galileo navigation constellations based on discrete VTWs and 0–1 variables. For large-scale range scheduling scenarios, Liu et al. [139] expressed the VTWs of TT&C tasks through 0–1 variables and determined the task beginning time within the corresponding VTWs based on conflict-avoidance rules. The above VTW allocation models not only determine the VTWs and resources of the TT&C tasks but also determine the specific task beginning time within the corresponding VTWs, which is suitable for the scheduling scenarios of TT&C tasks, including that of high-orbit satellites. Moreover, the priority sequencing model can also be used for the range scheduling problems of high-orbit satellites, but the too many heuristic

rules involved in the model are not conducive to the global optimization of the problems, so it is not discussed in this section.

In addition to the above models, 0–1 knapsack model [140], ontology and rule-based model [141–143], game theory model [144, 145], multi-agent model [146, 147], and machine learning model [148, 149], and so on have also been applied in the studies of satellite range scheduling, especially emergency task scheduling.

In the above studies, satellite range scheduling examples are usually categorized into two types: 1)simulation examples based on STK, where the number of tasks is usually less than 100 and the number of measurement stations is usually less than 10; and 2) the benchmark [150] released by the Air Force Satellite Control Network (AFSCN) for the scheduling of 7 high-orbit and low-orbit satellite joint TT&C tasks (the latest version was updated in 2003) with an average of 308 tasks and 19 stations daily. Considering the differences in task and resource scales, the difficulty of solving scheduling problems of the two types of examples is also very different. In fact, the task scales currently faced by domestic satellite control agencies have already exceeded that of the AFSCN benchmark. As a result, there are some limitations in many studies based on small-scale examples currently.

To summarize, this section lists the main models of studies on task scheduling of remote-sensing satellites, relay satellites, navigation satellites and satellite TT&C from the perspectives of task sequencing models, and VTW allocation models and analyzes the characteristics, forms of decision variables, and advantages and disadvantages of each type of model. Please refer to the summary in Table 1.1, in which, based on the author's experience in developing new satellite models, the practical operation constraints of domestic satellite task scheduling that are seldom considered by relevant literature are also summarized. These constraints greatly limit the transformation and application of the latest study results in satellite task scheduling. Therefore, how to incorporate these complex operation constraints into satellite task scheduling models in a standardized, modular, and concise manner becomes one of the key parts of this book.

From Table 1.1, it can be seen that the common models used in the study on task scheduling of remote-sensing satellites, relay satellites, navigation satellites, and satellite TT&C are quite similar, which is essentially due to the similarity of the satellite task scheduling problems mentioned above. Although satellites may carry different payloads and be assigned different tasks, the visibility of tasks and resources in time, space, and frequency domains is always the starting point and objective of problem modeling. Therefore, it is feasible to generalize the description and modeling of the above satellite scheduling problems. On the other hand, due to the variations in tasks and resources, there are many differences in the above-mentioned satellite task scheduling problems, especially the constraints, which are important factors for reflecting the characteristics of satellite task scheduling problems and distinguishing the types of satellite task scheduling.

Through the above summarization and analysis of scheduling models for remote-sensing satellites, relay satellites, navigation satellites, and satellite TT&C tasks, it can be seen that: 1)the similarity of satellite task scheduling problems is a prerequisite for creating general-purpose models for various types of problems; 2) to enhance the generality for various types of problems, the general-purpose scheduling model should closely focus on the visibility of the tasks and resources in the domains of time, space, and frequency; and 3) the general-purpose scheduling model also relies on the characteristic constraints of satellite task scheduling problems for better application in different problem scenarios.

Based on the analysis of the common characteristics and differences of various satellite task scheduling models mentioned above, the following summarizes three deficiencies of existing studies and gives the corresponding countermeasures:

(1) The above-mentioned models for satellite task scheduling problems are similar but not universal, and there is a lack of a unified modeling system and general-purpose modeling methodology. Through the analysis in this section, it is found that the various satellite task scheduling models have great similarities, and their principles of modeling based on the visibility of tasks and resources in time, space, and frequency domains are unified. However, these models also have certain differences due to the constraint characteristics of application scenarios, types of requirements, and solving rules. Therefore, designing a general-purpose satellite task scheduling modeling method to incorporate common satellite task scheduling problems into a set of unified modeling system is an important way to study the general-purpose satellite modeling and solving technology and meet the needs for joint and universal control of cross-agencies and cross-satellite models.

(2) Due to the oversimplification of many satellite task scheduling problems, some models have no practical application value in space operation control systems. Table 1.1 lists practical constraints that are rarely considered in many related studies. Although a certain degree of problem simplification can reduce the problem scale and improve the solution efficiency, some simplifications (e.g., without considering the constraint of memory erasure) have affected the usability of the problem modeling methods and the results of the algorithms, resulting in the fact that many of the study results cannot be transformed and applied to the actual satellite management systems. Therefore, summarizing the key constraints, general constraints, and characteristic constraints in various satellite task scheduling problems, restoring the real needs of various satellite operation management activities in the scheduling model as much as possible, and following the reality-based and application-oriented principle are the important prerequisites for guaranteeing the utility value of the relevant technologies.

(3) Effective mechanisms to reduce the scale of the satellite task scheduling problems and improve the efficiency of solving problems at the model level have not been emphasized. Scheduling algorithms have been the main object of study in satellite task scheduling problems, while scheduling models have received

Table 1.1: Summary of existing studies on satellite task scheduling models.

Scheduling model	Coding model	Coding process	Decoding process	Advantages and disadvantages of the model	Main constraints considered	Operational constraints not considered
Resource-oriented model	VRP model	The task resource execution sequence is expressed through decision variables	Beginning time is determined based on rules and constraints, and the decoding process produces feasible solutions	1. The task sequence is expressed, and is convenient for checking constraints of task sequence transition time 2. Imaging and downlink sequences are simultaneously determined under the working mode of "sequential playback" 3. It is not easy to describe the task scheduling problems with long VTWs and complex constraints 4. "Optimal sequence" is not necessarily equal to the "optimal result." The more complex the constraints are, the more likely that the decoding process loses high-quality solutions, and the higher the time cost of the decoding process is.	1. Constraints of uniqueness of task execution 2. Constraints of task visibility (e.g., VTWS of time domain, space domain, and frequency domain) 3. Constraints of task sequence (e.g., stripe splicing, 3D imaging, and sequential playback) 4. Constraints of the task working mode (e.g., constraints of recording, playback, real-time mode, and recording-playback ratio of remote-sensing satellites) 5. Resource capability constraints (e.g., satellite batteries, memory capacity, number of tasks; constraints of payload, operating frequency, duration of ground stations and downlink stations, etc.)	1. Constraints of nonlinear, functionalized, and process-dependent resource capability (e.g., for remote-sensing satellites, single-orbit imaging time is related to the number of orbit rolling, the minimum rolling angle, whether it is real time, etc.); 2. Constraints of resource capability and conversion time for distinguishing between continuous and discontinuous tasks (e.g., limitations on the number and duration of single-orbit continuous imaging of SuperView-1) 3. Constraints of resource capability and transition time for distinguishing sunlight conditions (e.g., the working time limit of the relay satellite in the Earth's shadow areas)
	Graph theory model					
	JSP model					
task sequencing model	Priority sequencing model	The task resource allocation sequence is expressed through decision variables				
	Priority model					

			6. Constraints of resource conversion time **7.** Constraints of resource preemptiblity and non-preemptibility (e.g., during TT&C tasks, other tasks can be performed while the phased array is idle)	**4.** Constraints of resource capability and conversion time for distinguishing between standby, work, and other operating states (e.g., payload transition and time limit for the performing tasks) **5.** Constraints of transition for single-payload resource mode (e.g., transition time between the time synchronization mode and the data downlink mode of ground-based stations of the Beidou Navigation Satellite System) **6.** Constraints of transition for multi-payload resource mode (e.g., transition time between CCD and hyperspectra of some satellites) **7.** Constraints of memory management of remote-sensing satellites based on erasing the entire memory
Task-oriented VTW allocation model	Rule-based allocation model	VTWs of tasks are expressed through 0–1 variables	Beginning time is determined based on rules and constraints, and decoded as feasible solutions	**1.** VTWs and beginning time of tasks are expressed directly and reflect the task sequence **2.** Not limited to "sequential playback," and imaging and downlink tasks can be scheduled simultaneously **3.** Task scheduling problems and complex constraints with long VTWs can be described **4.** Rule-based allocation model has a smaller solution space, but there is also a decoding process involved **5.** Multiple decision allocation model and discrete-VTW allocation model are more complete, guaranteeing diversity of solutions, but with a larger solution space
	Multiple decision allocation model	VTWs and beginning time are expressed, respectively, through 0–1 variables and integer variables	Constraints are not checked and may be decoded as infeasible solutions	
	Discrete-VTW allocation model	Discrete VTWs and beginning time are directly expressed through 0–1 variables		

far less attention. In fact, the key factors affecting the optimization efficiency such as the coding form, the neighborhood structure, and the search space of scheduling algorithms are largely determined by scheduling models. Therefore, in the study process of a general-purpose modeling method for satellite task scheduling problems, designing scientific scheduling models and modeling strategies is an inevitable requirement for reducing the problem scale, decreasing the coding complexity of algorithms, improving the solving efficiency of algorithms, and promoting the decoupling of models and algorithms.

1.3 Current status of studies on satellite task scheduling algorithms

Based on satellite task scheduling models, the task scheduling algorithms play a crucial role in model solving, profit optimization, and scheme output, which is important to instantiate problem models and determine the quality of model solving. Heuristic algorithms, exact solving algorithms, and meta-heuristic algorithms are three common scheduling algorithms that have been widely applied in the field of satellite task scheduling. Practice has shown that these three types of algorithms usually have unique performance advantages. For example, heuristic algorithms can quickly construct high-quality initial solutions, exact solving algorithms can provide the optimal solution in specific scenarios, and meta-heuristic algorithms have a good ability to find the optimal solution in complex solution space as well as compatibility with the first two types of algorithms. At the same time, the above algorithms tend to be closely related to the satellite task scheduling models and have practical problems such as insufficient generality and expandability, which urgently require more development and practical testing.

As a result, this section summarizes the commonly used algorithms in satellite task scheduling studies, including heuristic algorithms, exact solving algorithms, and meta-heuristic algorithms. Also, this section explores their coding forms, neighborhood structures, support strategies, and advantages and disadvantages, clarifies their applicable models, indicates the necessity of decoupling of algorithms and models and the in-depth integration of the algorithms, and provides a reference for the study of satellite task scheduling algorithms in this book.

1.3.1 Heuristic algorithms

1.3.1.1 Priority sequencing algorithms

Based on the priority sequencing model, priority sequencing algorithms are a class of algorithms that allocate resources and time to task sequences according to priority sequences and other rules. The algorithms have the advantages of clear logic, simple

structure, and fast operation speed, which is in line with human subjective experience. They are commonly used algorithms in satellite task scheduling studies and
practical scheduling systems. The sequencing rules in priority sequencing algorithms
are usually dominated by the task priority sequence [78, 79, 106, 107] and also include
some combined priority sequence related to time, length, quantity, and other attributes of VTWs [76, 77]. The commonly used resource allocation rules in the algorithms
include the earliest start principle [20, 36, 37, 43, 111, 112], the latest start principle [85],
the highest imaging quality principle [40, 41, 151], and the minimum possible conflict
principle [139]. As introduced above, since the algorithms are often tightly coupled
with sequencing models and the principles are relatively simple, this section will not
introduce them in detail again.

1.3.1.2 Conflict-avoidance algorithms (or de-conflicting algorithms)

Conflict-avoidance algorithms refer to a class of methods that reduce the degree of
conflict and provide conflict solutions during task scheduling by analyzing potential
conflicts between tasks and resources. Due to the special nature of conflict-avoidance
algorithms, they can also be regarded as priority sequencing algorithms based on
complex rules. In terms of their application in remote-sensing satellite task scheduling, Liu et al. [152] designed conflict-avoidance algorithms based on the imaging probability and non-mutually exclusive chains, respectively, for the task sequencing
model. Ran et al. [153] introduced a conflict judgment and conflict degree assessment
strategy as prior knowledge for crossover and mutation operations of a genetic algorithm (GA) in the process of electronic reconnaissance satellite task scheduling. Liu
et al. [154] achieved partial constraint conflict-avoidance by reducing the duration of
satellite imaging tasks. Chen et al. [49] constructed the initial task sequence in a way
that minimizes conflicts and also introduced a conflict-avoidance mechanism in the
subsequent iterative optimization process. In terms of their application in satellite
range scheduling, Jin et al. [155] designed a task resource allocation algorithm based
on the principle of minimum conflict. Yang et al. [156] calculated the potential conflict
set of satellite TT&C tasks and designed a conflict-based callback algorithm to realize
the "backtracking" mechanism in the resource allocation process. Tsatsoulis and Van
Dyne [157] proposed a variety of conflict-avoidance mechanisms based on case, rule,
and iteration. Luo et al. [133] calculated the irreconcilable conflict set of satellite TT&C
tasks and then designed a pre-scheduling and re-scheduling algorithm for satellite
TT&C tasks, which achieved the best results in the current AFSCN benchmark. In addition, the construction of edges in graph theory models often involves the idea of conflict recognition and avoidance. As can be seen, conflict-avoidance algorithms have
shown fast and effective optimization performance in many satellite task scheduling
studies, but they are rarely directly used as solving algorithms for task scheduling
problems. In most cases, the algorithms are used to generate initial scheduling solutions or assist other algorithms. In addition, conflict-avoidance algorithms mostly re-

volve around the constraint relationship between tasks and VTWs. The conflict-avoidance principles and methods between different algorithms are relatively similar, and they are tightly coupled with problem constraints, so it is difficult to make breakthroughs in algorithm theory.

1.3.1.3 Task allocation algorithms

Task allocation algorithms are a class of algorithms that pre-allocate satellite tasks based on certain experiences or rules to address the problems of large-scale and low optimization efficiency in satellite task scheduling. Like conflict-avoidance algorithms, task allocation algorithms are often used as support algorithms in the scheduling process. Task allocation algorithms help to alleviate the "combinatorial explosion" caused by the task scale in satellite task scheduling problems and play an important role in reducing the solution space and improving the optimization efficiency of task scheduling. Taking their applications in remote-sensing satellite task scheduling as examples, Lemaître et al. [158] designed the principle of fair allocation of remote-sensing satellite tasks, which evenly distributes tasks to different satellites based on task weights, satellite capabilities, and so on. Xu et al. [21] and Wang et al. [30] calculated the VTW overlap among tasks and assigned tasks to the satellite with the lowest VTW overlap. Du et al. [159] evaluated and predicted the probability of tasks being executed by satellites through various task characteristics such as VTW overlap and task priority and assigned tasks to the satellite with the highest probability. Zhou [160] introduced the "bidding, tendering, and evaluation" mechanism of contract net in the task allocation process, which improved the reliability of task allocation. Qiu et al. [161] and Sun et al. [162] decomposed the task scheduling problems into two subproblems: task allocation and task synthesis. He et al. [53] designed a task allocation and scheduling framework with a feedback mechanism, which achieved outperformance in large-scale agile satellite task scheduling scenarios by reallocating unscheduled tasks during the task scheduling process.

The above heuristic algorithms have effectively reduced the decision-making dimension and solving difficulty of the problems, but their application effectiveness largely depends on the rationality of the algorithm design. Meanwhile, most of the algorithms are tightly coupled with the characteristics of the problem scenarios, tasks, and resources, and they are not universal enough. In this regard, it is necessary to fully utilize the characteristics and experience of satellite task scheduling scenarios and design general-purpose and adaptive heuristic algorithms to solve large-scale and complex satellite task scheduling problems under the new normal.

1.3.2 Exact solving algorithms

Exact solving algorithms can obtain the global optima in small-scale satellite task scheduling problems and can also ensure the global optimality of solutions in dynamic or uncertain environments. Although it is generally difficult for exact solving algorithms to solve large-scale satellite task scheduling problems with complex constraints in limited time, the algorithm ideas are also of guiding significance for problem modeling and solution space optimization. This section briefly introduces two commonly used exact solving algorithms in related studies.

1.3.2.1 Branch and bound (B&B) algorithms

Branch and bound (B&B) algorithms are a class of exact solving algorithms proposed by Land et al. [163] in 1960 and formally named by Little et al. [164] in 1963. As one kind of the most commonly used algorithms for solving integer linear programming problems, B&B algorithms can reduce solution space by means of branching, bounding and pruning, and then search for the optimal solution through each branch. Due to the fact that satellite task scheduling problems are often simplified into linear programming problems, B&B algorithms have been widely applied in the field of satellite task scheduling [19, 38, 45, 50]. Moreover, to improve the efficiency of solving large-scale task scheduling problems, B&B algorithms are often used in conjunction with column generation algorithms [5, 30, 165], cutting plane algorithms [166, 167], and Lagrangian relaxation algorithms [22, 93, 168] for solving satellite task scheduling problems or their boundaries. In addition, through the mathematical programming solver CPLEX, B&B algorithms can usually complete design, improvement, and problem-solving work. Some branching and relaxation strategies in CPLEX also provide an important reference for designing B&B algorithms in the aerospace field.

1.3.2.2 Dynamic programming (DP) algorithms

Proposed by Bellman and Kalaba [169] in 1965, dynamic programming (DP) algorithm is a class of exact solving algorithm that searches for the optimal solution through problem decomposition and recursion. For the scheduling problems of remote-sensing satellite imaging tasks, Lemaître et al. [20] tried to solve the problems using a DP algorithm based on a graph theory model. Bai et al. [170] decomposed the problems into an optimal synthesis problem for single-orbit tasks. Damiani et al. [171] designed an evaluation vector that includes the current task, satellite battery level, and memory capacity and decomposed the problems into an optimal task combination problem for the evaluation vector. For the scheduling problems of remote-sensing satellite downlink tasks, Liu et al. [172] decomposed the problems into the optimal task combination problem for downlink tasks within VTW. Qin and Zhang [173] divided the problems into several stages by time. In addition, Peng et al. [174] decoded the imaging beginning time of agile remote-sensing satellites within VTW through DP during the iterative search process, replacing traditional heuristic decoding rules such as earliest start and imaging quality priority. The idea of problem decomposition adopted by the above DP algorithms is also informative for solving large-scale satellite task scheduling problems.

1.3.3 Meta-heuristic algorithms

1.3.3.1 Evolutionary algorithms (EAs)

Evolutionary algorithms (EAs) mainly refer to a class of meta-heuristic algorithms that iteratively optimize combinational optimization problems by simulating the evolution mechanism or group behavior of biotic populations in nature. EAs are widely used in the field of satellite task scheduling. This section selects three typical EAs, namely GA, ant colony optimization algorithm(ACO), and particle swarm optimization algorithm(PSO), and introduces their model bases, coding methods, and improvement strategies in the process of solving satellite task scheduling problems.

1.3.3.1.1 Genetic algorithms (GAs)

GA is a class of EAs proposed by Holland [175] in 1975 that simulate the principles of natural selection and genetic mechanisms in evolutionary theory. GAs are iteratively optimized in the form of population and probabilistic genetic mechanism and have been widely used in satellite task scheduling problems because of their implicit parallelism, diverse solution representations, and good global optimization ability. On the basis of the task sequencing model, to solve AFSCN benchmark problems, Parish [102] proposed the classical algorithm Genitor that utilizes gene segments to represent the resource allocation sequence of TT&C tasks, as shown in Figure 1.9(a), where the decoding process sequentially allocates resources for tasks 3, 5, 2, 4, 6, and 1. Zhou et al. [176] and Chen et al. [177] designed a GA that contains immune GA and learning strategy, showing good performance in large-scale multi-satellite and multi-station range

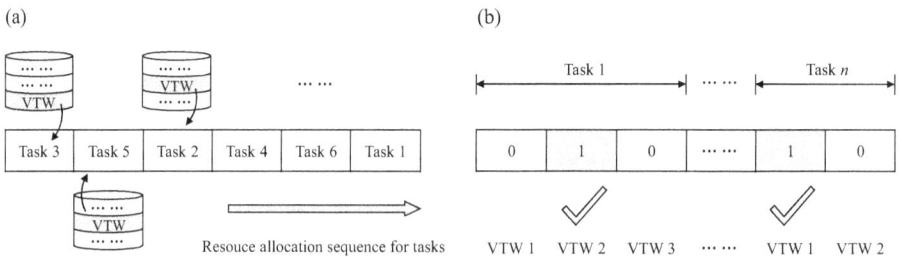

Figure 1.9: Example of coding form of genetic algorithm for solving satellite task scheduling problems: (a) a coding form of a task sequencing model and (b) a coding form of a VTW allocation model.

scheduling. The above genetic coding method fits well with the task sequencing model and has become a common coding method in satellite task scheduling algorithms. Sun et al. [25], Song et al. [39], Zhang et al. [107], Li and Wu [178], and Han et al. [179] have also effectively solved satellite task scheduling problems based on the coding form, as well as the strategies of imaging quality first, local search, depth-first search, conflict avoidance, and elitism selection.

Based on the VTW allocation model, Kim and Chang [42], Li et al. [54], and Niu et al. [180] expressed the VTWs of remote-sensing satellite tasks through 0–1 gene, as shown in Figure 1.9(b). The decoding process is to assign VTW 2 to task 1, and so on. Hosseinabadi et al. [181] expressed the satellite and VTW executing tasks through gene segments. Sun et al. [162] expressed the VTWs of remote-sensing satellite execut-ing tasks through integer gene and introduced the knowledge model and parameter learning strategies into the algorithm. Xhafa et al. [134, 135] used direct coding form to represent satellite TT&C tasks and designed a GA containing population competition and elimination strategies. Tang et al. [95] and Du et al. [182] directly expressed the discretized VTWs of navigation satellite tasks and TT&C tasks through gene and fur-ther designed the multi-objective optimization algorithms. Zhang and Li [40] also in-troduced the Metropolis criterion into the iterative mechanism of traditional GA and optimized the task scheduling problems of agile remote-sensing satellites by repre-senting VTWs and task working mode through genetic variables. It can be seen that based on the task sequencing model and VTW allocation model, GA can effectively represent solutions and solve problems. However, due to the limitations of the coding process of the task sequencing model itself, GA based on such model may lose high-quality solutions during the optimization process, while GA based on the VTW alloca-tion model may have too long encoding length and too many types of gene represen-tations, which may have a certain impact on algorithm iteration and search efficiency. Therefore, when using a GA to solve satellite task scheduling problems, a reasonable satellite task scheduling model and coding method, as well as targeted algorithm im-provement measures, are crucial.

1.3.3.1.2 Ant colony optimization (ACO) algorithms

Proposed by Colorni et al. [183] in 1991, ACOis a class of EA that simulates the path-finding behavior of ant colony. ACO continuously adjusts the path of ant colony through mechanisms of pheromone accumulation, feedback, volatilization, and com-munication during the pathfinding process of ant colony, demonstrating good asymp-totic convergence, robustness, implicit parallelism, and global optimization ability. In terms of their application in remote-sensing satellite task scheduling, Qiu et al. [184] and Geng et al. [185] solved the remote-sensing satellite task scheduling problems based on strategies such as ant colony system, max-min ants, and dynamic transition probability. Yan et al. [186] and Chen et al. [187] introduced knowledge learning and pheromone restriction strategies in traditional ACO. To address the problem of long optimization cycles and easy to fall into local optima of ACO, Zhu et al. [188] determined task sequences and VTWs based on comprehensive heuristic information, conflict avoidance, and perturbation strategies. In addition, Gao et al. [189] and Wu et al. [190] introduced a local search strategy into the framework of ACO and achieved good performance. In terms of their application in task scheduling of navigation satellites and satellite TT&C, Zhang and Feng [115] Zhang et al. [116–118] designed a cooperative communication mechanism between ant colonies to optimize the satellite range scheduling. Xing and Chen [191] and Yao and Xing [192] introduced the guided local search strategy and knowledge model in ACO, which improved the local and global optimization ability of the ACO in large-scale range scheduling problems. Chen and Wu [12] introduced pseudo-random effects in ant transition probability and reduced the search space of ACO based on conflict-avoidance strategies for TT&C tasks. To solve navigation satellite task scheduling problems, Huang et al. [193] designed a dynamic biased exploration probability based on the ant colony system to realize adaptive adjustment of the ant exploration ratio. Considering the pheromone scarcity at the early stage of ACO optimization, Li et al. [194] adopted GA and ACO together to achieve good optimization results. ACOs and their improvement strategies have been widely applied in satellite task scheduling problems, but due to the basic idea of ACO path optimization, related studies mainly use ACOs in task sequencing models to solve scheduling problems, where the execution sequence of satellite tasks is usually represented by ant paths. Therefore, affected by decoding strategies, ACOs may lead to the loss of high-quality solutions, which may be detrimental to global optimization in task scheduling scenarios with complex constraint logic.

1.3.3.1.3 Particle swarm optimization (PSO) algorithms

Proposed by Kennedy and Eberhart [195] in 1995, PSO is a class of EA that simulates the foraging behavior of birds. Compared with GA and ACO, PSO is easier to implement because of its simple structure. The algorithm is also characterized by fast convergence and good robustness. Initially, PSOs are mainly used for continuous optimization problems, and most of the PSOs used in satellite task scheduling studies at this stage are based on the discrete PSO (DPSO) algorithm proposed by Kennedy and Eberhart [196] in 1997. Considering that the traditional PSO cannot solve discrete optimization problems, Tang et al. [197] represented VTWs for executing tasks through particle position vectors and the number of VTWs through particle dimensions. Chang and Wu [198] represented the probability of task VTW selection through particles, completed the decoding based on probability sequence, and introduced an automatic regulation mechanism of particle quantity based on population diversity. Chen et al. [199] simultaneously represented both the VTWs and satellite operating modes for downlink tasks through particle vectors and introduced the quantum behavior and mutation operators to enhance the algorithm's global optimization ability. In addition, Chen et al. [200], Guo et al. [201], and Liu and Wang [202] expressed task resources with particle vectors and introduced mechanisms such as genetic, Tabu, and direction backtracking on task scheduling problems such as joint TT&C tasks of high-orbit and low-orbit satellites and area target observation of forest remote-sensing satellites. Due to the fact that the above-mentioned algorithms are usually based on discrete optimization-oriented DPSO, the scheduling models using these algorithms are also dominated by the VTW allocation model. However, PSOs themselves possess the dual characteristics of continuous optimization and discrete optimization, and excessive emphasis on the discrete coding form may not be conducive to the optimization of the algorithm. On the contrary, the coding methods with discrete optimization (to determine VTWs) and continuous optimization (to determine the beginning time of tasks) may be more suitable for solving satellite task scheduling problems with long VTWs at present.

1.3.3.2 Local search algorithms

Local search algorithms are a class of meta-heuristic algorithms that iteratively search the neighborhood solution space starting from the initial solution and selectively move toward high-quality solution space. Local search algorithms commonly used in satellite task scheduling studies at this stage include Tabu search (TS) algorithms and SA algorithms. In this section, the model bases, neighborhood structures, and search strategies of these two algorithms in the process of solving satellite task scheduling problems are introduced.

1.3.3.2.1 Tabu search (TS) algorithms

TS is a class of local search algorithm with a memory strategy proposed by Glover in 1986 [203, 204]. TS records local optimal solutions or the operations that generate local optimal solutions during the optimization process through a Tabu table, to avoid repeated local search and achieve the effect of skipping from local optima for exploring other high-quality solution spaces. As one of the earliest algorithms used for solving satellite task scheduling problems, TS is simple, versatile, and practical. On the basis of the task sequencing model, Cordeau and Laporte [3] and Cordeau et al. [4] utilized a TS to solve VRP problems for optimizing single-orbit task of remote-sensing satellites. He and Tan [205] designed a neighborhood structure based on the JSP scheduling model for inserting, moving, and exchanging machines of workpiece. Zuo and Wang [206] designed neighborhood structures for adding, deleting, and replacing tasks in task sequences. Chen et al. [207] and Li et al. [208] designed a variety of neighborhood structures based on task sequences and developed a TS based on a variable neighborhood strategy. Blöchliger and Zufferey [112] designed a neighborhood structure based on color selection in the graph coloring model and used a TS with variable Tabu length to solve satellite range scheduling problems. Based on the VTW allocation model, Xhafa et al. [137] constructed neighborhoods by perturbing VTWs and beginning time. Sarkheyli et al. [47] designed neighborhood structures for task addition and deletion, as well as conflict avoidance within VTWs, and selected the matching relationship between tasks and VTWs as Tabu objects. Habet et al. [209] designed a neighborhood evaluation mechanism to improve the neighborhood search efficiency of TS for executing remote-sensing satellite tasks. In addition, TS is also used to solve satellite task scheduling problems based on the knapsack model [57]. TSs have shown good generality and solving effectiveness, but due to the relatively simple mechanism of such algorithms, there have been relatively few studies in recent years on using TSs to directly solve satellite task scheduling problems. Therefore, TSs should be deeply integrated with the latest intelligent optimization technologies to better adapt to the current complex and diverse satellite task scheduling problems.

1.3.3.2.2 Simulated annealing (SA) algorithms

SA was proposed by Metropolis et al. [210] in 1953 and applied in the field of combinational optimization by Kirkpatrick et al. [211] in 1983. SA is a class of local search algorithm derived from the principle of solid annealing, which probabilistically accepts inferior solutions according to temperature changes, achieving the effect of skipping from local optima for exploring other high-quality solution spaces. Since SA is usually less affected by the initial solution quality and problem scale and has good asymptotic convergence, it has been widely applied in satellite task scheduling problems. Based on the task sequencing model, Huang and Zhang [212] constructed neighborhoods by inserting or exchanging mutually exclusive vertices in the task sequence and designed an SA algorithm containing secondary search and elitism strategy. On the basis of the VTW allocation model, He et al. [213] and Gao et al. [214] designed neighborhood structures for adding, deleting, merging, or decomposing tasks, and added strategies such as random perturbation and rescheduling to the SA algorithm. Xu et al. [215] and Du et al. [216] constructed the neighborhood of the SA algorithm by exchanging decision variables. Huang et al. [217] borrowed feedback characteristics from ACO and combined prior and posterior knowledge to design neighborhoods, and triggered disturbances when the algorithm fell into local optima. Wu et al. [43] designed neighborhood operations such as task addition, deletion, and migration based on adaptive probability and introduced adaptive temperature control and Tabu strategies into the SA algorithm. Lin [218] constructed neighborhoods by exchanging task VTWs and frequency bands and adopted SA and DP together to effectively solve the scheduling problems of remote-sensing satellite multi-payload tasks. In addition, in recent years, in response to the characteristics of satellite task scheduling problems that are huge in scale and easy to fall into local optima, local search algorithms such as variable neighborhood search [219] and adaptive large neighborhood search [41, 53] have been widely applied, showing good performance.

1.3.3.3 Other algorithms

1.3.3.3.1 Memetic algorithm (MA)

Proposed by Moscato [220] in 1989, MA is a hybrid meta-heuristic algorithm that combines EAs with local search algorithms. Derived from the word "meme" in Dawkins' book *The Selfish Gene* [221], the term "memetic" implies cultural genes; hence, MA is also known as the cultural gene algorithm. MA not only retains the optimization characteristics of EAs based on population evolution but also possesses the good local optimization ability of local search algorithms, offsetting the weaknesses by learning from the strong points of both algorithms. MA has also been widely applied in the field of satellite task scheduling. According to the hybrid form of EAs and local search algorithms, it can be mainly divided into the following three types: 1)MA based on the EA framework, that is, the earliest MA proposed by Moscato [220]. For example, Du et al. [182], Xing and Chen [191], Yao and Xing [192], and Liu et al. [202] introduced local search mechanisms into GA, ACO, and PSO for solving satellite task scheduling problems, respectively. As the most commonly used class of MA, this type of MA can be regarded as an EA based on individual improvement and repair strategies. However, the local search mechanism for individuals may increase the time of population iteration, so the design of reasonable local search frequency, number of iterations, and triggering conditions is the key to this type of MA. 2)MA based on local search framework. For example, He and Tan [205] and Huang et al. [217] introduced evolutionary strategies such as individual crossover and pheromone feedback into TS and SA, respectively. This type of MA can be regarded as a kind of local search algorithm that designs neighborhood structures and guides search directions through evolutionary strategies, but due to the lack of support of population, this type of MA is inferior to EAs in terms of solution diversity and implicit parallelism. 3)MA based on hierarchical framework. For example, to leverage the early optimization capabilities of GA and deep search capabilities of SA, Zhu et al. [52] designed a two-stage optimization algorithm based on GA and SA, which achieved outperformance in remote-sensing constellation task scheduling. However, this type of MA also has higher requirements on the generality of the coding form of the solution and the reasonableness of algorithm switching. It can be seen that MA is actually a framework that combines EAs with local search algorithms. A combination framework of collaborative and complementary algorithms plays a crucial role in solving complex and large-scale satellite task scheduling problems.

1.3.3.3.2 Decision algorithms based on machine learning

Decision algorithms based on machine learning refer to a class of methods that train satellite task scheduling decision models through supervised learning, reinforced learning, and other machine learning methods to schedule satellite tasks. In terms of supervised learning, some scholars use historical data of satellite task scheduling to train predictive models for satellite task schedulability and allocate resources to tasks one by one based on the prediction results and task sequencing models. For example, Li et al. [222] constructed a remote-sensing satellite task schedulability prediction model using robust decision trees, support vector machines, and artificial neural networks. Liu et al. [223] extracted task characteristics such as task priority, duration, and conflict degree from historical data of remote-sensing satellites and constructed a task schedulability prediction model based on a Back Propagation, also well known as BP neural network model with variable hidden layers. Considering uncertainty factors such as cloud cover, Xing et al. [64] extracted meteorological features and designed an autonomous task scheduling algorithm for remote-sensing satellites using a task schedulability prediction model. Du et al. [159] trained a task schedulability prediction model using evolutionary neural network algorithms to assign tasks to the satellite that is most likely to perform the tasks, significantly improving the efficiency of large-scale agile remote-sensing satellite task scheduling. In terms of reinforced learning, Wang et al. [65, 66] trained satellite selection strategies in action sets using a Q-learning algorithm and an evolutionary neural network algorithm for collaborative task scheduling problems of remote-sensing satellites. Wang et al. [67] trained a remote-sensing satellite task scheduling decision model by the A3C algorithm. When a new task is input, the model can determine whether to execute the task and assign the longest VTW to the task decided to be executed, providing a solution for autonomous scheduling of satellite tasks. It can be seen that decision algorithms based on machine learning, with the combined characteristics of the simplicity and rapidity of heuristic algorithms as well as the adaptive and self-learning nature of machine learning techniques, can be regarded as a class of algorithms that guide task sequencing and allocation based on advanced rules. At present, many satellite control agencies have accumulated a large amount of historical data on task scheduling, enabling extensive application of related technologies. On the other hand, although the above algorithms have realized autonomous scheduling of satellite tasks, they show the limitation of "making decision for each new task one by one," which is difficult to guarantee the global optimization of large-scale task scheduling. Moreover, the above algorithms have oversimplified the problems under study to extract features of tasks and resources for model training, but under the increasingly complex satellite task scheduling problems at the present stage, it may be difficult for the decision model based on simple features to achieve the ideal effect.

In summary, this section summarizes the commonly used algorithms in satellite task scheduling studies from three perspectives, such as heuristic algorithms, exact solving algorithms, and meta-heuristic algorithms, and analyzes the coding characteristics, neighborhood structures, and strengths and weaknesses of each type of algorithm. Please refer to Table 1.2 for the summary. The algorithms have successfully solved many satellite task scheduling problems and contributed important theoretical and practical experience, but there are also the following two deficiencies:

(1) The satellite task scheduling algorithms are tightly coupled with the models, and the algorithms are not universal and flexible enough. All the above-

Table 1.2: Summary of commonly used algorithms in satellite task scheduling studies.

Algorithm type	Algorithm name	Common coding model	Algorithm strengths		Algorithm weaknesses	
Heuristic algorithms	Priority sequencing algorithms	Task sequencing model	1.	Simple in structure and fast in computing speed	1.	Tightly coupled with problem characteristics and scheduling models, with low degree of generality
	Conflict-avoidance algorithms	Mainly task sequencing models	2.	Consistent with human subjective perceptions and supported by experience		
	Task allocation algorithms	Mainly VTW allocation models			2.	Poor in optimization quality, commonly used for assisting decision-making, and rarely used directly for generating scheduling plans
			3.	Helpful to reduce the problem solution space and increase the solution efficiency when used in combination with other algorithms		
Exact solving algorithms	Branch and bound algorithms	Task sequencing or VTW allocation models	1.	Possible to obtain the optimal solution, ensuring optimality even under uncertainty	1.	Only suitable for solving small-scale problems with simple constraints, and the ability to solve large-scale problems with complex constraints is limited
	Dynamic programming algorithms		2.	Informative for other algorithms, and good prospects when used in combination with other algorithms	2.	Usually require a significant degree of simplification of problems

Table 1.2 (continued)

Algorithm type	Algorithm name	Common coding model	Algorithm strengths	Algorithm weaknesses
Meta-heuristic algorithms	Evolutionary algorithms — Genetic algorithms	Task sequencing or VTW allocation models	1. Implicit parallelism and diverse solution representation capabilities; 2. Good global optimization ability 3. Good compatibility with other algorithms 4. Coding method of the task sequencing model is concise with simple operation	1. Limited local search capability 2. Coding method of the task sequencing model may lose high-quality solutions 3. Coding method of the VTW allocation model is complex and lengthy, which may affect the optimization efficiency
	Ant colony optimization algorithms	Mainly task sequencing models		
	Particle swarm optimization algorithms	Mainly VTW allocation models		
	Local search algorithms — Tabu search algorithms	Task sequencing or VTW allocation models	1. Simple principle, easy implementation, and good generality 2. Good local search capability 3. Certain capability to escape local optima	1. Limited global search capability and prone to falling into local optima 2. Relying on the neighborhood structure for design
	Simulated annealing algorithms	Mainly VTW allocation models		
	Other algorithms — Memetic algorithms	Task sequencing or VTW allocation models	Complementary to local search and evolutionary algorithms	Relying on the framework design and possibly reducing the iteration efficiency
	Decision algorithms based on machine learning	Mainly task sequencing models	Simple, fast, and adaptive with self-learning	"Making decision for each new task one by one" and insufficient global optimization

mentioned satellite task scheduling algorithms have their own applicable task scheduling models that vary depending on the problems, resulting in the tight coupling between satellite task scheduling models and the algorithms, and the insufficient generality of scheduling algorithms at this stage. Therefore, combined with the current study status of satellite task scheduling models, it is of great practical significance to study a satellite task scheduling architecture that decouples algorithms from models and uses flexible and general-purpose algorithms.

(2) **An effective mechanism for deep collaboration and reasonable matching between different algorithms has not yet formed.** The scheduling algorithms have different strengths, weaknesses, and generality. For example, heuristic algorithms are commonly used to assist in decision-making, exact solving algorithms have limited ability to solve large-scale problems, and EAs are strong at global optimization. Although the MAs provide a good approach for making algorithms collaborative, there is still room for improvement in terms of solving ability, generality, and other aspects at the current stage. Therefore, a reasonable combination of different types of algorithms to form a fully collaborative and complementary algorithm combination framework is a necessary way to solve large-scale and complex satellite task scheduling problems.

1.4 Development status of general-purpose solving techniques for satellite task scheduling

General-purpose solving techniques for satellite task scheduling refer to general-purpose solvers, toolboxes, algorithm packages, and so on that can establish general-purpose task scheduling models for different types of satellite tasks or use different task scheduling algorithms for model solving. General-purpose solving techniques for satellite task scheduling can solve practical problems such as satellite task scheduling in different scenarios and heterogeneous multi-satellite collaborative task scheduling. They can efficiently call diverse and expandable scheduling algorithms, providing practitioners with convenient, diverse, and effective problem modeling and solving methods. Therefore, general-purpose solving techniques for satellite task scheduling play an important role in improving the level of automation, intelligence, and integration of satellite comprehensive control. They are important criteria for measuring a country's comprehensive strength in spaceflight industry and key techniques that need to be urgently developed in the spaceflight sector.

At present, satellite task scheduling tools with general-purpose modeling and solving features include CPLEX, STK/Scheduler, Europa2 (the second generation of Extensible Universal Remote Operations Architecture), and the task scheduling subsystem of the commercial remote-sensing satellite SuperView-1. In the following, for the four general-purpose solving techniques mentioned above, this section introduces their modeling features, solving algorithms, and main functions, summarizes their strengths and weaknesses, and clarifies the necessity of developing a domestic general-purpose solver for satellite task scheduling independently and the new application ideas in the light of the development status of the satellite comprehensive control, providing a reference for the study of the general-purpose satellite task scheduling engine in this book.

1.4.1 Mathematical programming solver CPLEX

CPLEX is a mathematical programming solver developed by IBM of the United States for solving four types of mathematical programming problems, namely, linear programming, quadratic programming, mixed-integer linear programming, and mixed-integer quadratic programming [224]. In terms of modeling language, CPLEX is concise and easy to understand and can be compatible with mainstream programming languages such as C++, Java, Python, MATLAB, and .NET. In terms of solving the quality, CPLEX has a series of built-in exact solving algorithms such as simplex method, interior point method, and branch and bound algorithm to provide the optimal solutions to the problems, holds a series of optimal records in the benchmark of classic operations studies, and is therefore applied to solve satellite task scheduling problems.

Aiming at solving integrated scheduling problems of remote-sensing satellite imaging and downlink tasks, Xiao et al. [19] established a segmented FSP model to solve the problems through CPLEX. However, due to the NP-hard nature of the problems, only 20 tasks can be scheduled within the optimization time of 7,200 s. Considering that the time-dependent attitude transition time constraint of agile satellites cannot be represented by a linear function, Liu et al. [41] optimized the simplified problems through CPLEX, but only 12 tasks can be completed. To guarantee the optimality of task scheduling in uncertain environments, Valicka et al. [50] used CPLEX to schedule remote-sensing satellite tasks considering random cloud cover, and the optimization time was about 2,000–3,000 s. For the task scheduling problems of CBERS-3 remote-sensing satellite, Xu and Tan [225] completed the optimized scheduling of 11 tasks through CPLEX. Bensana et al. [56] successfully used CPLEX to optimize the scheduling of single-orbit satellite tasks but failed to complete the effective scheduling of multi-orbit satellite tasks.

Moreover, CPLEX can be used together with other algorithms or strategies, which to some extent can improve the efficiency of solving satellite task scheduling problems. For example, for the Galileo navigation constellations' task scheduling problems, Marinelli et al. [93] introduced Lagrangian relaxation and heuristic strategies in CPLEX, which can optimize the scheduling of 360 navigation satellite tasks in 18,000 s of optimization time. Meanwhile, Marinelli also pointed out that in the absence of other support strategies, the solving capability of CPLEX is only 120 tasks. Xu et al. [21] and Wang et al. [226] combined the local search algorithm with CPLEX to schedule 100 remote-sensing satellite tasks within 1,800 s.

The above studies indicate that although CPLEX performs well in solving small-scale satellite task scheduling problems, uncertain task scheduling problems, and problem boundaries, it is inefficient in solving large-scale task scheduling problems and cannot accurately describe nonlinear constraints and profits. Therefore, at the present stage when the scheduling scale is growing, it is not feasible to use CPLEX in satellite scheduling systems in reality since CPLEX is difficult to satisfy the dual requirements of satellite control agencies on the quality and speed of task scheduling.

1.4.2 General-purpose satellite scheduling software STK/Scheduler

STK/Scheduler is a commercial satellite task scheduling software developed by Orbit Logic of the United States [227, 228] and can be called directly from STK version 6.0 and above. STK/Scheduler can quickly establish task scheduling models based on STK models and data to provide task scheduling schemes through built-in algorithms with user-friendly interfaces, as shown in Figure 1.10. Since the satellite orbits and VTWs calculated by STK are very accurate, the software is often used as a tool for calculating VTWs in satellite task scheduling studies.

STK/Scheduler contains the following five scheduling algorithms: 1)one-pass algorithm, which determines task resources and beginning time based on the task priority sequence and built-in desirability functions, is a typical task sequencing algorithm; 2) sequential algorithm, which considers other factors such as task VTWs and duration on the basis of one-pass algorithm; 3) multi-pass algorithm, which runs the one-pass algorithm multiple times based on certain rules and adjusts the scheduling scheme when necessary, is a kind of algorithm combining task sequencing with conflict avoidance; 4) neural network algorithm, which assigns VTWs and beginning time to tasks based on certain rules and repairs infeasible solutions, is a kind of algorithm combining task allocation with conflict avoidance; 5) random algorithm, which is similar to the neural network algorithm but adopts a randomized strategy in allocating VTWs and beginning time.

Li et al. [229] implemented task scheduling for relay satellites based on STK/Scheduler, but the constraints involved are relatively simple, especially the constraints with the characteristics of relay satellites such as frequency band matching and transition time are not modeled. Li et al. [230] pointed out that STK/Scheduler took a long time to send commands, which was not conducive to large-scale task scheduling. Bai et al. [231] tested five built-in algorithms and the results showed that multi-pass algorithm and random algorithm had the best performance in a short period of time. Li et al. [232] introduced a task dynamic adjustment mechanism based on STK/Scheduler. Fisher and Herz [233] gave an introduction of customized algorithms for STK/Scheduler, but it was limited to the function of adjusting the sequencing rules, exchange strategies, and so on. In conclusion, although STK/Scheduler has made some achievements in the application of solving satellite task scheduling prob-

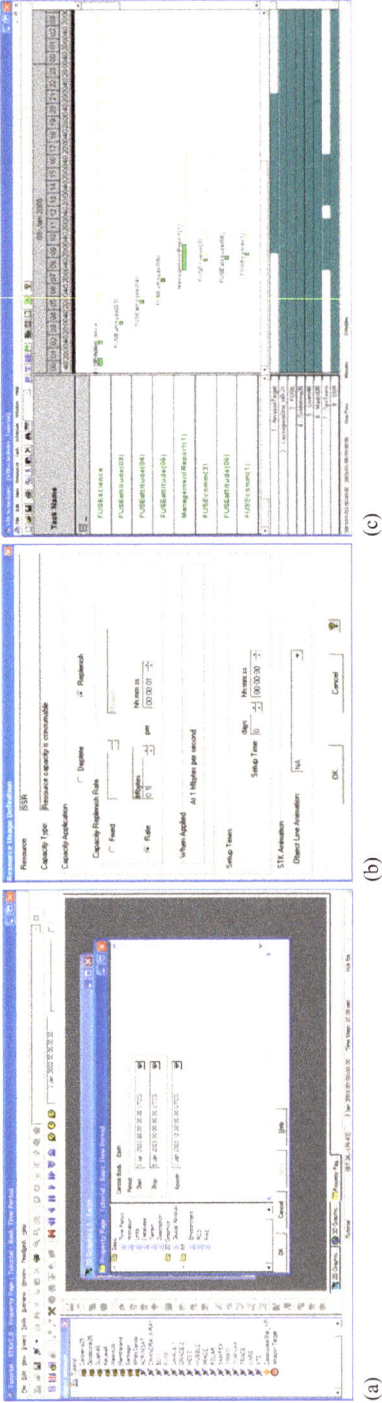

Figure 1.10: Example of STK/Scheduler operation interfaces: (a) scenario (target and resource) modeling interface; (b) resource attribute setting interface; and (c) scheduling result display interface.

lems, it has revealed some shortcomings in the current trend of increasing complexity and diversification of problems:

(1) Although STK/Scheduler provides user-friendly encapsulations for satellite tasks and resources, it greatly restricts the attributes and constraint formats of the tasks and resources, which is not conducive to solving problems containing complex tasks or constraints and makes secondary development very difficult.

(2) Due to the lack of dynamic adjustment functions for tasks, resources, profits, or constraints, STK/Scheduler is typically used to solve conventional satellite task scheduling problems.

(3) The main solving function of STK/Scheduler was last updated in 2006. Until now, it does not include meta-heuristic algorithms such as GA, ACO algorithm, TS algorithm, or SA algorithm, and relies on rules. With the increasing scale and complexity of satellite task scheduling problems in recent years, the built-in algorithms may not be effective in solving the problems.

Moreover, Herz et al. [234] used STK/Scheduler Online for solving satellite range scheduling problems in 2013, as shown in Figure 1.11, accessed the STK/Scheduler server through the Internet, and completed a series of functions such as scene modeling, parameter configuration, and optimal scheduling. This remote style of task scheduling gives users convenient and efficient access, helps with iterative updates of service and quick fixes of faults, and contributes to scheduling efficiency with the support of high-performance servers. Therefore, STK/Scheduler Online also provides a new approach for the design and deployment of satellite task scheduling systems.

1.4.3 Satellite task planning software Europa2

The Europa2 is developed by the National Aeronautics and Space Administration (NASA) for spacecraft task planning [235], as shown in Figure 1.12. Since 1998, Europa and Europa2 have been widely used in NASA's Hubble Space Telescope [236], DS-1 Autonomous Satellite [237], and other projects. Unlike the classic Planning Domain Description Language (PDDL) in the planning field, the New Domain Description Language (NDDL) on which Europa2 is based is a state-variable-based, object-oriented, and declarative language. Therefore, in Europa2, the description of a class of task scenarios is mainly accomplished by defining tags, transactions, objects, classes, timelines, and planning solutions, and a feasible solution is given by constraint planning techniques such as constraint propagation [238–241].

Since Europa2 is based on state variables, it is mainly for solving task planning problems, such as action logic and state transition, for satellites to perform tasks. Such problems describe the logical relationship of satellite tasks, give feasible solutions to satisfy the constraints, and require few resource scheduling. Whereas the task scheduling problems discussed in this book mainly solve the allocation of task

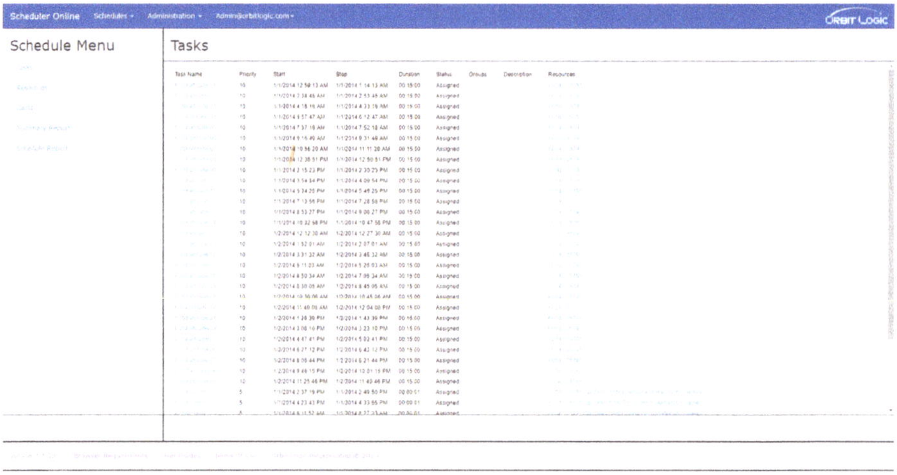

(a)

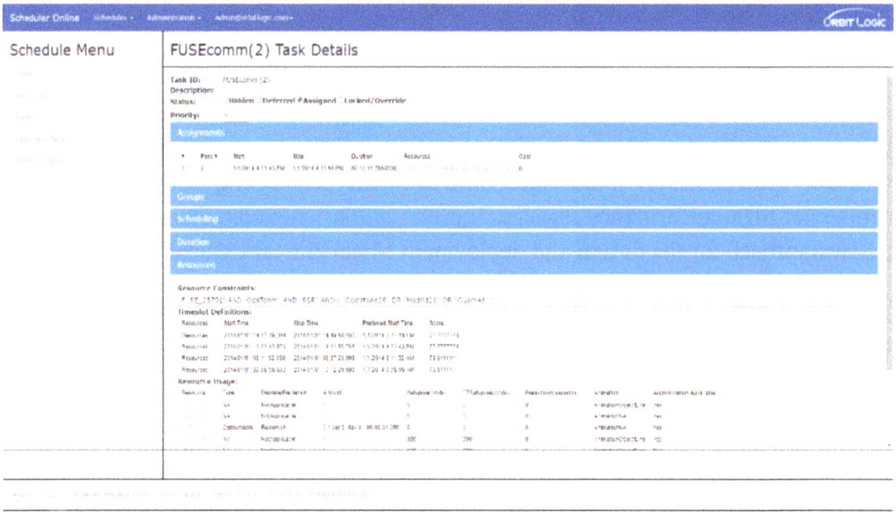

(b)

Figure 1.11: Example of STK/Scheduler online operation interface: (a) interface of remote task list and (b) interface of remote task attribute setting.

resources and time, such as VTWs for executing tasks, satellites and their battery level, memory, and other payload resources. Therefore, in response to the satellite task scheduling problems studied in this book, Europa2 has the following short-comings:

(1) Due to the lack of objective function and scheduling optimization mechanism, Europa2 can only give feasible satellite action sequences through constraint plan-

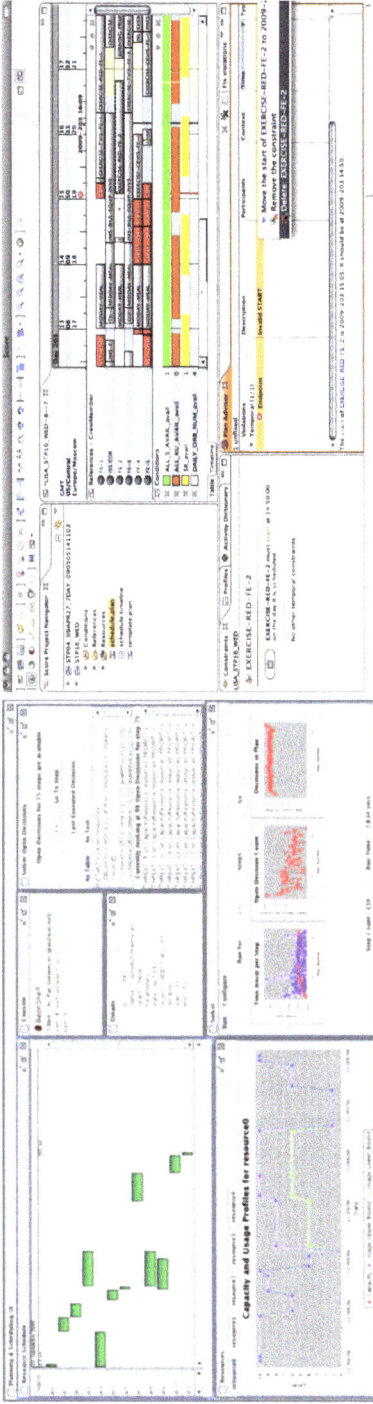

Figure 1.12: Example of Europa2 operation interface: (a) Europa 2.6 operation interface and (b) OpenSPIFe operation interface.

ning and other methods, but cannot give better scheduling solutions based on iterative search driven by the objective function.

(2) The constraint planning algorithm based on constraint propagation can only be applied to small-scale task scenarios. For example, the planning time of 20 single-satellite single-orbit tasks is very long. The latest version of Europa 2.6 released in 2011 [235] has been difficult to cope with the current needs of multi-satellite networked, large-scale, and complex satellite task scheduling.

However, the constraint propagation algorithm of Europa2 can quickly provide feasible solutions in the event of new tasks' inputs or changes in task or resource attributes, which is enlightening for the design of dynamic task scheduling algorithms and frameworks that respond quickly to demands. On top of that, although NASA has not continuously updated Europa2, it released the Open Scheduling and Planning Interface for Exploration (OpenSPIFe) based on Europa2 in 2015 [242]. OpenSPIFe has functions such as action planning, dynamic adjustment, and visual interface, but its applications in reality are still rare, and further study is needed.

1.4.4 SuperView-1 task scheduling subsystem

SuperView-1 is the first commercial agile remote-sensing constellations domestically. SuperView-1 constellations currently consist of four 0.5-m resolution optical imaging satellites in the sun-synchronous orbit. The first and second satellites were launched in December 2016, and the third and fourth satellites were launched in December 2017. The orbital period of each satellite is about 97 min. At present, these four satellites have formed a unique constellation configuration of "same orbit with different phase," that is, the four satellites phase 90 °from each other on the same orbit. As a result, Super-View-1 constellation is able to image any target on the Earth's surface in a single day and observe 80% of the Earth's area twice a day. According to the relevant plan, more than 20 agile remote-sensing satellites will be launched in succession to form a new SuperView-1 constellation network together with the existing four satellites. Given the commercial operation mode and the increasing number of satellites in the SuperView-1 constellations, efficiently scheduling satellite tasks and maximizing the economic benefits are critical concerns for the operation control agencies of SuperView-1.

At present, one of the scheduling models used by SuperView-1 task scheduling subsystem is the task sequencing model, which decodes the task sequences based on the principle of highest imaging quality (i.e., each task is executed at the midpoint of the corresponding VTW). Since the constraints of the task sequences are checked in the process of model decoding, all decoded sequences are feasible. Consequently, the scheduling model is universal, and complex constraints of the satellite tasks can also be addressed during this process.

On the basis of the task sequencing model, the scheduling algorithms used by SuperView-1 task scheduling subsystem are mainly task allocation algorithms and priority sequencing algorithms. For example, based on a series of task and resource attributes such as task profits, duration, and number of VTWs, the algorithms define the comprehensive priority of the tasks, build a sequence of single-satellite tasks according to the principle of descending priority, and allocate the resources to the tasks sequentially. Due to its simple mechanism, the SuperView-1 task scheduling subsystem can quickly provide single-satellite task scheduling results, but it also reveals some problems such as poor task scheduling effectiveness, scarce satellite synergy, and insufficient resource utilization. Meanwhile, when the task sequencing results are further optimized through manual adjustment, it is still necessary to redo the decoding and constraint checking. Due to the fact that the results of manual adjustment cannot get real-time feedback, there is a certain degree of blindness in manual adjustment, which may result in situations where the results of manual adjustment are not as good as the original scheduling. Therefore, although the models and algorithms can deal with complex satellite operation constraints, there is still room to improve the task scheduling efficiency and response speed.

In summary, this section summarizes the fundamentals, universality, and strengths and weaknesses of CPLEX, STK/Scheduler, Europa2, and SuperView-1 task scheduling subsystems. Table 1.3 summarizes the content of this section. Through the summary and analysis in this section, it can be concluded that:

(1) At this stage, there is no universal solver that can meet the needs of satellite task scheduling. Although the above-mentioned universal solvers for satellite task scheduling have achieved good solving results in some application scenarios, the limitations of each solver are also very obvious. For example, it is difficult for CPLEX to deal with large-scale and nonlinear problems; it is difficult for the STK/Scheduler to address complex constraints and make secondary development; Europa2 lacks optimization of the objective function; and the SuperView-1 task scheduling subsystem has limited optimization ability. Even though STK/Scheduler and Europa2 can be applied to simple satellite task scheduling problems, their built-in algorithms are relatively outdated, and they have not been updated for many years. At present, satellite task scheduling problems are rapidly growing in scale, complexity and emergency response frequency. The above-mentioned universal solvers for satellite task scheduling cannot meet the development needs.

(2) It is quite necessary to develop a universal solving technique for satellite task scheduling with independent intellectual property rights. Taking into account factors such as constraint description capability, solving effectiveness, copyright, and service support, CPLEX, STK/Scheduler, and Europa2 have not been applied in the domestic aerospace system. Perhaps the universal solvers for satellite task scheduling, such as STK/Scheduler and Europa2, have new versions incorporating advanced technologies, but they have not been released to the public. Therefore, it needs to develop a universal solving technique for satellite task scheduling that is

Table 1.3: Summary of general-purpose solving techniques for aerospace.

Universal solver	Common scheduling model	Solving algorithm	Solver strength	Solver weakness
CPLEX	Task sequencing model or VTW allocation model represented by linear programming model	Exact solving algorithms such as branch and bound algorithms	1. Concise language and good compatibility 2. Diverse built-in algorithms, reducing the difficulty of algorithm design, while opening up interfaces for algorithm design and improvement 3. Possible to obtain the optimal solution, ensuring optimality even under uncertainty 4. Large user base and continuously updated version	1. Difficult to describe nonlinear constraints and nonlinear profits and to linearize complex constraints 2. Difficult to solve large-scale task scheduling problems 3. Requiring a strong foundation in linear programming theory for further usage 4. Lack of specialization in the aerospace sector
STK/ Scheduler	Task sequencing model	Task sequencing algorithms and conflict-avoidance algorithms	1. Compatible with STK and simple modeling 2. Concise language and good compatibility 3. Remote service function available with STK/ Schedule Online	1. High degree of encapsulation, difficult to describe complex constraints, and difficult for secondary development 2. Lack of dynamic adjustment function 3. Limited algorithm optimization ability and not a continuously updated version
	VTW allocation model	Task allocation algorithms and conflict-avoidance algorithms		

Table 1.3 (continued)

Universal solver	Common scheduling model	Solving algorithm	Solver strength	Solver weakness
Europa2	The description of a planning model is mainly accomplished by defining tags, transactions, objects, classes, timelines, and planning solutions	Constraint propagation algorithms and other constraint programming algorithms	1. State variables and action sequences oriented with intuitive solving process 2. Facilitating the design of dynamic scheduling mechanisms with constraint propagation algorithms	1. Lack of objective function and scheduling optimization mechanism 2. Only applicable to small-scale (single-satellite and single-orbit) task scheduling, and not a continuously updated version
SuperView-1 task scheduling subsystem	Task sequencing model	Task sequencing algorithms based on the principle of highest imaging quality	1. Concise model and algorithms 2. All complex constraints considered	1. Scheduling quality to be improved 2. Mainly manual dynamic adjustment with no real-time feedback of results

suitable for national conditions and has independent intellectual property rights, so as to take the initiative in developing core technology and provide technical support for improving comprehensive strength in satellite control.

(3) **Remote satellite task scheduling service based on cloud computing and high-performance computing is a new concept of application.** STK/Scheduler Online provides a new online service mode for satellite task scheduling. This mode makes user access more convenient and efficient, helps the iterative updating of service functions and the rapid repair of faults, and also significantly improves efficiency with the support of high-performance servers. Therefore, utilizing the current technical advantages of cloud computing and high-performance computing, it is a novel concept of application to provide remote task scheduling services for satellite control agencies to improve the operation efficiency.

1.5 Main work of this book

1.5.1 Study content and organization structure

In response to the satellite control status quo of "one satellite, one system," the contradiction between the needs of conventional control and emergency response, and some countries' technological blockade, after studying the studies and development status of relevant models, algorithms, and universal solving techniques, this book conducts a study on the general-purpose scheduling engine for satellite tasks, focusing on the general-purpose modeling methods for satellite task scheduling, as well as conventional and emergency task scheduling algorithms. The main study content and organizational structure are as follows:

(1) A literature review is conducted in this chapter. In response to the new normal of large-scale and complex satellite task scheduling, as well as the new requirements of flexible networking and rapid response of satellites, this chapter summarizes the development status of satellite task scheduling models, algorithms, and universal solving technologies, reveals the common characteristics of the models, discusses the coding characteristics of the algorithms, and introduces the main functions of some universal solving tools such as CPLEX, STK/Scheduler, Europa2, and SuperView-1 task scheduling subsystem, providing a reference for the study of satellite task scheduling engine framework, models, and algorithms in this book.

(2) The top-level framework of the satellite task scheduling engine is designed in Chapter 2. Aiming at the problems of four main types of satellite task scheduling, namely remote-sensing satellite scheduling, relay satellite scheduling, navigation satellite scheduling, and satellite range scheduling, the scope of this book and the reality-based and application-oriented study principle are clarified, the definition of a satellite task scheduling engine is given, and the top-level framework of the engine for decoupling "model-conventional algorithm-emergency algorithm" is designed to provide a general-purpose and modularized modeling and new solving approach for the satellite task scheduling problems, guiding the specific directions for the study on models and algorithms in this book.

(3) A general-purpose modeling method for satellite task scheduling is proposed in Chapter 3. To solve the problems of insufficient generality of satellite task scheduling models and the satellite control status quo of "one satellite, one system," following the reality-based and application-oriented study principle, the satellite task scheduling problems are systematically and finely described, targeting at the practical requirements of satellite control. A general-purpose modeling method for satellite task scheduling that decouples "decision-constraints-profits" is proposed, and a general-purpose 0–1 mixed integer decision model, constraint model, and profit model are constructed in sequence, giving the key coupling point in the loosely coupled engine framework in this book. The method proposed in this chapter successfully integrates the problems of four main types of satellite task scheduling, including remote-sensing

satellite scheduling, relay satellite scheduling, navigation satellite scheduling, and satellite range scheduling, into a unified modeling system. It opens up a new approach
for mathematical and general-purpose modeling of satellite task scheduling problems
and provides an important general-purpose model support for the satellite task scheduling engine in this book.

(4) An adaptive parallel MA (APMA) for satellite conventional task scheduling is
proposed in Chapter 4. To satisfy the daily and weekly conventional task scheduling
needs of satellite control agencies, a general-purpose APMA is proposed based on the
algorithm's design requirements such as initial solution quality, local optimization
ability, global optimization ability, adaptability, generality, and time complexity. A
heuristic-based fast initial solution construction strategy, a parallel-search-based general-purpose local optimization strategy, a competition-based algorithm and operator
adaptive selection strategy, and a population-evolution-based global optimization
strategy are sequentially designed, forming a synergistic and complementary strategic
advantage. Then, APMA is tested through benchmark tests such as orientation problems and simplified remote-sensing satellite task scheduling problems, proving its
generality and effectiveness in solving standard problems. This provides a general-
purpose and efficient solution for satellite conventional task scheduling problems and
core algorithm for the satellite task scheduling engine.

(5) A distributed dynamic rolling optimization (DDRO) algorithm for satellite
emergency task scheduling is designed in Chapter 5. To meet the emergency task
scheduling needs of satellite control agencies under dynamic influence such as task
additions and deletions and satellite failures, a DDRO algorithm is designed based on
the design requirements of satellite emergency task scheduling algorithms. A task negotiation and allocation strategy based on dynamic contract net, a single-platform
task rescheduling strategy based on rolling time domain, a task rapid insertion strategy based on schedulability prediction, and a real-time conflict-avoidance strategy
based on constraint network are sequentially designed to implement dynamic and
real-time optimization on the current satellite task scheduling scheme, providing general-purpose and flexible solving methods for satellite emergency task scheduling
problems and another algorithm for the satellite task scheduling engine.

(6) The practical application of the satellite task scheduling engine is illustrated in
Chapter 6. Based on the general-purpose modeling method for satellite task scheduling mentioned above, as well as the conventional and emergency task scheduling algorithms such as APMA and DDRO, the satellite task scheduling engine completes the
application experiment by solving the satellite task scheduling problems closely related to engineering practice, such as the SuperView-1 commercial remote-sensing
constellations, the Tianlian-1 relay constellations, the Beidou-3 navigation satellite system, and the satellite range scheduling from the U.S. Air Force. This practices the general-purpose modeling method for satellite task scheduling in this book, tests the conventional and emergency task scheduling algorithms APMA and DDRO, introduces the
relevant application systems, and illustrates the feasibility and application prospects

of the satellite task scheduling engine, providing a practical basis for the engineering application of the engine.

(7) The entire book is summarized in Chapter 7. The book is summarized, conclusions are drawn, and ideas and perspectives for future study work are presented in view of the inadequacies in this book.

1.5.2 Technical route

The flowchart of the technical route of this book is shown in Figure 1.13.

1.5.3 Innovations

Aiming at the current situation and practical needs of satellite control, compared with current studies on satellite task scheduling, this book mainly shows the following four innovations:

(1) **A new method of general-purpose and refined modeling of satellite task scheduling decoupling "decision-constraints-profits" is proposed.** To address the issues of insufficient generality of satellite task scheduling models and current satellite control situation of "one satellite, one system," this book proposes a satellite task scheduling modeling method that decouples "decision-constraints-profits," reveals the common combinational optimization relationship in the problems of remote-sensing satellite scheduling, relay satellite scheduling, navigation satellite scheduling, and satellite range scheduling, and successfully incorporates these four types of satellite task scheduling problems into a scientific and unified modeling system to satisfy new demands for integrated and flexible control of satellites. The generality and feasibility of the modeling method have been practiced in task scheduling experiments for satellites such as SuperView-1, Tianlian-1, and Beidou-3. The method can fundamentally break through the status quo of "one satellite, one system" and provide an important model for the general-purpose application of the satellite task scheduling engine in this book.

(2) **A new adaptive satellite conventional task scheduling algorithm coordinating the "parallel-competition-evolution" strategies is proposed.** To satisfy the daily and weekly conventional task scheduling needs of satellite control agencies, this book proposes an APMA that coordinates multiple strategies. Through the improvement based on traditional MAs through strategies such as "parallel," "competition," and "evolution," the algorithm's optimization efficiency and adaptability are enhanced with a new feature of strategy coordination and complementary advantages, achieving outperformance in a series of experiments. As a general-purpose and efficient technique for solving satellite conventional task scheduling problems, APMA becomes the core algorithm for the application of the satellite task scheduling engine in this book.

Chapter 1 Introduction

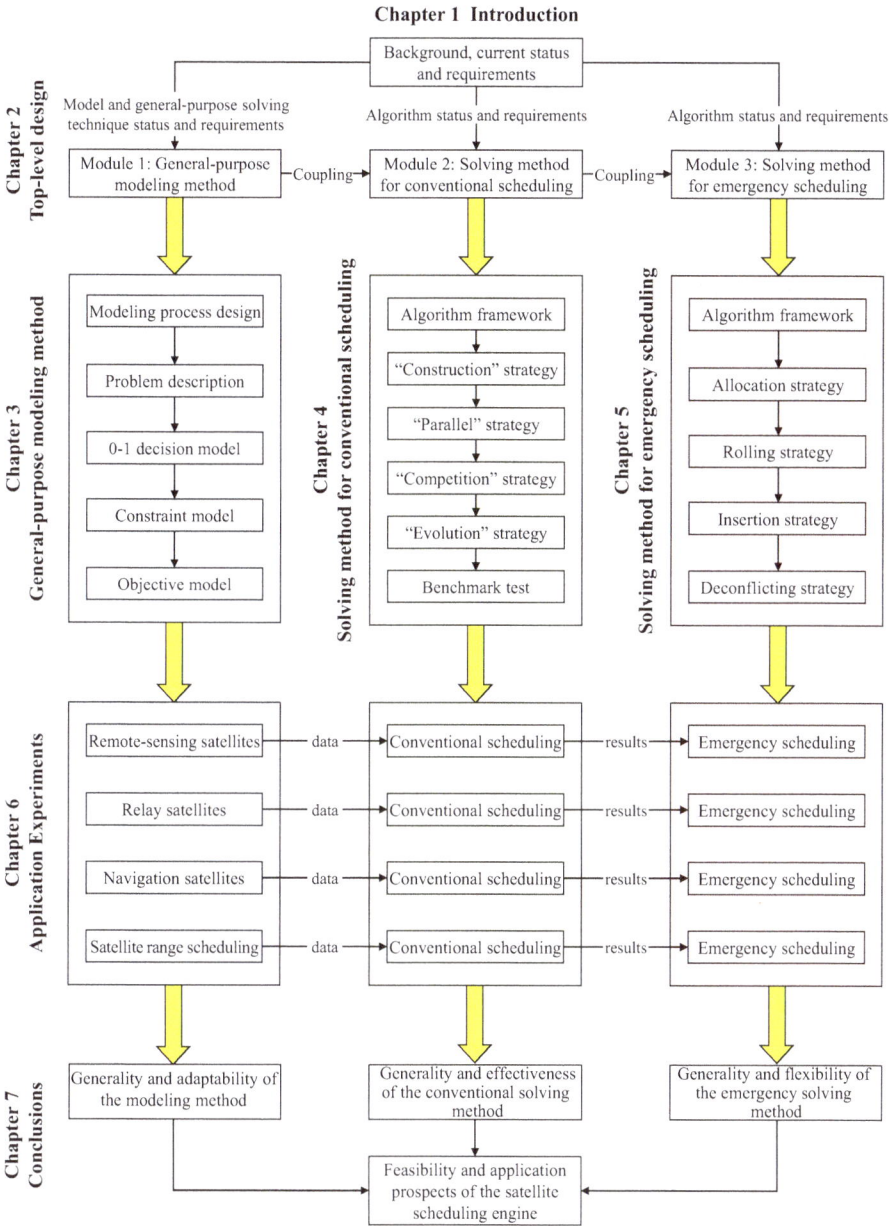

Figure 1.13: Flowchart of the technical route of this book.

(3) A new dynamic response satellite emergency task scheduling algorithm integrating "negotiation-rolling-insertion" mechanisms is proposed. To meet the emergency task scheduling needs under dynamic influence, this book designs a DDRO algorithm. By integrating commonly used emergency scheduling mechanisms such as "negotiation," "rolling," and "insertion," the algorithm dynamically and quickly optimizes the current satellite task scheduling scheme, achieving good application results in a series of experiments. As a general-purpose and flexible technique for solving satellite emergency task scheduling problems, DDRO becomes another important algorithm for the application of the satellite task scheduling engine in this book.

(4) A satellite task scheduling engine decoupling "model-conventional algorithm-emergency algorithm" is designed and implemented. This book designs a satellite task scheduling engine framework that decouples "model-conventional algorithm-emergency algorithm" and implements a satellite task scheduling engine to solve four main satellite task scheduling problems, including remote-sensing satellite scheduling, relay satellite scheduling, navigation satellite scheduling, and satellite range scheduling, while considering the dual needs of satellite conventional and emergency control. The engine can create a good environment for the studies, development, and application of satellite task scheduling models and algorithms, skillfully solve the contradiction between "conventional and emergency tasks," and promote the transformation of study results. It can be an important reference for the development of domestic general-purpose satellite task scheduling systems and related basic software.

Chapter 2
Top-level design of satellite task scheduling engine

The top-level design of the satellite task scheduling engine is an important basis for the study of this book, and it is the fundamental principle of technical study and application. In this regard, this chapter first defines four main types of satellite task scheduling problems, namely remote-sensing satellite scheduling, relay satellite scheduling, navigation satellite scheduling, and satellite range scheduling. The scope of this book and the reality-based and application-oriented study principle are clarified; then the definition of the satellite task scheduling engine is given and the functional requirements of the satellite task scheduling engine are specified; finally, a framework of the engine for decoupling "model-conventional algorithm-emergency algorithm" is designed, and the advantages of the framework are explained, which provides a general-purpose and modularized modeling and solving approach for the satellite task scheduling problems and guides the specific directions for the study on models and algorithms in this book.

2.1 Definitions of satellite task scheduling problems

For the four main satellite task scheduling problems, that is, remote-sensing satellite scheduling, relay satellite scheduling, navigation satellite scheduling, and satellite range scheduling, according to the actual situation of satellite control, this section defines the tasks and resources involved in the satellite task scheduling problems, gives specific examples, and clarifies the scope of this book. In addition, this section also analyzes the characteristics of the problems and emphasizes the reality-based and application-oriented study principle, which provides the basic premise for the study of this book.

2.1.1 Definitions of tasks

2.1.1.1 Description of tasks

In different studies of satellite task scheduling problems, the definitions of tasks are often different; some are macroscopic and some are concrete. At this point, to clarify the scope of this book and highlight the characteristics of combinational optimization in task scheduling problems, this section first gives the definitions of the following two kinds of tasks.

Definition 2.1 (Missions): They are a kind of macroscopic, subjective, and long-term tasks proposed by the satellite user agency to the control agency or spontaneously proposed by the control agency to realize the social, economic, or military value of satellites.

https://doi.org/10.1515/9783111537191-002

Definition 2.2 (Task): They are a kind of concrete, objective, and short-term tasks created by the satellite control agency in the form of a list after planning and breaking down missions, and need to be scheduled for the execution of satellites.

Based on the above definitions, to illustrate the relationship between the two tasks more intuitively, this section provides an example as shown in Figure 2.1. In the figure, a remote-sensing satellite user agency proposes the mission of imaging the 10 famous lakes, and for planning purposes, the control agency breaks down the mission into some tasks: imaging West Lake in Hangzhou, Slender West Lake in Yangzhou, Xuanwu Lake in Nanjing, and so on. Moreover, due to the large area of Taihu Lake in Wuxi, it is necessary to image the lake several times. Accordingly, tasks of imaging areas A, B, and C of Taihu Lake in Wuxi are created. It can be seen that a mission is a collection of tasks; the former is macroscopic and subjective, while the latter is concrete and objective.

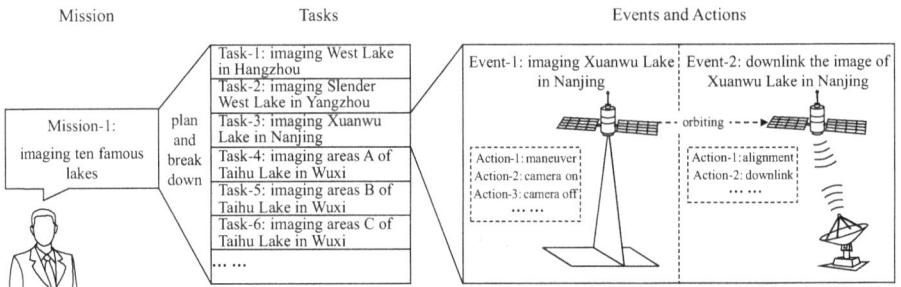

Figure 2.1: Example to show the relationship between missions, tasks, events, and actions of a remote-sensing satellite.

From the perspective of combinational optimization, the tasks in the satellite task scheduling problems should be concrete and objective. Therefore, the satellite task scheduling problems studied in this book only refer to the scheduling problems of tasks rather than those of missions. This book does not examine the planning and breaking down of missions, which usually require specialized studies based on other requirements such as market, business, tactics, and strategies.

2.1.1.2 Task composition

Considering the complexity of satellite task scheduling problems, some researchers have defined more concrete task types, such as meta-task, sub-task, or atomic task, to represent the components or necessary conditions of satellite tasks. In this regard, further definitions are provided in this section to clarify the elements of the tasks in this study:

Definition 2.3 (Event of task): Referred to as an event, it is an uninterrupted, indivisible, and objective activity that a satellite or ground station is required to carry out to accomplish a certain task.

Definition 2.4 (Action): An instantaneous and predefined command-level operation triggered by a satellite or a ground station to accomplish an event.

For example, in Figure 2.1, the task of imaging Xuanwu Lake in Nanjing includes two objective events: 1) imaging Xuanwu Lake in Nanjing by the satellite and 2) downlinking the image of Xuanwu Lake in Nanjing by the satellite to the ground station. Both of these events are necessary conditions for the completion of the task, and none is dispensable.

This would appear to indicate that events have the following characteristics: 1) Objectivity: reflecting the objective prerequisites for the completion of a task; 2) uninterruptedness: quantifying and distinguishing different activities carried out by satellites or ground stations; and 3) non-divisibility: clarifying the smallest unit of task composition and determining the degree of concreteness in the modeling of task scheduling problems. In view of this, how to objectively and accurately describe the tasks and their included events involved in the satellite task scheduling problems is the first priority of this book.

In addition, for the two events in Figure 2.1, the satellite and the ground station need to trigger a series of command-level actions, such as attitude maneuver, camera switching on and off, and antenna alignment. It should be noted that in the process of satellite management and control, the compilation of command-level actions is performed after the task scheduling studied in this book. Meanwhile, the constraints of command-level actions have been incorporated into the constraints of task scheduling in advance. For example, the constraints of transition time between imaging and downlink events of "SuperView-1" actually include the constraints of command-level actions such as switching-on and switching-off times of the camera and memory. Therefore, this book only needs to consider the scheduling problems of tasks and events. There are other specialized subsystems that compile the scheduled tasks and events into command-level actions, and this book does not need to consider the actions.

2.1.1.3 Examples of tasks

Based on the above definitions and descriptions, according to the actual situation of satellite control, Figure 2.2 and Table 2.1 give specific examples of missions, tasks, and events for the four main types of satellite task scheduling problems, namely, remote-sensing satellite scheduling, relay satellite scheduling, navigation satellite scheduling, and satellite range scheduling, and further explain the actual meanings of various types of satellite tasks and events, defining the study scope of satellite task scheduling problems in this book.

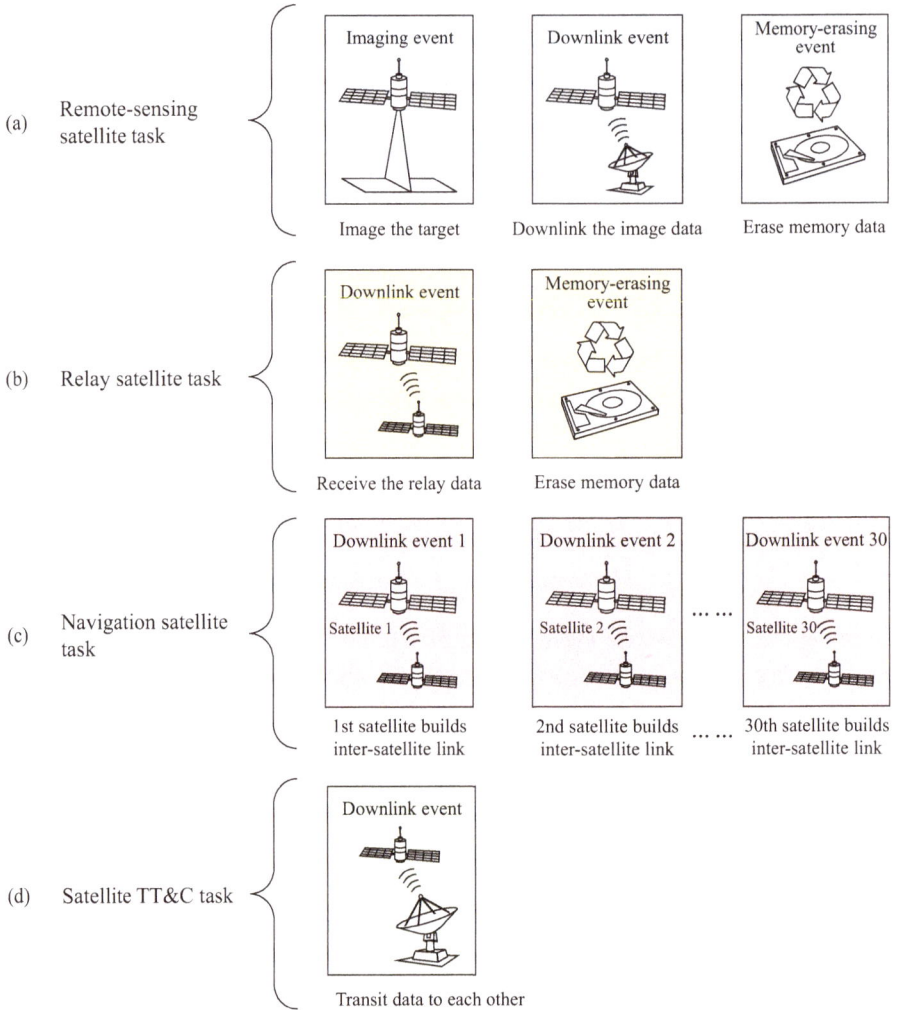

Figure 2.2: Examples of tasks in various satellite task scheduling problems: (a) remote-sensing satellite tasks; (b) relay satellite tasks; (c) navigation satellite tasks; and (d) satellite range scheduling tasks.

2.1.1.3.1 Remote-sensing satellite tasks

In remote-sensing satellite task scheduling problems, for "the mission of imaging the 10 famous lakes" in Table 2.1, the ith task to be scheduled is "to image the ith target." Further, this task contains an imaging event, a downlink event, and a memory-erasing event, whose actual meanings are that the satellite images the target, the satellite downlinks the target image data to the ground station, and the satellite erases memory data, respectively, as shown in Figure 2.2(a). It should be noted that most of the current remote-sensing satellites must periodically empty their on-board memory data to realize the reuse of on-board memory resources due to the influence of on-board memory technology, command templates, and operation requirements, and this event is called a memory-erasing event. However, only imaging events are usually considered in related studies, and a few studies have considered both imaging and downlink events, while very few studies have considered the memory-erasing event. Although these studies have reduced the complexity of remote-sensing satellite task scheduling and achieved rich theoretical study results, the study results are often "derailed" from the actual situation and difficult to apply in engineering because of the neglect of the downlink and memory-erasing events and many related constraints. Therefore, to objectively and completely describe remote-sensing satellite task scheduling problems and provide practical and usable task scheduling modeling and solving methods for satellite control, this book includes memory-erasing events as one of the events of remote-sensing satellite tasks.

Table 2.1: Examples of missions, tasks, and events in satellite task scheduling problems.

Satellite type	Mission	ith task	Events and their actual meanings
Remote-sensing satellite	Imaging the 10 famous lakes	Imaging the ith target	Imaging: the satellite images the target
			Downlink: the satellite downlinks the target image data to the ground station
			Memory-erasing: the satellite erases memory data
Relay satellite	Providing relay service	Receiving the ith relay request	Downlink: the satellite receives relay data
			Memory-erasing: the satellite erases memory data
Navigation satellite	Reducing system delay and ensuring navigation accuracy.	Constructing the inter-satellite link network in the ith navigation timeslot	Downlink: satellite 01 builds an inter-satellite link with other satellites
			Downlink: satellite 02 establishes an inter-satellite link with other satellites
			Downlink: . . .
			Downlink: satellite 30 builds an inter-satellite link with other satellites
Satellite TT&C	Track, telemetry, and command of in-orbit satellites	Executing the ith TT&C request within a certain satellite range	Downlink: the ground station and satellite transmit TT&C data to each other

2.1.1.3.2 Relay satellite tasks

In relay satellite task scheduling problems, for "the mission of providing relay service" in Table 2.1, the ith task to be scheduled is "to receive the ith relay request" (usually from other satellites). Further, this task contains a downlink event and a memory-erasing event, whose actual meanings are that the satellite receives the relay data and the satellite memory-erasing data, respectively, as shown in Figure 2.2(b). It should be noted that the relay satellites involved in this book mainly refer to high-orbit (geostationary orbit) relay satellites, which can basically realize real-time mutual downlink with the ground, so the satellite memory-erasing events are not considered in the following for the time being. However, the modeling method and model interface of the events are retained to provide support for future studies of low-orbit relay scheduling, lunar relay scheduling, deep-space relay scheduling, and other related tasks.

2.1.1.3.3 Navigation satellite tasks

In navigation satellite task scheduling problems, for "the mission of reducing system delay and ensuring navigation accuracy," this section first introduces the current time division system of the navigation system, as shown in Figure 2.3. As shown in the figure, the time axis is divided into consecutive superframes (1 min per superframe), and then 1 superframe is divided into 20 timeslots (3 s per timeslot). The navigation system maintains an inter-satellite link network structure in each timeslot. As a result, every time the timeslot changes, the inter-satellite link of the navigation system switches once, forming a dynamic network structure that evolves over timeslots. It should be noted that the satellite control agency specifies one superframe as one cycle of the navigation satellite task scheduling problem. In other words, a navigation satellite task scheduling problem is an optimization problem of the inter-satellite link network within 20 timeslots.

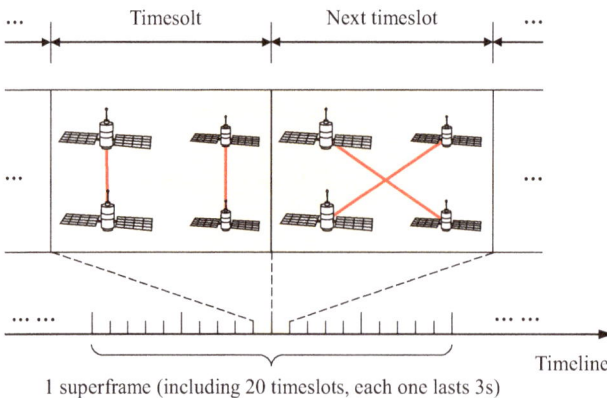

Figure 2.3: Illustration of the time division system of the current navigation system and examples of superframes and timeslots.

Under the above-mentioned time division system and satellite control requirements, the ith task to be scheduled in the navigation satellite task scheduling problems is "to construct the inter-satellite link network in the ith navigation timeslot." Further, this task contains the event that each satellite in the navigation system establishes an inter-satellite link with other satellites, as shown in Figure 2.2(c). Taking the Beidou-3 navigation system as an example, the system has 30 navigation satellites, so a navigation satellite task scheduling problem contains 20 tasks (timeslots), and each task contains 30 downlink events.

In addition to the above-mentioned inter-satellite links of navigation satellite task scheduling problems, the navigation system also involves another kind of task scheduling problem, that is, satellite-ground time synchronization. Since such kind of problem can be regarded as a type of (high-orbit) satellite range scheduling problem, it will not be studied separately in this book.

2.1.1.3.4 Satellite range scheduling

In satellite range scheduling problems, for "the mission of track, telemetry, and command of in-orbit satellite," the ith task is "to execute the ith request of range scheduling." The task only contains one downlink event; that is, the ground stations and the satellite transmit TT&C data to each other, as shown in Figure 2.2(d). Here, the stations not only include traditional ground-based ground stations but also include space-based ground stations constructed with relay satellites. Therefore, the satellite range scheduling problems and the relay satellite task scheduling problems have greater similarity. The main difference between the two is that the former mainly involves ground-based ground stations and has fewer constraints, while the latter involves relay satellites and should consider satellite-related constraints.

In summary, based on the actual situations of satellite control, after defining tasks and events, this section elaborates on the specific meanings of the tasks and events in the four major types of satellite task scheduling problems, that is, remote-sensing satellite scheduling, relay satellite scheduling, navigation satellite scheduling, and satellite range scheduling, defines the study scope of this book, and provides the prerequisite for the later study on the model and algorithm of satellite task scheduling problems.

2.1.2 Definitions of resources

Based on the above definitions and examples of tasks, this section further defines the types of resources involved in the study of satellite task scheduling problems in this book, as shown in Table 2.2.

As shown in the table, the actual satellite control involves many resources and is a kind of complex combinational optimization problem. In this regard, to describe the satellite task scheduling problems as objectively and systematically as possible, practically solve the actual needs of the control agency, and overcome the problem that traditional studies cannot be utilized in actual application due to problem oversimplification, the satellite task scheduling problems studied in this book consider the resources of platform, payload/equipment, and windows in Table 2.2. In short, this study follows the "reality-based and application-oriented" principle, aiming to solve the real needs of the control agency and provide practical satellite task scheduling technology.

Table 2.2: Types and examples of resources involved in the actual satellite control process.

Resource category	Resource subcategory	Example	Whether considered in this book
Platform resources	Remote-sensing satellites	01–04 satellites of "SuperView-1"	Considered
	Relay satellites	01–04 satellites of "TianLian-1"	
	Navigation satellites	01–30 satellites of "BDS-3"	
	Ground station	Beijing Station, Kashi Station, and Jiamusi Station	
Payload/equipment resources	Satellite camera	0.5 m resolution CCD optical remote-sensing camera	Considered
	Satellite memory	2,000 G on-board memory	
	Satellite battery	On-board solar battery	
	Satellite antenna	X-band downlink antenna	
	Ground station antenna	X-band TT&C antenna	
Window resources	Time windows	Time windows for imaging and downlink	Considered
	Satellite orbits	01–30 orbits	
	Date	April 12, 2020	
Human and post resources	Posts of ground stations	Day shift, night shift, image processing post, logistics	Not considered
	Ground station staff	TT&C engineer, image processing specialist, and physicians	
	Ground station vehicles	TT&C vehicles, patrol vehicles, and shuttle buses	

On the other hand, the human and post resources involved in the actual satellite control, as well as the related scheduling problems such as matching and scheduling of humans, posts, and vehicles, are not within the scope of this study and are not considered in this book. In other words, this book assumes that there are sufficient human and post resources for satellite task scheduling. This assumption is also in line with the general practice in the current satellite control process.

2.1.3 Characteristics of the problems

(1) Basic characteristics of combinational optimization problems. From the above definitions and examples, it can be seen that the satellite task scheduling problems studied in this book belong to a kind of combinational optimization problems and have the basic characteristics of route planning, resource scheduling and other combinational optimization problems. For example, there is a combinational optimization relationship between the tasks and the resources with certain constraints, clear optimization objectives, and NP-hardness of the problem [2]. In view of this, the study on satellite task scheduling problems in this book also follows the basic process of combinational optimization problem study, that is, constructing a task scheduling model, designing an optimization algorithm, and testing the generality and effectiveness of the model and algorithm. Meanwhile, in the long-term process of satellite control, the satellite task scheduling problems are changing, which requires dynamic and emergent optimization, and have the relevant characteristics of dynamic and emergent combinational optimization problems. However, in addition to the basic characteristics of combinational optimization problems, the satellite task scheduling problems studied in this book have two important characteristics that are not present in other combinational optimization problems, as described below, which increases the complexity of satellite task scheduling problems and has an important impact on the problem modeling methodology and algorithm design.

　　(2) Few simplifications, few uncertain factors, and high requirements for refinement of the optimization scheme. In many traditional combinational optimization problems, there exists a certain degree of problem simplification, which does not require a high degree of refinement of the optimization scheme. For example, in vehicle routing problem, by assuming ideal road conditions and vehicle driving conditions, it will not consider vehicle acceleration, braking, traffic lights, and other too fine factors in the process of scheme optimization. Practice has also shown that such a refined scheme optimization has no practical significance due to the influence of uncertainty factors such as road conditions and vehicle driving dynamics. However, for satellite task scheduling problems, satellites are a class of precision space instruments, controlled by a set of strict internal systems, and maintain close and frequent communication with the ground control center. In other words, the satellites' working status is closely monitored, and there are relatively few uncertainty factors. In this case, the optimization scheme of satellite task scheduling must reach the degree of refinement of its internal system and ground control to meet the fine optimization needs and constraints. For example, if its internal system provides that the minimum interval time between satellite imaging and downlink is 100 s, then the optimization scheme must explicitly give the specific time of satellite imaging, downlink, corresponding payload transition, and attitude conversion while meeting the interval time constraint of 100 s. Otherwise, the internal system of the satellites will trigger the self-protection procedure, which will make the task scheme impossible to be implemented and even cause irreversible failures.

This characteristic of the satellite task scheduling problems has an important impact on its problem modeling and algorithm design. Many idealized and simplified traditional combinational optimization models and algorithms are often insufficiently refined to meet the refined satellite control requirements and are not good enough to be applied to actual satellite operation systems. As a result, constructing satellite task scheduling optimization models more precisely and accurately, designing practical optimization algorithms, restoring the original characteristics of satellite task scheduling problems as much as possible, and achieving the objective requirements of the reality-based and application-oriented principle have become important contents of this book's study on satellite task scheduling modeling methodology.

(3) Coexistence of the objectivity of computer optimization and the subjectivity of human decision. Usually, when studying the modeling and algorithm design methods for combinational optimization problems, the optimization process is realized by computers instead of humans to achieve the optimization goal. The objectivity of computer optimization is presented in the case of given optimization objectives, and satellite task scheduling problems are no exception. However, the process of abstracting the complex satellite task scheduling problems into a computer-recognizable mathematical model is somewhat subjective and fuzzy, and there may be some discrepancies with the complex needs of the satellite control agency as well as the subjective habits and experiences of the control personnel. On the other hand, the process of satellite control is usually accompanied by repeated communication and coordination with the user agency. For example, after a satellite has begun to implement the task scheduling program, the user agency requests the addition (or deletion) of a certain task, and the control agency has to make corresponding adjustments. It can be seen that the subjective and personalized task requirements sometimes raised by the user agency often require human intervention by the control personnel. Therefore, the satellite task scheduling problems exhibit the coexistence of the objectivity of computer optimization and the subjectivity of human optimization. How to properly deal with this contradiction, provide the required optimization models and algorithms, and satisfy the practical requirements of the control agency have become another important part of the study on the modeling and solving methods of the satellite task scheduling problems in this book.

In summary, according to the actual situation of satellite control, this section defines and describes the tasks and resources involved in the four main satellite task scheduling problems studied in this book, analyzes the characteristics of the problems, clarifies the study scope of this book, and emphasizes the reality-based and application-oriented study principle, which provides the basic premise for the study work of this book.

2.2 Functional positioning of the satellite task scheduling engine

Based on the above definitions of the problems, this section clearly provides the definition of the satellite task scheduling engine, analyzes the characteristics and functional requirements of the engine, and clarifies the basic logic of the study on the combinational optimization model and related algorithms in this book, which provides a reference basis for the framework design of the satellite task scheduling engine in this chapter and the study work of this book.

2.2.1 Basic definition

In the computer field, an engine refers to the core component of a computer platform for developing a program or system. By using an engine, developers can quickly build and lay out the functionality required by a program or system. Common engines include game engines, search engines, simulation engines, and antivirus engines [243].

Scheduling engine, also known as a scheduler, refers to a generic solver for solving certain types of scheduling problems, such as the distributed task scheduling engine TBSchedule of Alibaba and the high-performance task scheduling engine Volcano of Huawei. Moreover, there are many general-purpose mathematical planning and combinational optimization solvers that can be used to solve scheduling problems in related fields, such as the mathematical planning solver CPLEX of IBM, the mathematical planning solver Gurobi, the operations study optimization solver OR-Tools of Google, the combinational optimization solver OptaPlanner of Red Hat, and the mathematical optimization solver COPT of Cardinal Operations. Since there is no authoritative definition of a "satellite task scheduling engine," this section draws on Orbit Logic's definition of its general-purpose satellite scheduling software STK/Scheduler to define the satellite task scheduling engine in this book.

Orbit Logic defines its general-purpose satellite scheduling software, STK/Scheduler, as a solver "for any type of space system scheduling problem" and attributes the software's strengths to its "unique approach to task and resource definition along with the powerful algorithm implementation" [227, 228]. It is thus clear that the core technology of STK/Scheduler is mainly embodied in its general-purpose task and resource modeling approach for solving various kinds of scheduling problems and the corresponding optimization algorithms. In view of this, this book defines the satellite task scheduling engine as follows:

Definition 2.5 (Satellite task scheduler/scheduling engine): A general-purpose solver that supports the development of scheduling systems by solving any type of satellite task scheduling problem through general-purpose combinational optimization models and algorithms.

As can be seen from the definition, the satellite task scheduling engine mainly has the following three features: 1) Generality: the engine can solve any type of satellite task scheduling problem to support the development of any type of satellite scheduling system; 2) Combinational optimization: the engine takes combinational optimization models and algorithms as tools to solve scheduling problems; 3) Practicality: the overall practicality of the engine's model and algorithms reaches the basic level of a solver for fundamental application.

Based on the above definition and features, this chapter is going to further analyze the functional requirements of the satellite task scheduling engine, design the framework of the satellite task scheduling engine, and consequently study the general-purpose modeling method of satellite task scheduling (i.e., construction of the combinational optimization model), as well as the conventional and emergency scheduling algorithms, so as to provide the required technology for the satellite task scheduling engine.

2.2.2 Functional requirements

This book aims at solving practical problems for the actual needs of the satellite control agency. According to the above definition and characteristics, the functional requirements of the satellite task scheduling engine can be summarized as follows:

(1) Requirements of optimization. These are fundamental to the satellite task scheduling engine and fundamentally reflect the fact that satellite task scheduling problems are a class of combinational optimization problems. Therefore, the related contents of the satellite task scheduling engine should serve the optimization requirements of the satellite control agency and meet the fundamental needs of maximizing the social, economic, and military benefits of satellites. In other words, the study of the satellite task scheduling engine, its model, and algorithms in this book is a study of combinational optimization technology.

(2) Requirements of general-purpose application of the model and algorithms. These are fundamental to the satellite scheduling engine. The introduction to this book has stated that the models and algorithms for satellite task scheduling are not sufficiently generalized, resulting in the current "one satellite, one system" for the development and use of satellite control systems. Therefore, when designing the framework of the satellite task scheduling engine, the generality of the engine's model and algorithms should be fully considered to overcome the shortcomings of the current "one satellite, one system" for satellite control. Specifically, the generality of the model and algorithms includes the following three aspects: 1) model generality, that is, the model needs to meet the modeling requirements of remote-sensing satellite scheduling, relay satellite scheduling, navigation satellite scheduling, satellite range scheduling, and other types of satellite task scheduling problems; 2) algorithm generality, that is, on the basis of model generality, the algorithms need to meet the solving requirements of any type of satellite task scheduling problems; 3) generality of the interface between the model and algorithms, that is, the data and process interface of the model and algorithms need to work in a "loosely coupled" way, and eliminate the limitation that "an algorithm is designed only for a model, and a model can only be solved by one algorithm" in many related studies and in-service systems. As a result, the satellite task scheduling engine should solve the above technical difficulties of models and algorithms for the general-purpose application of the engine.

(3) **Dual application requirements of conventional scheduling and emergency scheduling.** On the one hand, the satellite control agency needs to deal with conventional task scheduling requirements on a daily and weekly basis, optimize the task scheduling plan within a certain period of time, and formulate daily and weekly task scheduling plans. On the other hand, in the long-term process of satellite control, due to the frequent occurrence of dynamic events such as adding or deleting tasks, and satellite failures, the satellite control agency has to repeatedly optimize the task scheduling scheme and formulate an emergency task plan in a very short period of time. As a result, the satellite task scheduling engine also needs to be applicable in both conventional and emergency scheduling. It is worth noting that the complexity of satellite conventional task scheduling problems often conflicts with the timeliness of emergency task scheduling, that is, the "conventional-emergency" contradiction, which is also very common in the studies and application of many combinational optimization problems. In this case, the satellite task scheduling engine framework needs to take into account both the need to meet the overall optimization requirements of the complex satellite conventional task scheduling scenarios and the need to meet the rapid response requirements of the satellite emergency task scheduling scenarios so as to make a good balance between the "conventional-emergency" application requirements. In addition, the satellite payload technology has undergone profound changes, with more and more satellites equipped with hardware capable of autonomous computation and response, and autonomous satellite task scheduling within a certain range has become particularly important in the context of the dynamic and changing space environment and the need for rapid response.

(4) **Management requirements of flexible model adjustment in the long-term satellite control process.** The design life of satellites ranges from several years to decades, while the actual service life is often longer than the design life. Satellite missions, constraints, and networking with other satellites are not static in the long-term process of satellite control. At the same time, the relevant control requirements are changed in special cases. For example, to protect the satellite and its payloads, the

maximum number of single-orbit imaging of a satellite is 10 times, but when responding to major emergencies such as natural disasters and military operations, the maximum number of imaging can be relaxed to 20 times. Meanwhile, the satellite forms a temporary constellation with several other satellites to implement joint task scheduling. Therefore, to meet the flexible and autonomous control needs of the satellite control agency, the satellite task scheduling engine should enable the agency to adjust and manage the task scheduling model and constraints to a certain extent. Meanwhile, under the background of integrated satellite system control, good model and system generality of the satellite task scheduling engine are also conducive to flexible model adjustment and management in the long-term satellite control process. In addition, the satellite task scheduling system can also be iterated and upgraded through this process to further create a good model and system-compatible environment, resulting in a new paradigm of "incremental" system development and utilization.

(5) **Subjective application requirements of flexible manual adjustment of task scheduling results.** As mentioned above, there is a contradiction between the objectivity of computer optimization and the subjectivity of human optimization in satellite task scheduling. Although most satellite control schemes can be generated by a computer (scheduling system), the control agency always requires the system to retain the function of manually adjusting the task scheduling results. It is worth noting that after the task scheduling scheme is adjusted manually, the original scheme may violate the constraints, leading to a decrease in profit. Therefore, to quickly implement the overall adjustment of the scheme to ensure the normal operation of the satellites, this requirement also includes the emergency scheduling and rapid scheduling requirements mentioned above, which raises higher requirements for the framework and algorithm design of the satellite task scheduling engine.

It is not difficult to find that the above-mentioned functional requirements of the satellite task scheduling engine always revolve around the satellite task scheduling model, as well as optimization algorithms under different scheduling requirements, which essentially have not been separated from the framework of combinational optimization technology. Therefore, in the following framework design of the satellite task scheduling engine and subsequent study, this book will still follow the general logic of combinational optimization model and algorithm study, and carry out the corresponding design and study work around the goals of general-purpose design and application.

To sum up, this section clearly provides the definition of the satellite task scheduling engine, analyzes the characteristics and functional requirements of the engine, and clarifies the basic logic of the study on the combinational optimization model and related algorithms in this book. As a result, a reference basis is provided for the framework design of the satellite task scheduling engine in the next section and the study of this book.

2.3 Framework design of the satellite task scheduling engine

After defining the problems and clarifying the functional requirements in the previous section, this section first presents the design ideas of the satellite task scheduling engine framework, and then designs a top-level framework decoupling "model-conventional algorithm-emergency algorithm" for modeling and solving satellite task scheduling problems. It also explains the advantages and feasibility of the framework, providing a new way of thinking for general-purpose and modularized satellite task scheduling problem modeling and solving, and pointing out the specific direction for the study of the model and algorithms in this book.

2.3.1 Design ideas

According to the functional requirements of the satellite task scheduling engine and the basic "reality-based and application-oriented" principle, this section first presents the design ideas of the satellite task scheduling engine framework, as shown in Table 2.3.

Table 2.3: Functional requirements and framework design ideas of satellite task scheduling engine.

No.	Requirements	Design idea
1	Requirements of optimization	1. Abstracting the problems as a combinational optimization problem by satellite task scheduling modeling methods
		2. Meeting the requirements of combinational optimization by satellite task scheduling optimization methods
2	Requirements for the general-purpose application of the model and algorithms	1. Constructing a general-purpose modeling and solving framework decoupling "model-algorithms"
		2. Designing a general-purpose modeling method for remote-sensing satellite task scheduling, relay satellite task scheduling, navigation satellite task scheduling, and satellite range scheduling, so as to improve the generality of the model
		3. Designing general-purpose model-solving algorithms to improve the generality of the algorithms

Table 2.3 (continued)

No.	Requirements	Design idea
3	Dual requirements of conventional scheduling and emergency scheduling	1. Further decoupling of "conventional algorithm-emergency algorithm" in the framework 2. Designing high-performance optimization algorithms for conventional satellite task scheduling problems 3. Designing flexible and effective dynamic optimization algorithms for satellite emergency task scheduling
4	Requirements of flexible model adjustment in the long-term satellite control process	1. Realizing the loose coupling of "decision-constraints-profits" on the basis of general-purpose modeling 2. Designing constraint templates and opening functional interfaces further enhance the flexibility of the model.
5	Requirements of flexible manual adjustment of task scheduling results	1. Further opening the functional interfaces for model adjustment 2. Supporting the rapid implementation of secondary optimization after adjustment through the emergency scheduling algorithm

The above ideas can be summarized as follows: 1) At the framework level, the satellite task scheduling engine needs to have certain flexibility and decouple the model and algorithms; 2) at the model level, the satellite task scheduling model needs to be universal and flexible, and decouple the decision, constraints, and profits; 3) at the algorithm level, it is necessary to design satellite conventional and emergency task scheduling algorithms, respectively. The decoupling of "model-conventional algorithm-emergency algorithm" is thus realized to meet the functional requirements of the engine.

2.3.2 Engine framework

Based on the design ideas in the previous section, this section presents the engine framework decoupling the "model-conventional algorithm-emergency algorithm," as shown in Figure 2.4. In this figure, the satellite task scheduling engine framework contains: 1) a general-purpose modeling method for satellite task scheduling problems; 2) a general-purpose solving method for satellite conventional task scheduling; and 3) three "loosely coupled" modules, such as a general-purpose solving method for satellite emergency task scheduling. The main contents of each module are as follows:

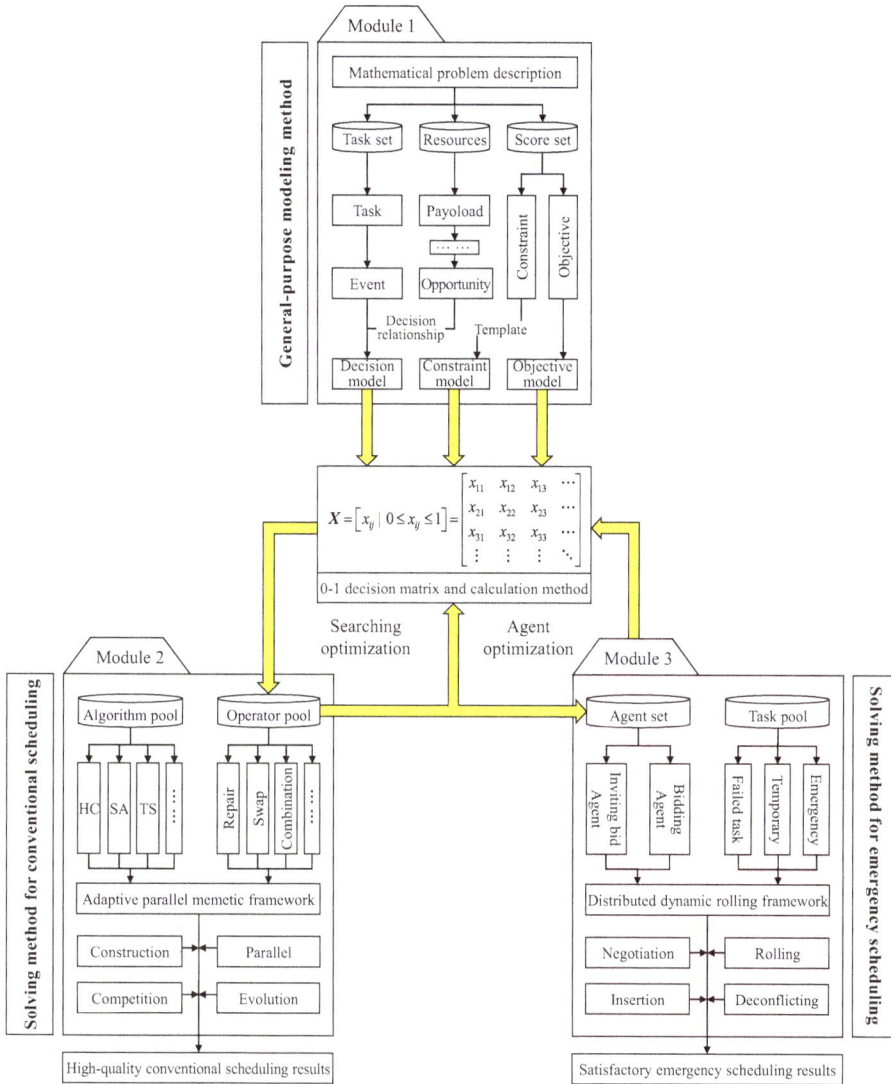

Figure 2.4: Satellite task scheduling engine framework designed in this book.

(1) Module 1: General-purpose modeling method for satellite task scheduling problems The general-purpose modeling method of satellite task scheduling is the first module of the satellite task scheduling engine. The module contains the following main contents: 1) Mathematical description of problems: describing the task set, resource set, and score set (including constraint set and profit set) of task scheduling problems; 2) construction of decision model: organizing the decision-relation between "task set-resource set" (i.e., "event-event executable opportunity") in satellite task scheduling problems, and constructing a general-purpose 0–1 mixed integer decision model, which provides a general-purpose model interface for the subsequent satellite task scheduling algorithms; 3) construction of constraint model: providing a constraint value calculation interface for subsequent satellite task scheduling algorithms; and 4) construction of profit model: providing a profit calculation interface for subsequent satellite task scheduling algorithms. The details of this module will be described in Chapter 3. Here, the general-purpose 0–1 mixed integer decision model and the calculation methods of constraints and profits are the "key coupling points" between the model and the algorithms in this framework, forming a "loosely coupled" data and process connection method. Moreover, the module also realizes the decoupling of "decision-constraints-profits" in the process of constructing the satellite task scheduling model, and provides functional interfaces for decision variable modification, constraint calculation, and profit calculation for subsequent algorithms, which meets the needs of universal construction and flexible management of satellite task scheduling models and provides a general-purpose and flexible model for subsequent algorithms.

(2) **Module 2: General-purpose solving method for satellite conventional task scheduling** The general-purpose solving method for satellite conventional task scheduling is the first problem-solving module of the satellite task scheduling engine. Based on the general-purpose modeling method for satellite task scheduling problems and the interfaces provided by the model, this module iteratively updates the above-mentioned 0–1 decision matrix through general-purpose algorithms and operators to optimize the satellite task scheduling scheme. This module builds a general-purpose adaptive parallel memetic algorithm framework, integrates a series of optimization algorithm strategies such as construction, parallelism, competition, and evolution, provides a general-purpose and efficient solving method for satellite conventional task scheduling problems, and outputs high-quality scheduling schemes to satisfy the basic optimization needs of the control agency for daily and weekly satellite task scheduling plans under conventional circumstances. The details of this module will be described in Chapter 4.

(3) **Module 3: General-purpose solving method for satellite emergency task scheduling** The general-purpose solving method for satellite emergency task scheduling is the second problem-solving module of the satellite task scheduling engine, and the module is specially designed to solve satellite emergency task scheduling problems. Although module 2 can solve satellite conventional task scheduling problems, in the long-term satellite control process, dynamic events such as adding or deleting tasks and satellite failures occur frequently, and the solving efficiency of module 2 is insufficient. As a result, for the satellite emergency task scheduling problem, the module provides feasible scheduling schemes for emergency tasks through real-time emergency optimization of the current task scheduling scheme by the agent. This module builds a distributed real-time rolling general-purpose algorithm framework, integrates emergency optimization algorithm strategies such as negotiation, rolling, insertion, and conflict avoidance, provides a general-purpose and flexible solving method for satellite emer-

gency task scheduling problems, and outputs feasible and satisfactory emergency scheduling schemes to satisfy the needs of the control agency for emergency optimization of the scheduling scheme under special circumstances and the needs of some satellites for self-optimization of the scheduling scheme during autonomous operation. The details of this module will be described in Chapter 5.

2.3.3 Advantage analysis

Based on the process presented in the previous section for solving satellite conventional and emergency task scheduling problems, this section summarizes the three advantages of the satellite task scheduling engine framework as follows:

(1) At the overall framework level, the decoupling of "model-algorithms" is realized, which gets rid of the limitation of the design idea that "an algorithm is designed only for a model" and contributes to the flexible management and expansion of the model and algorithms. In this framework, modeling and optimization are clearly divided into two independent stages, and decision matrix and calculation methods of constraints and profits are used as loosely coupled interfaces between the model and algorithms, which gets rid of the limitation that "an algorithm is designed only for a model, and a model can only be solved by one algorithm" in many related studies and in-service systems. In this way, during the long-term application of this framework, the control agency can flexibly and spontaneously select, modify, and update the model or algorithms, realizing flexible management and flexible expansion to meet the needs of application, maintenance, and research and development. This is of great positive significance for improving the long-term management and control level of satellite systems, promoting the transformation of study results of satellite task scheduling algorithms, and the "incremental" research and development of in-service systems.

(2) At the model level, the decoupling of "decision-constraints-profits" is realized, which highlights the flexibility and practicality of model management, especially constraint management. Satellites have a long in-orbit operation cycle, and with the passage of time or under the background of major requirements, the missions, constraints, and network organization of the satellites may change. In this regard, based on the decoupling of the model and algorithms, this framework further breaks down the general-purpose modeling method into decision modeling, constraints, and profit modeling, and realizes the decoupling of "decision-constraints-profits," which highlights the flexibility and practicality of model management, especially constraint management, and meets the actual needs of the control agency. It is more conducive to the long-term control of satellite systems and the "incremental" development of related technologies.

(3) At the algorithmic level, the unity of "conventional-emergency" optimization is realized, taking into account the design purpose and application scenarios of conventional control and emergency response, and guaranteeing the feasibility and effectiveness in different scenarios. Satellite task scheduling involves both conventional scheduling and emergency scheduling, with the former requiring the fullest possible high-quality optimization, and the latter requiring fast and flexible instant optimization. In this regard, this book does not attempt to provide a "one-trick pony" solution but gives a reasonable optimization mechanism in which conventional scheduling works first and emergency scheduling works later to directly solve the contradiction between the needs of satellite conventional and emergency task scheduling, so as to realize that the organic combination and unity of "conventional-emergency" give full play to the application advantages of the satellite task scheduling algorithms according to circumstances, and meet the dual optimization requirements of satellite conventional and emergency task scheduling, which is of great positive significance to enhance the level of satellite system control under different scenarios and to promote the transformation of study results of satellite task scheduling algorithms.

In summary, this section designs a satellite task scheduling engine framework decoupling the "model-conventional algorithm-emergency algorithm" and explains the advantages and feasibility of the framework. Therefore, a general-purpose and modularized modeling and solving method is provided for various satellite task scheduling problems, and the direction for the study of the model and algorithms in this book is determined.

2.4 Chapter summary

Aiming at the four main types of satellite task scheduling problems, namely, remote-sensing satellite scheduling, relay satellite scheduling, navigation satellite scheduling, and satellite range scheduling, this chapter completes the top-level design of the satellite task scheduling engine, which includes the following main contents:

(1) The tasks and resources involved in the various satellite task scheduling problems are defined, the study scope and problem characteristics are clarified, and the "reality-based and application-oriented" study principle is highlighted, which provides the basic premise for subsequent study in this book.

(2) The definition of the satellite task scheduling engine is clearly given, the functional requirements of the satellite task scheduling engine are listed, and the basic logic of the combinational optimization model and algorithm study in this book is revealed, which provides a reference basis for the design of the satellite task scheduling modeling and solving framework in the next step.

(3) A satellite task scheduling engine framework decoupling "model-conventional algorithm-emergency algorithm" is designed, and the advantages and feasibility of

the framework are explained, which provides a general-purpose and modularized modeling and solving approach for satellite task scheduling problems and guides the specific directions for the study on models and algorithms in this book.

The framework of the satellite task scheduling engine designed in this chapter breaks through the traditional design idea of "an algorithm is designed only for a model, and a model can only be solved by one algorithm," and realizes the combination of "conventional-emergency" solving methods, which provides a brand-new and feasible application idea for deploying the satellite task scheduling engine under the current rapid development of space systems and utilizing the advantages of satellite task scheduling technology according to circumstances.

Chapter 3
General-purpose modeling method for satellite task scheduling problems

To address the issues of insufficient generality and flexibility of satellite task schedul-ing model and current management situation of "one satellite, one system," following the reality-based and application-oriented principle, this chapter puts forward a gen-eral-purpose modeling method of satellite task scheduling that decouples the decision, constraints, and profits, and can properly incorporate modeling of the aforemen-tioned four kinds of satellite task scheduling problems into a general-purpose man-ner. This manner opens up a new idea of general-purpose and refined modeling of satellite task scheduling problems, supporting the required general-purpose models for the satellite task scheduling engine presented in this book.

In this chapter, firstly, the satellite task scheduling problem is systematically and hierarchically generalized into a task set, a resource set, a score set, and a decision matrix. Secondly, the decision relationship between the task set and the resource set, specifically between the satellite event and its executable opportunity, is creatively explained. With this relation, the general-purpose 0–1 mixed integer decision model is then built with examples based on four satellite scheduling problems including re-mote-sensing satellite scheduling, relay satellite scheduling, navigation satellite sched-uling, and satellite range scheduling to play the key coupling points in the loosely cou-pled engine framework in this book. Thirdly, the main constraints of the satellite task scheduling problems are analyzed and summarized, and a general-purpose constraint template is designed through the constraint objects, constraint thresholds and con-straint relationships, and a constraint value calculation method based on the con-straint network is designed. Finally, the objective functions of various satellite task scheduling problems is formulated to illustrate optimization direction of satellite task scheduling.

3.1 General-purpose modeling process for satellite task scheduling

3.1.1 Requirements analysis

The first step to address satellite task scheduling is to build a systematic and complete task scheduling model that clarifies the combinational optimization relationship in the problems. The previous chapter has comprehensively summarized the models commonly used in satellite task scheduling studies, and emphasized that problem modeling should follow the reality-based and application-oriented principle. Specifi-

https://doi.org/10.1515/9783111537191-003

cally, at present, the main requirements for the problem modeling of satellite task scheduling are as follows:

(1) Requirements for systematicness, authenticity, and completeness. It has been explained in the preceding text that the models in satellite task scheduling studies are too simple and such study results are difficult to meet satellite task scheduling requirements in reality. At present, with the increasing scale and diversified functions of satellites, the complexity of task scheduling increases sharply. Therefore, the primary requirement for satellite scheduling model is to systematically, truthfully, and completely describe the actual problems of the scheduling. The model simplification in related studies mainly includes: 1) Simplification of decision relations. For example, most remote-sensing satellite task scheduling studies only decided imaging events, while a few studies decided imaging and downlink events at the same time, and both of them rarely considered memory-erasing events. Such simplified models cannot truly and accurately cover the actual application requirements of remote-sensing satellite task scheduling. 2) Simplification of constraints. On the one hand, with the simplification of decision relations, related constraints were ignored; On the other hand, a large number of tedious and complex satellite operation constraints were simplified. Such study results cannot be utilized in actual application since they lose contact with reality. Therefore, it is necessary to systematically, truthfully, and completely describe the decision relations and constraints of satellite task scheduling to provide practical basis for satellite task scheduling problem modeling.

(2) Requirements for generality and expandability. Under the current satellite management system, various satellites with different modes and functions are operated at the same time, and sometimes it is even necessary to perform cross-agency and cross-model operation. Therefore, it is crucial for satellite task scheduling model to be compatible with multiple types of satellites and adapt to various task scheduling scenarios. It is important to clarify that generality does not mean that all satellite task scheduling problems are described by one model, but refers to the unity of modeling ideas and methods in different satellite task scheduling problems; In other words, the main satellite task scheduling problems can be uniformly described by a set of general modeling methods. Specifically, to address the four main satellite task scheduling problems, it is necessary to design a set of general modeling methods to describe the combinational optimization problems in a unified way. Based on the methods, cross-agency and cross-model satellites can be flexibly and integrally scheduled to meet the new requirements of flexible networking and quick response of satellite management, conducive to fostering incremental development and flexible expansion of satellite task scheduling system, and advancing the course of integrated management of satellite systems.

(3) Requirements for flexibility of constraint modeling and management. The flexibility of constraint modeling and management is a key factor of the generality of models. The constraints in different satellite task scheduling problems may be different, which is the main reason that the task scheduling models of different satellites are incompatible and the development cycle of new satellite task scheduling system is long. Moreover, some special and personalized constraints for some satellites, tasks and special periods may be proposed. Therefore, there is absolutely a need for a flexible and independent manner of constraint modeling and management. Experience has shown that there are some common features in constraints of various satellite task scheduling problems, which can be described uniformly through reasonable classification and induction to reduce modeling complexity for higher efficiency. At the same time, it is also indispensable to have an open and user-oriented constraint modeling environment, which provides a solid support for the long-term management of satellite systems.

(4) **Requirements for reasonability of objective functions.** The optimization requirements from satellite control agency may be vague and inaccurate. It is also very important to transform the subjective optimization requirements into objective and quantifiable objective functions in modeling satellite task scheduling problems. For example, while maximizing the number of satellite images, the remote-sensing satellite control agency also seeks to improve the imaging quality to have a comprehensive performance of imaging quantity and quality. While minimizing the average delay of inter-satellite links, the navigation satellite control agency also seeks to diversify the topology of links in different timeslots and increase the frequency of inter-satellite ranging. Therefore, it is necessary to fully understand and analyze the optimization requirements of the management agency and build a reasonable and accurate objective function model to illustrate optimization direction of satellite task scheduling model.

3.1.2 Process design

Based on the above analysis of requirements, this section puts forward a general-purpose modeling process of satellite task scheduling that decouples the decision, constraints, and profits, as shown in Figure 3.1. The process includes:

(1) **The satellite task scheduling problems are mathematically and hierarchically generalized to meet the requirements of systematicness, authenticity, and completeness of the satellite task scheduling model.** To address the problem that the models in conventional studies are too simple, the satellite task scheduling problem is systematically and mathematically generalized to a task set, a resource set, a score set, and a decision matrix. Especially, to address problems of complexity and diversity of resources, a "bottom-up" resource structure is established based on the logic that "event executable opportunity" is the underlying resource in satellite task scheduling, providing objective basis for building decision, constraint and profit models.

(2) **A general-purpose 0–1 mixed integer decision model is built to explain the decision-relation between the task set and the resource set and meet the requirements for generality and expandability of satellite task scheduling model.** To address the problem of insufficient generality of conventional satellite task sched-

uling models, based on the mathematical description of satellite task scheduling problems, the decision-relation between the task set and the resource set, specifically between the satellite event and its executable opportunity, is creatively explained. With this relation, the general-purpose 0–1 mixed integer decision model is then built with examples based on four satellite scheduling problems including remote-sensing satellite scheduling, relay satellite scheduling, navigation satellite scheduling and satellite range scheduling, providing a new and general decision model for various satellite task scheduling problems, laying an important foundation for subsequent general-purpose modeling of constraints and objectives and algorithm design, and playing the key coupling point in the loosely coupled engine framework in this book.

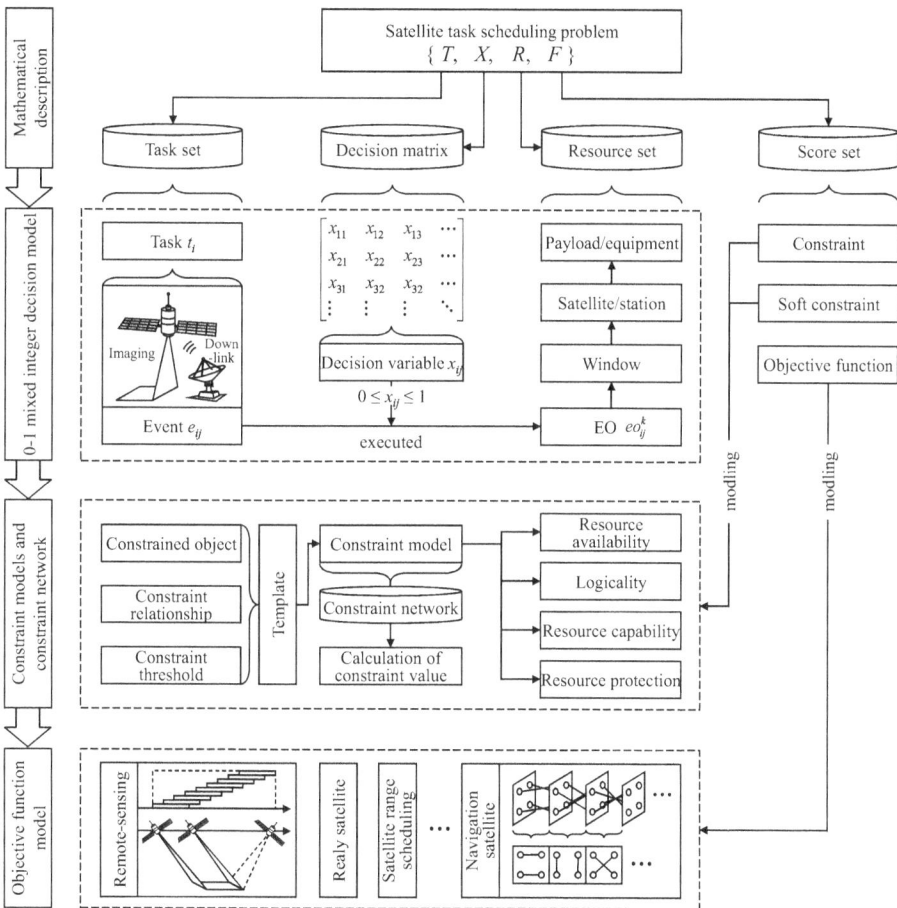

Figure 3.1: General-purpose modeling process of satellite task scheduling.

(3) Constraint template is designed to build constraint model and constraint network for providing general, objective and flexible manners to describe constraints to meet the requirements of constraint modeling and management of satellite task scheduling. Based on the general-purpose 0–1 mixed integer decision model, a general-purpose constraint template composed of an object, a threshold and a relationship is designed to uniformly describe the constraints of satellite task scheduling. Detailed examples and mathematical models are given. Then, the constraint network formed by the above constraints is provided, and the constraint value calculation method for the constraint network is further designed, providing a general-purpose constraint evaluation basis and an efficient constraint value calculation method for satellite task scheduling scheme, and providing a practical method for application and flexible expansion of related constraint models.

(4) **An objective model of satellite task scheduling is built to optimize satellite control.** For the purpose of optimizing actual satellite operation, the objective function models of remote-sensing satellites, relay satellites, navigation satellites and satellite range scheduling are established respectively, and the calculation methods of the objectives of agile remote-sensing satellites regarding the imaging quality as important and the time delay of navigation satellites are especially introduced, providing evaluation basis for satellite task scheduling schemes to guide the optimization direction for follow-up algorithms.

In summary, the decision model, constraint model and objective model jointly form the general-purpose satellite task scheduling model for the satellite task scheduling engine of this book that decouples the decision, constraints, and profits, creating a new idea for improving the generality and expandability of existing satellite task scheduling models to promote the integrated management process of satellite systems.

3.2 Mathematical description for satellite task scheduling problems

Task scheduling problems are typical combinational optimization problems that usually include tasks, resources, constraints, objectives and decision relationships. In this section, the satellite task scheduling problems are systematically, hierarchically and mathematically generalized to a task set, a resource set, a score set, and a decision matrix. Especially, to address problems of complexity and diversity of resources, a "bottom-up" resource structure is established in this section based on the logic that "event executable opportunity" is the underlying resource in satellite task scheduling, providing objective basis for building decision, constraint and objective models. Meanwhile, the instantiated and object-oriented modeling manner in this section may benefit the design and application of satellite task scheduling system based on the study results of this book.

3.2.1 Quadruple

Satellite task scheduling problems belong to combinational optimization problems. To organize the key elements of satellite task scheduling problems to clearly show their combinational optimization characteristics, this section first describes the satellite task scheduling problems as a quadruple as follows:

$$STSP = \{T, R, X, F\} \tag{3.1}$$

where STSP is the satellite task scheduling problem, T is the task set, R is the resource set, X is the decision matrix, and F is the score set.

It can be seen that the quadruple contains four elements, including task set, resource set, decision matrix and score set. The meanings of each element are as follows:

(1) Task set
As one of the inputs of satellite task scheduling problems, task set is a set of tasks in the problems to be scheduled and their properties. In this study, the task set is directly and completely provided by the satellite control agency, that is, the proposers of satellite task scheduling problems. The task set is described in detail in Section 3.2.2.

(2) Resource set
As another important input of satellite task scheduling problems, resource set is the set of all levels and all kinds of resources in the problems and their properties. In this study, the resource set is also directly and completely provided by the satellite control agency (pre-processed and calculated). In this set, resources are divided into different levels and types, including concrete and top-level resources such as satellites and ground stations, and abstract and bottom-level resources such as satellite orbits, time windows and event executable opportunities. The scale of task set and resource set together determines the solution space scale of satellite task scheduling problems, commonly known as problem scale, which has a significant impact on the complexity and solution efficiency of task scheduling problem. Each level and type of resources of this set is described in detail in Section 3.2.3.

(3) Decision matrix
Decision matrix is the key element linking task set and resource set in satellite task scheduling problems. It is used to describe the decision-relation between the task set and the resource set, that is, to establish the combinational optimization relations between the two sets. Decision matrix provides a mathematical description manner for the scheme (solution) of satellite task scheduling problems. In other words, on the basis of task set and resource set, a decision matrix represents a satellite task scheduling scheme, namely, a solution of satellite task scheduling problems as the object of problem solving and outputting. On the other hand, decision matrix is also an object of iterative search and optimization of the algorithms in the process of problem solving. As an important basis for the design of algorithms, operators and strategies, deci-

sion matrix provides a general-purpose coding mode and neighborhood structure. Thus, it is an important bridge for connecting models and algorithms under the framework of satellite task scheduling engine in this book. The decision matrix and decision relations are described in detail in Section 3.3 to build a general-purpose satellite task scheduling decision model.

(4) Score set

The score set includes the constraints (including soft constraints) of satellite task scheduling problems and the objective function, as well as the constraints and objectives of the current scheduling scheme calculated from the above task set, resource set and decision matrix. Among them, the constraints and the objective function are collectively called score function, which is another important input of satellite task scheduling problems. Constraints and objectives are collectively referred to as score values, which is another output of the problems. Section 3.2.4 elaborates on the relevant definitions and classifications; Sections 3.4 and 3.5 systematically build the constraint model and objective model of satellite task scheduling problems.

3.2.2 Task set

For the task set in the quadruple of satellite task scheduling problems, this section describes the elements of the tasks truly and accurately according to the actual situation of satellite management to provide an objective basis for the modeling of satellite task scheduling problems.

First, the task set T and the task t_i in the satellite task scheduling quadruple are expressed by eqs. (3.2a) and (3.2b), respectively:

$$T = \{t_i | i = 1, 2, 3, \ldots\} \tag{3.2a}$$

$$t_i = \{E_i, P_i^C, P_i^V\} \tag{3.2b}$$

where t_i is the ith task, i is the task number, and $i = 1, 2, \ldots$, E_i is the event set of task t_i, P_i^C is the property set (constant) of task t_i, and P_i^V is the variable set of task t_i.

To clearly explain the task elements of event set E_i, property set P_i^C and variable set P_i^V, based on the task scheduling problems of SuperView-1 remote-sensing satellite, this section gives an example of satellite task composition and their relations as shown in Figure 3.2, and further clarifies the specific meanings of event set, property set and variable set.

3.2.2.1 Event set

The event set is the set of events that a task contains. The event set E_i of the task t_i can is expressed as follows:

$$E_i = \{e_{ij} | j = 1, 2, \ldots\}, i \le |T| \tag{3.3}$$

where e_{ij} is the jth event of task t_i, j is the event number (starting from 1), and $|T|$ is norm of task set T.

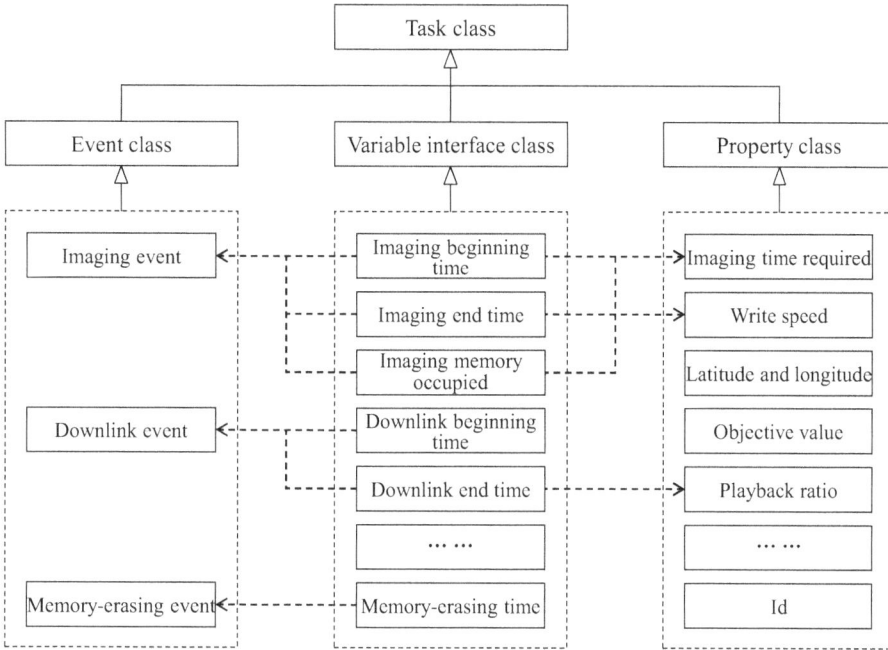

Figure 3.2: Example of the composition and relations of task set in SuperView-1 task scheduling problems.

In Figure 3.2, the event set E_i of the remote-sensing satellite task t_i contains imaging event e_{i1}, downlink event e_{i2}, memory-erasing event e_{i3}, etc. It is worth noting that most studies on remote-sensing satellite task scheduling only consider imaging events, while a few studies consider imaging and downlink events at the same time, and both of them rarely consider memory-erasing events. Although such studies have reduced the complexity of satellite task scheduling problems, the study results cannot be utilized in actual application since they have ignored downlink events, memory-erasing events and related constraints. Therefore, to objectively, truly and completely describe the task scheduling problems of remote-sensing satellites, and provide practical and usable task scheduling modeling and solving methods for satellite management, this section includes downlink events and memory-erasing events into the tasks of remote-sensing satellites.

3.2.2.2 Property set

The property set is the set of inherent properties (constants) given to the tasks by the satellite control agency in advance. In Figure 3.2, the property set P_i^C of remote-sensing satellite task t_i contains task number i, objective value, target latitude and longitude, writing rate, recording-playback ratio, imaging time and other elements. The elements contained in the property set may vary for different problems. This section does not explain them in detail.

3.2.2.3 Variable set

In the satellite task scheduling problems, there are some variables determined by property variables and task events. For example, in Figure 3.2, the imaging event e_{i1} of the remote-sensing satellite task determines the imaging beginning time $b(e_{i1})$, which together with the imaging required time determines the imaging end time $e(e_{i1})$, and together with the writing rate determines the imaging required memory $m(e_{i1})$; The downlink event e_{i2} determines the downlink beginning time $b(e_{i2})$, which together with the imaging required time, the recording-playback ratio, and so on determines the downlink end time $e(e_{i2})$. The memory-erasing event e_{i3} determines the memory erasure time $b(e_{i3})$ and the like. It is thus clear that the values of the above variables change with the combination of different task events, reflecting the characteristics of combinational optimization of task scheduling problems. Therefore, this section classifies the above variables into variable set to provide an open access interface and calculation approach in the process of task scheduling to get enough data for the evaluation of task scheduling schemes.

3.2.3 Resource set

The essence of task scheduling problems is to allocate resources for tasks. Clear and systematic description of various types of resources is crucial for modeling and solving task scheduling problems. For the resource set R in the quadruple of satellite task scheduling problems, this section divides it into platform set, payload/equipment set, window set and event executable opportunity set as follows:

$$R = \{P, Q, W, EO\} \tag{3.4}$$

where P is the platform set, Q is the payload/equipment set, W is the window set, and executable opportunity (EO) is the event executable opportunity set.

Based on the task scheduling problems of SuperView-1 remote-sensing satellite, Figure 3.3 shows the composition and mutual relations of resource set R, and this section illustrates the platform set P, payload/equipment set Q, window set W, event executable opportunity set EO and the specific meanings and internal relations of each element in Figure 3.3 in turn.

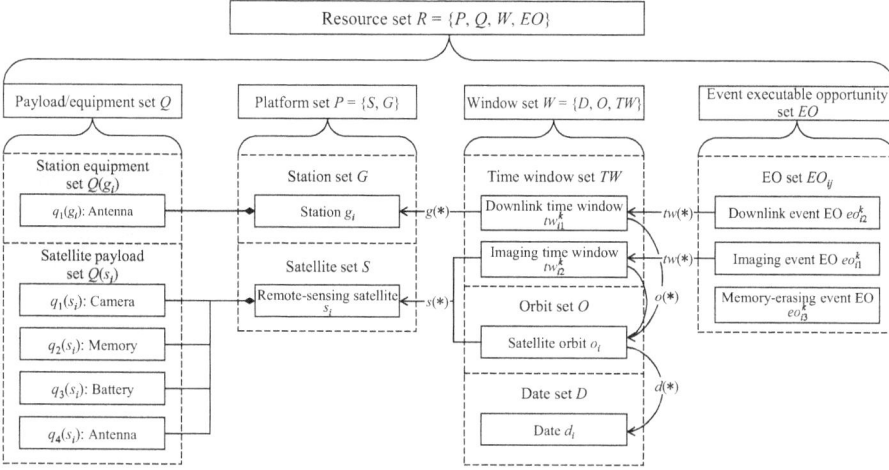

Figure 3.3: Example of the composition and relations of resource set in SuperView-1 task scheduling problems.

3.2.3.1 Platform set

Denoted as P, the platform set is a set of platform resources. In this book, as the main carriers for executing tasks, platform resources refer to satellites and ground stations (including relay satellites acting as ground stations). To facilitate the modeling in this book, the platform set P is divided into satellite set S and ground station set G as follows:

$$P = \{S, G\} \tag{3.5}$$

where S is the satellite set and G is the ground station set.

Then, satellite set S and ground station set G can be expressed by the following equations, respectively:

$$S = \{s_i | i = 1, 2, 3, \ldots\} \tag{3.6a}$$

$$G = \{g_i | i = 1, 2, 3, \ldots\} \tag{3.6b}$$

$$s_i = \{Q(s_i)\}, \ i \le |S| \tag{3.7a}$$

$$g_i = \{G(g_i)\}, \ i \le |G| \tag{3.7b}$$

where s_i is the ith satellite and i is the ground station number ($i = 1, 2, \ldots$), g_i is the ith ground station and i is the ground station number ($i = 1, 2, \ldots$), $Q(s_i)$ is the payload/equipment set carried by satellite s_i, and $G(g_i)$ is the payload/equipment set of station g_i.

For example, in Figure 3.3, in the task scheduling problems of SuperView-1 remote-sensing satellite, the platform set includes remote-sensing satellite and the station. Usually, as a kind of integrated resources covering other resources to perform satellite tasks, platform resources are the top-level resources in satellite task scheduling problems.

3.2.3.2 Payload/equipment set

Denoted as Q, the payload set is a set of payload/equipment resources. In this book, as important components of platform resources and tools for executing tasks, payload/ equipment resources refer to instruments, facilities, subsystems, and so on carried/ equipped by platform resources. Payload/equipment set Q of the satellite task scheduling problems and the payload/equipment sets $Q(s_i)$ and $Q(g_i)$ of satellite s_i and station g_i can be, respectively, expressed as

$$Q = \bigcup_{i \le |S|} Q(s_i) + \bigcup_{i \le |G|} Q(g_i) \tag{3.8a}$$

$$Q(s_i) = \{q_j(s_i) | j = 1, 2, 3, \ldots\}, \ i \le |S| \tag{3.8b}$$

$$Q(g_i) = \{q_j(g_i) | j = 1, 2, 3, \ldots\}, \ i \le |G| \tag{3.8c}$$

where $q_j(s_i)$ is the jth payload carried by satellite s_i, and j is the payload number (starting from 1). $q_j(g_i)$ is the jth equipment of station g_i, and j is the payload number (starting from 1).

For example, in Figure 3.3, the SuperView-1 satellite contains four types of payload resources, including camera, battery, memory, and downlink antenna, which respectively undertake functions of imaging, power supply, storage, and downlink. The station contains downlink antenna and other equipment resources for receiving data from the remote-sensing satellite. It's important to note that the satellite and the station also contain many other payloads and equipment, such as standby battery, backup memory, and temperature sensors, but such payloads and equipment are not considered in the modeling and solving process of task scheduling problems in this book since they are not directly related to the problems.

3.2.3.3 Window set

Denoted as W, window set is a set of window resources. In this book, as one of the abstract resources in satellite task scheduling problems, window resources refer to instantiated time intervals containing information such as beginning time, end time, platform resource or parent window. In existing studies of satellite task scheduling problems, the time window of task in time, space and frequency domain is often the basis of modeling and solving. However, related studies usually only describe a time window with the beginning and end time, which is not suitable for systematic modeling of complex satellite task scheduling problems since it simplifies the internal relations between the window and other types of resources.

In this regard, considering that in the actual situation of satellite management, days and orbits are taken as time units, this section further divides window set W into time window set TW, orbit set O and date set D as follows:

$$W = \{D, O, TW\} \tag{3.9}$$

Then, the date set D can be expressed as:

$$D = \{d_i | i = 1, 2, 3, \ldots\} \tag{3.10}$$

where d_i is the ith day in satellite task scheduling cycle (starting from 1).

It can be seen that compared to conventional satellite task scheduling studies only covering one orbit and one day, the scheduling cycles studied in this book are more comprehensive, providing an applicable basis for modeling and calculating relevant constraints in actual satellite task scheduling problems based on days and orbits as time units.

Then, the orbit set O can be expressed as:

$$O = \{o_i | i = 1, 2, 3, \ldots\} \tag{3.11a}$$

$$o_i = \{b(o_i), e(o_i), d(o_i), s(o_i)\} \tag{3.11b}$$

where o_i is the ith satellite orbit and i is the orbit number, $i_n = 1, 2, \ldots,$ $b(o_i)$ is the beginning time of orbit o_i; $e(o_i)$ is the end time of orbit o_i, $d(o_i)$ is the date of orbit o_i, $d(o_i) \in D$, and $s(o_i)$ is the satellite in orbit o_i, $s(o_i) \in S$.

Here, the orbit o_i includes its beginning time $b(o_i)$ and its end time $e(o_i)$. Similarly, in the following text, $b(*)$ and $e(*)$ are used to represent the beginning and end time of the relevant objects. In addition, as applicable to actual situation, the orbit o_i also includes its date $d(o_i)$ and its satellite $s(o_i)$ in modeling to provide an access channel from the orbit set O to the date set D and the satellite set S, building an important way for subsequent constraint and objective calculation involving different levels and types of resources.

Then, the time window set TW can be expressed as

$$TW = \bigcup_{i \le |T|} \bigcup_{j \le |E_i|} TW_{ij} \tag{3.12a}$$

$$TW_{ij} = \left\{ tw_{ij}^k | k = 0, 1, 2, 3, \ldots \right\}, i \le |T|, j \le |E_i| \tag{3.12b}$$

$$tw_{ij}^k = \left\{ b\left(tw_{ij}^k\right), e\left(tw_{ij}^k\right), o\left(tw_{ij}^k\right), g\left(tw_{ij}^k\right) \right\}, i \le |T|, j \le |E_i|, k \le |TW_{ij}| \tag{3.12c}$$

where TW_{ij} is the time window set of event $e_{ij}.tw_{ij}^k$ is the kth time window of event e_{ij}, and k is the window number (starting from 1).$b(tw_{ij}^k)$ is the beginning time of time window $tw_{ij}^k.e(tw_{ij}^k)$ is the end time of time window $tw_{ij}^k.o(tw_{ij}^k)$ is the satellite orbit corresponding to time window tw_{ij}^k, $o(tw_{ij}^k) \in O.g(tw_{ij}^k)$ is the ground station corresponding to time window tw_{ij}^k (if any), $g(tw_{ij}^k) \in G$.

Taking the task scheduling problems of SuperView-1 in Figure 3.3 as an example, the satellite flies around the Earth 15 circles a day, that is, there are 15 satellite orbits/ day (each orbit takes about 96 min); Task targets can be seen in some orbits, creating time windows (tens of seconds to hundreds of seconds) for imaging events. Similarly, orbits and stations create time windows (about hundreds of seconds) for downlink events. Therefore, the window resources in this section not only include basic elements such as beginning and end time, but also include instantiated and hierarchical information such as the satellite, the station and parent window, forming a bottom-up data access channel from lower resources to upper resources, which enables the modeling in this book to be consistent with actual complex situation and facilitates systematic description of satellite task scheduling problems.

3.2.3.4 Event executable opportunity set

Although the time window reflects the opportunity that satellite event can be executed, in satellite task scheduling problems such as range scheduling of agile remote-sensing satellites and high-orbit satellites, within the time window of an event, there is more than one executable opportunity for the event, meaning that the event may happen at any time within the time window. Therefore, it is impossible to determine the specific time when the event begins and ends only through the time window. In this section, the time window is further discrete and abstracted as a series of event executable opportunities as the direct basis for deciding the beginning and end time of events.

Definition 3.1 (Event executable opportunity (EO)): This refers to an opportunity for an event in a satellite task to be executed within its time window, abbreviated as executable opportunity.

Taking remote-sensing satellite as an example, based on the time allowed for imaging and downlink events of the satellite, 1 s is set as the minimum unit of time window (at present, the minimum unit of task control of remote-sensing satellite is usually 1 s), the imaging and downlink time windows are further discrete into a series of event executable opportunities, as shown in Figure 3.4. It can be seen that the event executable opportunity includes not only the specific time when the event begins and ends, but also the corresponding time window. Event executable opportunities are the underlying resources to be scheduled in satellite task scheduling problems since they can show the actual situation of the event execution more directly and accurately.

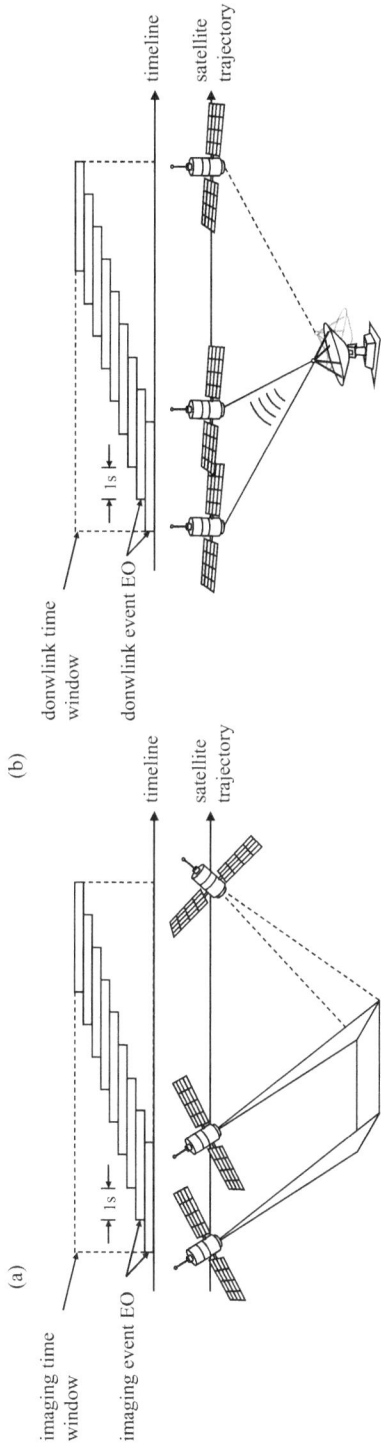

Figure 3.4: Example of the composition and relations of task set in SuperView-1 task scheduling problems: (a) imaging event and (b) downlink event.

In view of this, the set of event executable opportunities is called event executable opportunity set denoted as EO and expressed by the following equations:

$$EO = \bigcup_{i \le |T|} \bigcup_{j \le |E_i|} EO_{ij} \tag{3.13a}$$

$$EO_{ij} = \left\{ eo_{ij}^k | k = 1, 2, 3, \dots \right\}, \ i \le |T|, \ j \le |E_i| \tag{3.13b}$$

$$eo_{ij}^k = \left\{ b\left(eo_{ij}^k\right), \ tw\left(eo_{ij}^k\right), \ A\left(eo_{ij}^k\right) \right\}, \ i \le |T|, \ j \le |E_i|, \ k \le |EO_{ij}| \tag{3.13c}$$

where EO_{ij} is the executable opportunity set of event $e_{ij}.eo_{ij}^k$ is the kth executable opportunity of event e_{ij}, and k is the executable opportunity number (starting from 1).$b(eo_{ij}^k)$ is the beginning time of executable opportunity $eo_{ij}^k.tw(eo_{ij}^k)$ is the time window corresponding to executable opportunity eo_{ij}^k, $tw(eo_{ij}^k) \in TW_{ij}.A(eo_{ij}^k)$ is Euler angle $\{\theta, \varphi, \psi\}$ of the satellite attitude or ground station antenna corresponding to executable opportunity eo_{ij}^k.

As shown in Figure 3.3, taking satellite task scheduling problems of SuperView-1 as an example, this section systematically and hierarchically describes all levels and all kinds of resources in satellite task scheduling problems. It is important to note that Figure 3.3 also represents the connections between resources to create data channels for the bottom resources (event executable opportunities) to access the top resources (platforms and payloads). Further explanation is given as follows:

$$eo_{i1}^k \xrightarrow{tw(*)} tw_{i1}^k \xrightarrow{o(*)} o_i \xrightarrow{s(*)} s_i \rightarrow Q(s_i) \tag{3.14a}$$

$$eo_{i2}^k \xrightarrow{tw(*)} tw_{i2}^k \xrightarrow{g(*)} g_i \rightarrow Q(g_i) \tag{3.14b}$$

In eq. (3.14a), the time window tw_{i1}^k of imaging event executable opportunity eo_{i1}^k in time window set TW can be accessed correspondingly; Then, the corresponding orbit o_i in orbit set O can be accessed based on time window tw_{i1}^k; The corresponding satellite s_i in satellite set S can be accessed based on the orbit o_i; At the end, the payload set $Q(s_i)$ carried by satellite s_i can be accessed. The letter i only represents the resource numbers and the i of different resources may vary.

Similarly, in eq. (3.14b), the time window tw_{i2}^k of downlink event executable opportunity eo_{i2}^k in time window set TW can be accessed correspondingly. Then, the corresponding station g_i in station set G can be accessed based on the time window tw_{i2}^k; At the end, the payload set $Q(g_i)$ carried by station g_i can be accessed.

Therefore, this section systematically and hierarchically describes the resource sets in the quadruple of satellite task scheduling problems, clarifies the internal relations among the element sets and the elements in the resource set and establishes a bottom-up data access channel to fully show the complexity and diversity of resources in satellite task scheduling problems, laying a solid basis for the modeling of satellite task scheduling problems. Moreover, since this section fully utilizes the idea of instantiation and object-oriented (C++, Java, etc.) commonly used in data management and program design, the study results of this book can benefit the design of satellite task scheduling system.

3.2.4 Score set

In combinational optimization problems, constraints and objective functions are important basis for evaluating the quality of a solution. In the optimization process, as the constraints usually have higher priority than objective functions, constraints are always satisfied first before optimizing objective functions. In some ways, objective functions can be regarded as constraints with lower priority. On the contrary, constraints can also be regarded as objective functions with higher priority. In some related studies, there are hard constraints and soft constraints to show the constraints or objectives with different priorities.

In view of this, for the score set F in the quadruple of satellite task scheduling problems, this section further divides it into constraint set, objective function and soft constraint set as follows:

$$F = \{F^H, f, F^S\} \tag{3.15}$$

where F^H is the constraint set, f is the objective function, and F^S is the soft constraint set.

3.2.4.1 Constraint set

Denoted as F^H, constraint set is a set of hard constraints and can be expressed by the following equations:

$$F^H = \{f_i^H | i = 1, 2, 3, \ldots\} \tag{3.16a}$$

$$F^H(X) = \sum_{i=1}^{|F^H|} f_i^H(X) \tag{3.16b}$$

where f_i^H is the ith constraint, and i is the constraint number (starting from 1). X is the decision matrix. $f_i^H(X)$, abbreviated as constraint value, is the value (≤ 0) of the current scheduling scheme that is calculated based on the decision matrix X and violates the constraint $f_i^H.F^H(X)$, called as constraint value of the current scheduling scheme, is the total value (≤ 0) of the current scheduling scheme that is calculated based on the decision matrix X and violates the constraints.

Taking the constraint f_i^H of remote-sensing satellite task scheduling as an example, when the constraint requires that the total number of imaging events executed by the satellite in a single orbit is 10 times, and if the total number of imaging events executed by the satellite in the orbit is 12 times in the current satellite task scheduling scheme, the constraint value $f_i^H(X) = -2$. If the current scheme does not violate any constraint, the constraint value $F^H(X) = 0$, and the scheme is feasible.

3.2.4.2 Objective function

Denoted as f, objective function is the optimized objective function of satellite task scheduling problems. Denoted as $f(X)$, the objective function value of the current scheduling scheme calculated based on the decision matrix X is referred to as objective value. The objective function of common satellite task scheduling problems includes the number, duration and priority of tasks scheduled, as well as the weighted values of such indicators.

It's important to note that all the optimization problems involved in this book are single-objective problems, that is, there is only one objective function. This book does not study multi-objective optimization problems because: 1) In terms of requests, the satellite control agency basically only needs one satellite task scheduling scheme for implementation, rather than a multi-objective Pareto scheme set. This book only studies the modeling and solving means of single-objective optimization problems according to the actual requests; 2) in terms of algorithm, the multi-objective optimization algorithm based on EA has high time complexity and low constraint optimization efficiency. Under the background of complex satellite task scheduling constraints, it is difficult to meet the demand of the management agency on improving optimization quality and efficiency at the same time.

3.2.4.3 Soft constraint set

Similar to the constraint set, denoted as F^S, soft constraint set is a set of soft constraints and can be expressed by the following equations:

$$F^S = \{f_i^S \mid i = 1, 2, 3, \ldots\} \tag{3.17a}$$

$$F^S(X) = \sum_{i=1}^{|F^S|} f_i^S(X) \tag{3.17b}$$

where f_i^S is the ith soft constraint, and i is the soft constraint number (starting from 1). $f_i^S(X)$, abbreviated as soft constraint value, is the value (≤ 0) of the current scheduling scheme that is calculated based on the decision matrix X and violates the soft constraint $f_i^S(X).F^S(X)$, called as soft constraint value of the current scheduling scheme, is the total value (≤ 0) of the current scheduling scheme that is calculated based on the decision matrix X and violates the soft constraints.

Soft constraints are usually used to represent soft constraint requirements, secondary objectives or other optimization objectives of auxiliary algorithm. In the process of combinational optimization, as an indirect role in resolving constraints and optimizing objectives, soft constraints cannot directly change the constraint value or objective value, but can guide the search direction of the algorithm to a certain extent. For example, in SuperView-1 task scheduling problems, the management agency requires to shorten the storage time of satellite imaging data in on-board memory to improve the imaging and downloading efficiency. Under the influence of this soft constraint, the violation probability of memory-related constraints will be reduced, and the optimization efficiency of the algorithm can be improved.

3.2.4.4 Comparison method

Based on the above constraint value $F^H(X)$, objective value $f(X)$ and soft constraint value $F^S(X)$, $F(X)$ shown in eq. (3.18) is denoted as the score value that serves as the evaluation basis of satellite task scheduling scheme (solution) in this book:

$$F(X) = \{F^H(X), f(X), F^S(X)\} \tag{3.18}$$

The decision matrices of two given satellite task scheduling schemes are X_1 and X_2 respectively. In the process of comparing the score values $F(X_1)$ and $F(X_2)$ of the two schemes, the constraint value has the highest priority, the objective value has the secondary priority, and the soft constraint value has the lowest priority. Assuming that the satellite task scheduling problems in this book are all maximization problems, the comparison method of score values $F(X_1)$ and $F(X_2)$ is as follows:

Algorithm 3.1: Comparison method of score value.

Input: Score values $F(X_1)$ and $F(X_2)$

Output: $F(X_1) > F(X_2)$ or $F(X_1) = F(X_2)$ or $F(X_1) < F(X_2)$

```
1    if F^H(X_1) > F^H(X_2) then
2    |      return F(X_1) > F(X_2)
3    else if F^H(X_1) = F^H(X_2) & f(X_1) > f(X_2) then
4    |      return F(X_1) > F(X_2)
5    else if F^H(X_1) = F^H(X_2) & f(X_1) = f(X_2) & F^S(X_1) > F^S(X_2) then
6    |      return F(X_1) > F(X_2)
7    else if F^H(X_1) = F^H(X_2) & f(X_1) = f(X_2) & F^S(X_1) = F^S(X_2) then
8    |      return F(X_1) = F(X_2)
9    else
10   |      return F(X_1) < F(X_2)
11   end
```

To sum up, to meet the requirements of systematicness, authenticity and completeness of the satellite task scheduling model, this section describes the satellite task scheduling problems hierarchically and mathematically through a quadruple composed of task set, resource set, score set and decision matrix. Especially, to address problems of complexity and diversity of resources, a "bottom-up" resource structure is established in this section based on the logic that "event executable opportunity" is the underlying resource in satellite task scheduling, providing objective basis for building decision, constraint and objective models. Moreover, since this section fully utilizes the idea of instantiation and object-oriented (C++, Java, etc.) modeling, the study results of this book can benefit the data management and the design of satellite task scheduling system.

3.3 General-purpose 0–1 mixed integer decision model for satellite task scheduling

For the decision matrix in the quadruple of satellite task scheduling problems, based on the description of task set and resource set above, through a 0–1 mixed integer decision variable, the decision-relation between the task set and the resource set, specifically between the satellite event and its executable opportunity, is creatively explained in this section. With this relation, the general-purpose 0–1 mixed integer decision model is then built with examples based on four satellite scheduling problems including remote-sensing satellite scheduling, relay satellite scheduling, navigation satellite scheduling and satellite range scheduling. It provides a new and general decision model for various satellite task scheduling problems, lays an important foundation for subsequent general-purpose modeling of constraints, objectives and algorithm design, and plays the key coupling point in the loosely coupled engine framework in this book.

3.3.1 Decision variables and decision matrix

Based on task set T and resource set R, this section firstly illustrates the combinational optimization relation (i.e. decision relation) between them through a 0–1 integer variable. Furthermore, a 0–1 mixed integer decision variable and decision matrix are built for the establishment of general-purpose 0–1 mixed integer decision model.

3.3.1.1 Decision relation

For task t_i in task set T, there are a plurality of events to be executed, and each event e_{ij} can be executed at any one of the executable opportunity eo_{ij}^k in the executable opportunity set $EO_{ij}(EO_{ij} \subseteq R)$ (regardless of constraints), as shown in Figure 3.5. Then, this section first describes the relation by a 0–1 integer variable x_{ij}^k as follows:

$$x_{ij}^k = \begin{cases} 1, \text{ if event } e_{ij} \text{ is excecuted at EO } eo_{ij}^k \\ 0, \text{ other} \end{cases} \quad \forall i \leq |T|, j \leq |E_i|, k \leq |EO_{ij}| \qquad (3.19)$$

where T is the task set, E_i is the event set of task I, e_{ij} is the jth event of task i, EO_{ij} is the executable opportunity set of event e_{ij}, and eo_{ij}^k is the kth executable opportunity in executable opportunity set EO_{ij}.

The variable x_{ij}^k of the above equation reflects the combinational optimization relation (i.e. decision relation) between the event e_{ij} and its executable opportunity eo_{ij}^k.

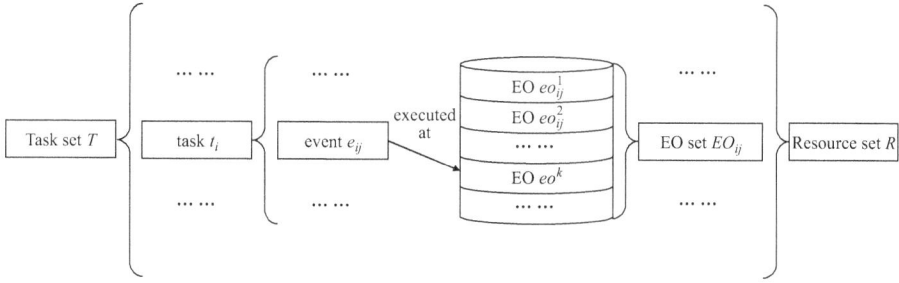

Figure 3.5: Relation between task set and resource set in satellite task scheduling problems.

Meanwhile, in the satellite task scheduling problems, since the event e_{ij} can only be executed once, the variable x_{ij}^k must satisfy the inherent conditions as follows:

$$\sum_{i=1}^{EO_{ij}} x_{ij}^k \leq 1, \ \forall i \leq |T|, j \leq |E_i| \qquad (3.20)$$

3.3.1.2 0–1 mixed integer decision variable

On the basis of the above decision relation and 0–1 integer variable x_{ij}^k, to further unify the variable range, simplify the form and facilitate generation of neighborhood for algorithm, this section standardizes the integer variable and transforms it into a 0–1 mixed integer variable $x_{ij}(0 \leq x_{ij} \leq 1)$ used as the decision variable of satellite task scheduling problems in this book, as shown in eq. (3.21). The decision variable naturally satisfies the inherent conditions of eq. (3.20).

$$x_{ij} = \begin{cases} \dfrac{k}{|EO_{ij}|}, & \text{if event } e_{ij} \text{ is excecuted at EO } eo_{ij}^k, \\ 0, & \text{if event } e_{ij} \text{ is not executed,} \end{cases} \quad \forall i \leq |T|, j \leq |E_i| \qquad (3.21)$$

In the equation, since the executable opportunity number k starts from 1, that is, $1 \leq k \leq EO_{ij}$, this book stipulates that $x_{ij} = 0$ when the event e_{ij} is not executed, so the range of decision variable x_{ij} can be standardized to [0, 1]. Conversely, the executable opportunity number k can also be calculated from the decision variable x_{ij} as follows:

$$k = x_{ij} \cdot |EO_{ij}|, \ \forall i \leq |T|, j \leq |E_i|, 0 < x_{ij} \leq 1 \qquad (3.22)$$

where $\lceil \cdot \rceil$ indicates upward rounding. Therefore, the decision relation between task set T and resource set R in satellite task scheduling problems, specifically between event e_{ij} and event executable opportunity eo_{ij}^k, can be clearly expressed as

$$e_{ij} \xrightarrow{k = x_{ij} \cdot |EO_{ij}|} eo_{ij}^k, \ 0 < x_{ij} \leq 1 \qquad (3.23)$$

3.3.1.3 Decision matrix

Then, the decision matrix X of satellite task scheduling problems can be expressed by eq. (3.24). As one of the quadruples of satellite task scheduling problems, the decision matrix X can determine a satellite task scheduling scheme:

$$
X = \left(x_{ij} \right) = \begin{bmatrix} x_{11} & x_{12} & \cdots & x_{1|E_1|} \\ x_{21} & x_{22} & \cdots & x_{2|E_2|} \\ \vdots & \vdots & & \vdots \\ x_{|T|1} & x_{|T|2} & \cdots & x_{|T||E_{|T|}|} \end{bmatrix} \tag{3.24}
$$

From the description of satellite task scheduling problems and 0–1 mixed integer decision model, it can be seen that task set T contains $|T|$ tasks, in which task t_i contains $|E_i|$ events, and each event e_{ij} corresponds to $|EO_{ij}|$ event executable opportunities. In other words, each decision variable x_{ij} in the decision matrix X may have $|EO_{ij}|$ values. Thus, the total number of possible values of each decision variable is

$$
\prod_{i=1}^{|T|} \prod_{j=1}^{|E_i|} |EO_{ij}| \tag{3.25}
$$

The above equation is the solution space scale of satellite task scheduling problems, where $|T|$ is the task scale, $|E_i|$ and $|EO_{ij}|$ are finite constants. Therefore, time complexity $O(2n)$ becomes exponential time when the brute force search is used to address the satellite task scheduling problems.

On the other hand, although the number $|EO_{ij}|$ of event executable opportunities in the above equation is not affected by the task scale, with the increase of the number of resources or the expansion of the time window in the satellite task scheduling problems, $|EO_{ij}|$ will also proliferate, and its influence on the solution space scale may even exceed the task scale. Accordingly, to reduce the scale of solution space and ensure the effectiveness of the model, in the process of discretizing time window into event executable opportunities based on the method in this chapter, it is necessary to choose the appropriate discretization accuracy, and screen and merge reasonably for preprocessing.

In conclusion, this section builds 0–1 mixed integer decision variables and decision matrix of satellite task scheduling, and based on the idea of generalization and loose coupling, establishes a concise general-purpose 0–1 mixed integer decision model with unified form and value range of decision variables, providing a general data interface for the following modeling of constraints and objectives, and benefiting the neighborhood design of subsequent algorithms.

3.3.1.4 Example of decision model

To further illustrate the general-purpose 0–1 mixed integer decision model and visually present the specific decision relations in different satellite task scheduling problems, this section takes satellite task scheduling problems including remote-sensing satellite scheduling, relay satellite scheduling, navigation satellite scheduling and satellite range scheduling as examples, and provides examples of 0–1 mixed integer decision model, as shown in Table 3.1 and Figure 3.6.

(1) Remote-sensing satellite task

In Table 3.1, the decision variables are expressed as x_{i1}, x_{i2} and x_{i3} for the remote-sensing satellite task t_i and its imaging event e_{i1}, downlink event e_{i2} and memory-erasing event e_{i3}. According to the decision variable in eqs. (3.21) and (3.22), the imaging event e_{i1}, the downlink event e_{i2} and the memory-erasing event e_{i3} of the remote-sensing satellite task t_i will be executed at the executable opportunities eo_{i1}^k, eo_{i2}^k and eo_{i3}^k respectively, as shown in Figure 3.6(a). It can be seen that the above decision variables clearly show the decision relations in the task scheduling problems of remote-sensing satellite, reflecting the characteristics of combinational optimization of task scheduling problems.

Table 3.1: Functional requirements and framework design ideas of the satellite task scheduling engine.

Task t_i	Event e_{ij}	Decision variable x_{ij}	Actual meaning
Remote-sensing satellite task t_i	Imaging event e_{i1}	x_{i1}	The event is executed at the EO eo_{i1}^k
	Downlink event e_{i2}	x_{i2}	The event is executed at the EO eo_{i2}^k
	Memory-erasing event e_{i3}	x_{i3}	The event is executed at the EO eo_{i3}^k
Relay satellite task t_i	Downlink event e_{i1}	x_{i1}	The event is executed at the EO eo_{i1}^k
	Memory-erasing event e_{i2}	x_{i2}	The event is executed at the EO eo_{i2}^k
Navigation satellite task t_i	Downlink event e_{i1}	x_{i1}	The event is executed at the EO eo_{i1}^k
	Downlink event e_{i2}	x_{i2}	The event is executed at the EO eo_{i2}^k
	. . .	. . .	. . . . . .
	Downlink event e_{i30}	x_{i30}	The event is executed at the EO eo_{i30}^k
Satellite range task t_i	Downlink event e_{i1}	x_{i1}	The event is executed at the EO eo_{i1}^k

Note: According to eq. (3.22), the EO number in the table is $k = x_{ij} \cdot |EO_{ij}|$.

(2) Relay satellite task

Similarly, in Table 3.1, for the relay satellite task t_i and its downlink event e_{i1} and memory-erasing event e_{i2}, the decision variables are expressed as x_{i1} and x_{i2}. According to the decision variable equation, the downlink event e_{i1} and the memory-erasing

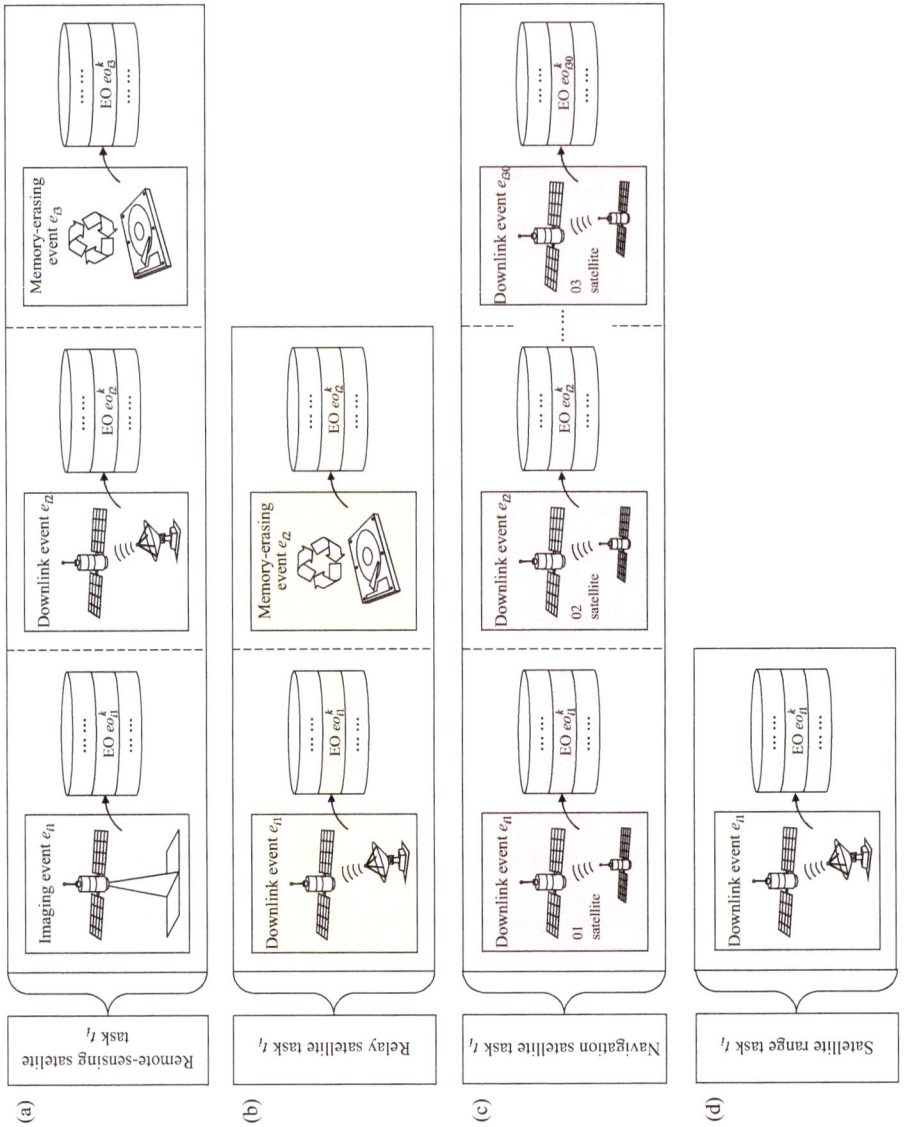

Figure 3.6: Illustrations of decision relations between events and executable opportunities in various satellite task scheduling problems: (a) remote-sensing satellite task; (b) relay satellite task; (c) navigation satellite task; and (d) satellite range task.

event e_{i2} of the relay satellite task t_i will be executed at the executable opportunities eo_{i1}^k and eo_{i2}^k respectively, as shown in Figure 3.6(b).

(3) Navigation satellite tasks

According to the problem defined in Section 2.1, the navigation satellite task t_i is to "build an inter-satellite link network in the ith navigation timeslot," and the downlink event e_{ij} of the satellite is to "establish an inter-satellite link between the jth satellite and other satellites." For 30 navigation satellites in BDS-3 navigation system, the decision variables are expressed as x_{i1} to x_{i30} in Table 3.1. According to the decision variable equation, the downlink events e_{i1} to e_{i30} of the navigation satellite task t_i will be executed at executable opportunities eo_{i1}^k to eo_{i30}^k respectively, as shown in Figure 3.6(c).

(4) Satellite range scheduling

Finally, for the satellite range task t_i and its downlink event e_{i1}, the decision variable is expressed as x_{i1} in Table 3.1, indicating that the satellite range task t_i will be executed at the EO eo_{i1}^k, as shown in Figure 3.6(d).

It can be seen that the general-purpose 0–1 mixed integer decision model in this section can clearly describe the decision relations in different satellite task scheduling problems, reflect the essential characteristics of combinational optimization in task scheduling problems, and meet the actual needs of satellite task scheduling modeling. Moreover, among the decision models of different satellite task scheduling problems, remote-sensing satellite tasks involve the most events (imaging, downlink and memory erasure), which increases the complexity of the problems to a certain extent. Since satellite range tasks involve the fewest events (only downlink events), the solution of task scheduling problems of satellite range are generally less difficult than that of other satellite task scheduling problems.

3.3.2 Advantage analysis

Compared to the conventional studies on satellite task scheduling problems, the decision model built in this section has the following two advantages:

(1) Since there is no subjective coding and decoding in the model following the reality-based and application-oriented principle of modeling in this book, the model explains the decision-relation more objectively. Taking remote-sensing satellite task scheduling as an example, on the one hand, as conventional studies usually only decide satellite imaging events, and seldom considers downlink events or memory-erasing events, such studies cannot truly and objectively reflect the decision-relation in practical problems and lose contact with reality. On the other hand, as mentioned in the introduction of this book, for task scheduling problems of agile remote-sensing satellites, conventional studies usually decide the specific beginning time of imaging events in the time window according to subjective principles such as "imaging quality first" and "imaging time first," resulting in a subjective coding and decoding process with high computational cost, which affects the objectivity of the decision-relation to a certain extent. On the basis of mathematical description of problems, the decision model in this section truly and objectively explains the decision-relation in the actual satellite task scheduling problems. In particular, in the model, the event beginning time is directly decided by decision-relation of event-event decision opportunity to avoid subjective encoding and decoding, opening up a new idea for objective description of satellite task scheduling decision-relation.

(2) **Covering four main satellite task scheduling problems, the decision model built in this section based on the modeling principle of generalization and loose-coupling is more general, which forms a unified decision system.** The decision model in this section is applicable to four main satellite task scheduling problems, including remote-sensing satellite scheduling, relay satellite scheduling, navigation satellite scheduling and satellite range scheduling. Compared with the non-universal decision models in conventional studies on satellite task scheduling problems, to fully meet the requirements of integrated management of satellite systems and address the "one satellite, one system" shortcoming to accelerate the integrative development of the systems, the decision model in this section has more generality and expandability for a wider scope of application. Moreover, in the aspect of model representation, this section standardizes the value range of decision variables to [0, 1] to construct a general-purpose 0–1 mixed integer decision matrix in a unified way. As a result, this section provides a general data interface for the follow-up modeling of constraints and objectives, facilitates the subsequent design of algorithm neighborhood, and plays the key coupling point for the satellite task scheduling engine framework in this book that loosely decouples the model – conventional algorithm – emergency algorithm and decision – constraints – objectives.

To sum up, in this section, with examples based on four satellite scheduling problems including remote-sensing satellite scheduling, relay satellite scheduling, navigation satellite scheduling and satellite range scheduling, the decision-relation between the task set and the resource set, specifically between the satellite event and its EO, is creatively explained. Also, a general-purpose 0–1 mixed integer decision model is built, providing a new and general decision model for various satellite task scheduling problems, laying an important foundation for subsequent general-purpose modeling of constraints and objectives and algorithm design, and playing the key coupling point in the loosely coupled Engine framework in this book.

3.4 Constraint models and constraint network for satellite task scheduling

To meet the actual requirements of the satellite constraint modeling and management, based on the above-mentioned general-purpose 0–1 mixed integer decision model, this section first analyzes four common constraints in satellite task scheduling problems including resource availability, logicality, resource capability and resource protection. Then, a general-purpose constraint template composed of an object, a threshold and a relationship is designed to uniformly describe the constraints of satellite task scheduling. Detailed examples and mathematical models are given. On this basis, this section illustrates the constraint network formed by the above constraints and designs the constraint value calculation method correspondingly. As a result, this section provides a general constraint evaluation basis and an efficient constraint value calculation method for satellite task scheduling schemes, and a practical method for application and flexible expansion of related constraint models, as well as a default and convenient constraint modeling method for satellite control agencies.

3.4.1 Constraint analysis and classification

There are many complex constraints in satellite task scheduling problems. For example, there are more than 100 constraints in the SuperView-1 task scheduling problems. In fact, since these constraints are usually formed due to different causes and some seemingly unreasonable and irregular constraints also have logical causes, reasonable analysis and classification of constraints and their causes are helpful to systematically build the constraint model of satellite task scheduling problems to understand the constraint system.

Practice shows that the main causes of constraints in satellite task scheduling problems include: 1) To ensure that resources such as satellites and stations are visible and available; 2) To ensure the rational logic of satellites and stations in the process of executing tasks; 3) To show the actual ability of satellites and stations in executing tasks; 4) To protect satellites and their payloads to prolong their service life. In view of this, this section classifies the main constraints in the current satellite task scheduling problems into the following four categories:

(1) Resource availability constraint
This kind of constraint is mainly to ensure that resources such as satellites and stations are visible and available. According to Definition 3.1, event EO is an opportunity for a satellite event to be executed within its time window. Therefore, in satellite task scheduling problems, an event only can be executed when the visible and available resources can provide the EO. In other words, this kind of constraint is the basis for the generation and screening of event EO in the modeling process of satellite task scheduling problems. For any satellite event, the EO in the EO set must satisfy this kind of constraint, and the EO that does not satisfy the constraint will be eliminated. For this reason, only such constraints can be pre-processed in satellite task scheduling problems. Among such constraints, the most typical one is the visible time window constraint. For example, the imaging events of remote-sensing satellite tasks can only be executed within the visible time window of "objective-satellite," and the downlink

events can only be executed within the visible time window of "satellite-station"; Events in navigation satellite tasks, relay satellite tasks and satellite range tasks also need to satisfy the visible time window constraints correspondingly.

(2) Logicality constraints

This kind of constraint is mainly to ensure the rational logic of task execution. In the process of task scheduling, any scheduling scheme should follow the basic logic of facts and the business process of satellite management. Therefore, this kind of constraint is the primary basis for the feasibility evaluation of satellite task scheduling scheme. For example, since the remote-sensing satellite can only downlink the acquired image data of the object to the station after it has imaged the task object first, the downlink event of the same task must be later than the imaging event; otherwise, it violates the basic logic of the fact.

(3) Resource capability constraints

In the satellite task scheduling problems, the ability of resources is usually limited. Therefore, as bottleneck constraints that limit the upper limit of task scheduling, resource capability constraints reflect the ability of resources to perform tasks. Under such constraints, how to maximize the utilization and ability of resources is a common topic in the study of various combinational optimization problems. In satellite task scheduling problems, the scope of resource capability constraints is extensive. For example, in the task scheduling problems of remote-sensing satellite, the resource capability constraints involve not only the battery and memory capacity of the satellite, but also the transition time constraint between its continuous imaging and downlink events (including attitude transition, working mode transition, payload transition, etc.). Another example is that, in the task scheduling problems of SuperView-1 satellite, there are more than 50 resource capability constraints, which are summarized and classified into three kinds of constraints through the constraint template below in this book. Therefore, in constraint modeling of satellite task scheduling problems, it is crucial to summarize and classify resource capability constraints based on knowledge and the characteristics of combinational optimization problems of satellites.

(4) Resource protection constraints

This kind of constraint is mainly to protect satellites and their payloads to prolong their service life. Because of the long service period of satellites in orbit, and the difficulty and high cost of fault maintenance, to protect satellites and their payloads, the satellite control agencies usually set some conservative constraints to avoid the satellites working under limited conditions for a long time. For example, in the task scheduling problems of remote-sensing satellite, to prevent satellite cameras, antennas and other payloads from working for too long with high frequency, which will affect their service life, the maximum number of times and time of imaging and downlink events for each orbit in a single day are usually constrained. It is worth noting that the resource protection constraints can be regarded as resource capability constraint set intentionally, which objectively reflects the constraint on the resource capability to perform tasks. Therefore, in the process of modeling satellite task scheduling constraints, this kind of constraint can be included in the category of resource capability constraints. At the same time, because such constraints are set intentionally, they are usually flexible and variable in special periods or unexpected situations. For example, with the rapid increase in the scale of satellite tasks and the limited task cycles, the related constraints such as the number of times and cycles of satellite tasks in each orbit may be appropriately compromised. In other words, as the constraints of satellite task scheduling problems are not static in the long-term management process, it is important to build an open and flexible constraint environment for constraint modeling.

In addition to the above four kinds of constraints, in satellite task scheduling problems, there may be some other constraints for meeting the individual needs of satellite control (or use) agencies. On the one hand, some individual constraints can be included in the above four kinds of constraints, for example, the special time, special resources, and special angles of task execution can be regarded as resource availability constraints, and special task order can be regarded as a logicality constraint. On the other hand, it is necessary to build special constraint models for other exclusive and distinctive constraints that cannot be included in the four types of constraints.

Based on the above constraint analysis and classification, a set of general constraint template for satellite task scheduling is designed below with examples and mathematical models of the four kinds of constraints, which provides a practical method for building related constraint models and their flexible expansion, as well as a default and convenient constraint modeling method for satellite control agencies.

3.4.2 Constraint template with examples

3.4.2.1 Template composition

Constraint template is helpful to describe complex and diverse constraints in satellite task scheduling problems in a standardized and intuitive way. Consequently, it contributes to building general constraint models conveniently, configuring constraint parameters flexibly, and improving the expandability of constraint models. Therefore, based on the above constraint analysis and classification, this section template the constraints in satellite task scheduling problems into constrained objects, constraint thresholds and constraint relationships, as defined below:

Definition 3.2 (Constrained objects): These refer to instantiated objects or functionalized indicators that are constrained in constraints.

Definition 3.3 (Constraint thresholds): These refer to the thresholds of the constrained objects, that is, the values when the constrained objects just meet the constraints.

Definition 3.4 (Constraint relationships): These refer to the object matching relationship or indicator value relationship that the constrained objects and the constraint thresholds must satisfy in constraints.

Thus, in the constraint set F^H of satellite task scheduling problems, any constraint f_i^H can be templated as:

$$f_i^H = \{c_i^H, y_i^H, \omega_i^H\}, \ i \leq |F^H| \tag{3.26}$$

where c_i^H is the constrained object of constraint f_i^H, y_i^H is the constraint threshold of constraint f_i^H, and ω_i^H is the constraint relationship of the constraint f_i^H.

Based on this template and the above constraint analysis and classification, taking the constraints of SuperView-1 remote-sensing satellite task scheduling problem as an example, as listed in Table 3.2, this section gives examples and corresponding explanations of 12 subcategories of the four main categories of constraints including resource availability constraints, logicality constraints, resource capability constraints and resource protection constraints, as well as the mathematical models of each constraint.

3.4.2.2 Example of resource availability constraints

The resource availability constraints listed in Table 3.2 can be expressed by the following equation:

$$0 \leq x_{ij} \leq 1, \ \forall i \leq |T|, \ j \leq |E_i| \tag{3.27}$$

where x_{ij} is the decision variable for the jth event of the ith task, $x_{ij} \in X$.

This constraint condition limits the value range of the decision variable x_{ij}. In other words, the EO at which an event is executed must belong to the EO set of the event. This constraint is consistent with the eqs. (3.21) and (3.22) to define decision variable. It is important to note that when the decision variable $x_{ij} = 0$, it indicates that the corresponding event is not executable. The situation when the decision variable $x_{ij} = 0$ is not considered in the following constraints and objective models.

3.4.2.3 Example of logicality constraints

The three subcategories of logicality constraints listed in Table 3.2 can be expressed by the following equation:

$$e(e_{ij}) \leq b\left(e_{ij'}\right), \ \text{if Seq}(e_{ij}) > \text{Seq}\left(e_{ij'}\right), \ \forall i, i' \leq |T|, j \leq |E_i|, j' \leq |E_{i'}| \tag{3.28}$$

$$s(e_{i1}) = s(e_{i2}), \ \forall i \leq |T| \tag{3.29}$$

$$b(e_{i2}) \leq b\left(e_{i3}\right), \ \text{if } b(e_{i1}) \leq b\left(e_{i3}\right) \& s(e_{i1}) = s\left(e_{i'1}\right), \ \forall i, i' \leq |T| \tag{3.30}$$

Table 3.2: Example and explanation of task scheduling constraints of SuperView-1 remote-sensing satellite based on the constraint template.

No.	Sector	Subcategory	Constraints			
			Constrained object	Constraint relationship	Constraint threshold	Equation
1		Resource availability constraint	An EO when a task event is executed	Belong to	EO set of the task event	(3.27)
2	Logicality	Event order constraint	End time of a task event	Less than or equal to	Beginning time of an event after of the event based on the predetermined order	(3.28)
3		Data consistency constraint	A satellite that performs an imaging event of a task	Equal to	Satellite that performs the downlink event of the same task	(3.29)
4		Memory erasure constraint	The end time of a downlink event of the task to which the imaging event before the satellite's arbitrary memory-erasing event corresponds	Less than or equal to	Beginning time of memory-erasing event	(3.30)
5	Resource capability	Memory capacity constraint	The total amount of data generated by an imaging event between two memory-erasing events of a satellite	Less than or equal to	Memory threshold of the satellite	(3.31)
6		Electricity consumption constraint	Total power consumption for executing events in a single orbit	Less than or equal to	Power threshold of the satellite	(3.32)
7		Event execution transition time constraint	The interval time between two events performed by a satellite or station	Less than or equal to	Transition time required for the satellite or station to perform the two events	(3.33)

Table 3.2 (continued)

No.	Sector	Subcategory	Constraints			
			Constrained object	Constraint relationship	Constraint threshold	Equation
8	Resource protection	Number of times and time allowed for satellite payloads to execute events in a single orbit	Total number of times and time allowed for satellite payloads to execute events in a single orbit	Less than or equal to	Maximum number of times and time allowed for satellite payloads to execute events in a single orbit	(3.35)
9	Resource protection	Number of times and time allowed for satellite payloads to execute events in a single day	Total number of times and time allowed for satellite payloads to execute events in a single day	Less than or equal to	Maximum number of times and time allowed for satellite payloads to execute events in a single day	(3.36)
10		Number of times and time allowed for satellite platform to maneuver in a single orbit	Total number of times and time allowed for satellite platform to maneuver in a single orbit	Less than or equal to	Maximum number of times and time allowed for satellite platform to maneuver in a single orbit	(3.37)
11		Number of times and time allowed for satellite platform to maneuver in a single day	Total number of times and time allowed for satellite platform to maneuver in a single day	Less than or equal to	Maximum number of times and time allowed for satellite platform to maneuver in a single day	(3.38)
12		Number of times and time allowed for satellite platform to maneuver continuously	Total number of times and time allowed for satellite platform to maneuver continuously	Less than or equal to	Maximum number of times and time allowed for satellite platform to maneuver continuously	(3.39)

where e_{ij} is j^{th} event of the ith task (in remote-sensing satellite task scheduling problems, e_{i1}, e_{i2}, and e_{i3} are imaging, downlink and memory-erasing events respectively), $e_{ij} \in E_i$. $b(e_{ij})$ is the beginning time of event e_{ij}.$e(e_{ij})$ is the end time of event $e(e_{ij})$. Seq (e_{ij}) is the predetermined start sequence of event e_{ij}. $s(e_{ij})$ is the satellite executing event e_{ij}, based on eqs. (3.11b), (3.12c), and (3.13c): $s(e_{ij}) = s(o(\text{tw}(e_{ij}))) \in S$.

In summary, the logic of constraint (3.28) is that the event must be executed in the predetermined sequence.

The logic of constraint (3.29) is that the data generated by the imaging event e_{i1} will be stored on the satellite $s(e_{i1})$ executing the event, and the corresponding downlink event I_2 can be performed by the satellite only.

The logic of the constraint (3.30) is that to prevent the data of the imaging event e_{i1} stored on the satellite from being mistakenly erased by the memory-erasing event $e_{i'3}$ before the downlink event e_{i2}, the end time $e(e_{i2})$ of the downlink event must be earlier than the time of memory-erasing event $b(e_{i3})$.

3.4.2.4 Example of resource capability constraints

The three subcategories of resource capability constraints listed in Table 3.2 can be expressed by the following equation:

$$\sum_{e_{i1} \in \left\{ e_{i1} | b(e_{i'3}) \le b(e_{i1}) \le b(e_{i''3}) \right\}} m(e_{i1}) \le M(s(e_{i1})), \text{ if } s(e_{i1}) = s\left(e_{i'1}\right) = s\left(e_{i''1}\right), \forall i, i', i'' \le |T|$$

$$(3.31)$$

$$\sum_{e_{ij} \in \left\{ e_{ij} | o(e_{ij}) = o_k \right\}} \varepsilon\left(e_{ij}\right) \le E(o_k), \forall i \le |T|, j \le |E_i|, k \le |O| \tag{3.32}$$

$$\delta\left(e_{ij}, e_{i'j'}\right) \ge \Delta\left(e_{ij}, e_{i'j'}\right), \text{ if } o\left(e_{ij}\right) = o\left(e_{i'j'}\right) || g\left(e_{ij}\right) = g\left(e_{i'j'}\right), \forall i, i' \le |T|, j \le |E_i|, j' \le |E_{i'}| \tag{3.33}$$

where $s(e_{ij})$ is the satellite executing event e_{ij}.$m(e_{ij})$ is the amount of data generated by event e_{ij}. $M(s(e_{ij}))$ is the memory threshold of satellite $s(e_{ij})$ for executing event e_{ij}.$o(e_{ij})$ is the orbit of the satellite executing event e_{ij}, based on eqs. (3.12c) and (3.13c): $o(e_{ij}) = o(tw(e_{ij})) \in O.\varepsilon(e_{ij})$ is the power consumption of event e_{ij}.$E(o(e_{ij}))$ is the electricity threshold of satellite orbit $o(e_{ij})$ executing event e_{ij}.$g(e_{ij})$ is the ground station executing event e_{ij}, based on eqs. $(e_{i'j'})$ is the actual interval between events e_{ij} and $e_{i'j'}$. $\Delta(e_{ij}, e_{i'j'})$ is the transition time required between events e_{ij} and $e_{i'j'}$.

In summary, constraint eq. (3.31) reflects the data storage capacity of satellite resources. The total amount of data generated by imaging events executed between the two memory-erasing events $e_{i'3}$ and $e_{i''3}$ shall not exceed the memory threshold of the satellite.

Constraint eq. (3.32) reflects the power supply capacity of satellite orbital resources. Under normal circumstances, since the satellite is not planned to execute tasks above high-latitude areas such as the North and South Poles, and the on-board batteries are charged during this period to supply the electricity required by the satellite to execute tasks in the next orbit, orbit is taken as the unit in the constraints modeling.

Constraint eq. (3.33) reflects the capability of resources such as the satellite and the station to execute two different events in succession. For example, the satellite executes attitude transition, working mode transition, and payload startup for two imaging events; and the station executes antenna elevation transition for two downlink events. In this section, the required transition time between two different events e_{ij}

and $e_{i'j'}$ is uniformly denoted as $\Delta\,(e_{ij},\,e_{i'j'})$, which can be calculated based on the actual situation; The actual interval time between the two events is denoted as $\delta(e_{ij},\,e_{i'j'})$, which can be calculated as follows:

$$\delta\left(e_{ij},e_{i'j'}\right) = \begin{cases} b\left(e_{ij}\right)e\left(e_{i'j'}\right), & \text{if } b\left(e_{ij}\right) \geq e\left(e_{i'j'}\right), \\ b\left(e_{i'j'}\right)e\left(e_{ij}\right), & \text{if } b\left(e_{i'j'}\right) \geq e\left(e_{ij}\right), \ \forall i,i' \leq |T|, j \leq |E_i|, j' \leq |E_i'| \\ 0, & \text{otherwise} \end{cases} \tag{3.34}$$

3.4.2.5 Example of resource protection constraints

The five subcategories of resource protection constraints listed in Table 3.2 can be expressed by the following equation, where constraints (3.35) and (3.36) constrain the number of times and time of events executed by the satellite payloads and station equipment to protect them:

$$\sum_{e_{ij} \in \left\{ e_{ij}|o(e_{ij}) = o_k, \, q\left(e_{ij}\right) = q_m \right\}} \begin{bmatrix} 1 \\ l\left(e_{ij}\right) \end{bmatrix} \leq \begin{bmatrix} n^E_{\max}(o_k, \, q_m) \\ l^E_{\max}(o_k, \, q_m) \end{bmatrix}, \, \forall i \leq |T|, \, j \leq |E_i|, \, k \leq |O|, \, m \leq |Q| \tag{3.35}$$

$$\sum_{e_{ij} \in \left\{ e_{ij}|d(e_{ij}) = d_k, \, q\left(e_{ij}\right) = q_m \right\}} \begin{bmatrix} 1 \\ l\left(e_{ij}\right) \end{bmatrix} \leq \begin{bmatrix} n^E_{\max}(d_k, \, q_m) \\ l^E_{\max}(d_k, \, q_m) \end{bmatrix}, \, \forall i \leq |T|, \, j \leq |E_i|, \, k \leq |D|, \, m \leq |Q| \tag{3.36}$$

where $o(e_{ij})$ is the satellite orbit in which event e_{ij} is executed, $q(e_{ij})$ is the payloads/ equipment required to execute event e_{ij}, $q(e_{ij}) \in Q.l(e_{ij})$ is the duration of event e_{ij}, $l(e_{ij}) = e(e_{ij})-b(e_{ij}).n^E_{\max}(o_k, q_m)$ is the maximum number of times of events executed by payload q_m in orbit o_k, $o_k \in O$, $q_m \in Q.l^E_{\max}(o_k, q_m)$ is the maximum time allowed by payload q_m in orbit o_k to execute an event. $d(e_{ij})$ is the date on which event e_{ij} is executed, based on eq. (3.11b): $d(e_{ij}) = d(o(e_{ij})) \in D.n^E_{\max}(d_k, q_m)$ is the maximum number of times the payload q_m executes an event on date d_k, $d_k \in D.l^E_{\max}(d_k, q_m)$ is the maximum time allowed by the payload q_m to execute an event on the date d_k.

Similarly, constraints (3.37) and (3.38) protect the satellite platform by constraining the number of times and timing of its maneuvers (i.e., the transition between two consecutive events):

$$\sum_{e_{ij} \in \left\{ e_{ij}|o(e_{ij}) = o_k \right\}} \begin{bmatrix} n^\Delta\left(e_{ij}\right) \\ \Delta\left(e_{ij}, \text{next}\left(e_{ij}\right)\right) \end{bmatrix} \leq \begin{bmatrix} n^\Delta_{\max}(o_k) \\ l^\Delta_{\max}(o_k) \end{bmatrix}, \, \forall i \leq |T|, \, j \leq |E_i|, \, k \leq |O| \tag{3.37}$$

$$\sum_{e_{ij} \in \left\{ e_{ij} | d(e_{ij}) = d_k, \, s\left(e_{ij}\right) = s_m \right\}} \begin{bmatrix} n^\Delta\left(e_{ij}\right) \\ \Delta\left(e_{ij}, \text{next}\left(e_{ij}\right)\right) \end{bmatrix} \le \begin{bmatrix} n^\Delta_{\max}(d_k, s_m) \\ l^\Delta_{\max}(d_k, s_m) \end{bmatrix},$$

$$\forall i \le |T|, \; j \le |E_i|, \; k \le |D|, \; m \le |S| \tag{3.38}$$

where $n^\Delta(e_{ij})$ is the number of satellite maneuvers required to execute event e_{ij}. Usually, it is 1 or $0.n^\Delta_{\max}(o_k)$ is the maximum number of maneuvers allowed for the satellite in orbit $o_k.l^\Delta_{\max}(o_k)$ is the maximum time allowed for the satellite in orbit o_k to maneuver.next(e_{ij}) is the next event that will be executed by the satellite executing event $e_{ij}.n^\Delta_{\max}(d_k, s_m)$ is the maximum number of maneuvers allowed for the satellite s_m on date d_k, $s_m \in S. \, l^\Delta_{\max}(d_k, s_m)$ is the maximum time allowed for the satellite s_m to maneuver on the date d_k.

In addition, constraint (3.39) further protects the satellite payloads and station equipment by preventing the satellite from maneuvering continuously too frequently in a short time:

$$\sum_{e_{ij} \in \left\{ e_{ij} | \, o\left(e_{ij}\right) = o_k, \Delta\left(e_{ij}, \text{next}\left(e_{ij}\right)\right) \Delta_C(o_k) \right\}} \begin{bmatrix} n^\Delta\left(e_{ij}\right) \\ \Delta\left(e_{ij}, \text{next}\left(e_{ij}\right)\right) \end{bmatrix} \le \begin{bmatrix} n^{\Delta_C}_{\max}(o_k) \\ l^{\Delta_C}_{\max}(o_k) \end{bmatrix},$$

$$\forall i \le |T|, \; j \le |E_i|, \; k \le |O| \tag{3.39}$$

where $\Delta_C(o_k)$ is the time threshold of a continuous maneuver in orbit $o_k.n^{\Delta_C}_{\max}(o_k)$ is the maximum number of continuous maneuvers allowed for the satellite in orbit $o_k. \, l^{\Delta_C}_{\max}(o_k)$ is the maximum time of continuous maneuvers allowed for the satellite in orbit o_k.

In summary, this section builds a general constraint template for satellite task scheduling problems and gives mathematical models of 12 subcategories of the four main categories of constraints including resource availability constraints, logicality constraints, resource capability constraints and resource protection constraints with examples of SuperView-1 remote-sensing satellite task scheduling problems. Based on the constraint template in this section, the constraint models can be built intuitively and conveniently, and the constraint parameters can be configured flexibly, which provides a standardized and general modeling method for complex and diverse constraints in satellite task scheduling problems.

In addition, the scheduling problems of relay satellite scheduling, navigation satellite scheduling and satellite range scheduling also involve some of the above constraints, as listed in Table 3.3. It is important to note that resource availability constraint and event execution transition time constraint, as the most common and important constraints in management of various satellites, are common to the four main satellite task scheduling. Since the logic of the constraints list in Table 3.3 is consistent with the constraint examples above, this section does not explain them in detail.

Table 3.3: Summary of constraint categories involved in various satellite task scheduling.

No.	Sector	Subcategory	Satellite task scheduling			
			Remote satellite task scheduling	Relay satellite task scheduling	Navigation satellite task scheduling	Satellite range scheduling
1	–	Resource availability constraint	●	●	●	●
2	Logicality	Event order constraint	●	●		●
3		Data consistency constraint	●		●	
4		Memory erasure constraint	●	●		
5	Resource capability	Memory capacity constraint	●	●		
6		Electricity consumption constraint	●	●		
7		Event execution transition time constraint	●	●	●	●
8	Resource protection	Number of times and time allowed for satellite payloads to execute events in a single orbit	●	●		
9		Number of times and time allowed for satellite payloads to execute events in a single day	●	●		
10		Number of times and time allowed for satellite platform to maneuver in a single orbit	●	●		
11		Number of times and time allowed for satellite platform to maneuver in a single day	●	●		
12		Number of times and time allowed for satellite platform to maneuver continuously	●	●		

3.4.3 Constraint network and calculation of constraint values

3.4.3.1 Constraint network

Based on the above examples of the template and the constraints, it can be seen that: 1) There are many kinds of constraints in satellite task scheduling problems, and the calculation is complex. Especially in the process of subsequent iterative search for the algorithm, the calculation of constraints may take up a lot of time, reducing the iterative efficiency of the algorithm. 2) Constrained objects and thresholds in all kinds of constraints are closely related to task events e_{ij}, i.e. decision variables x_{ij}, which naturally form a constraint network with decision variables x_{ij} as nodes, providing a new idea for the calculation of complex constraints. In this section first, the constraint network is clearly explained, and then a constraint value calculation method is designed based on the constraint network, becoming an important premise for incremental and efficient constraint calculation in the iterative search process of follow-up algorithms.

Now, taking the 2×2 decision matrix X shown in Figure 3.7(a) as an example, under the action of constraints (3.33), the combination of any two events e_{ij} may be involved, that is, the combination of any two decision variables x_{ij} in the decision matrix X may be involved, as shown in Figure 3.7(b).

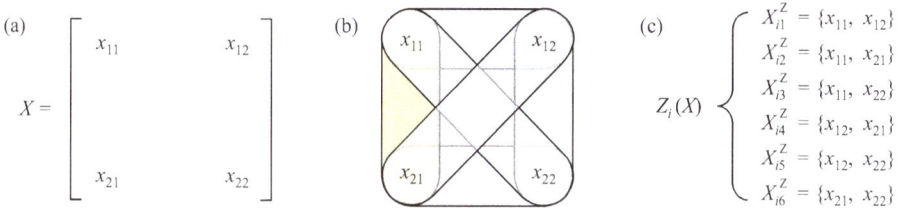

Figure 3.7: Example of constraint network for satellite task scheduling problems: (a) decision matrix X; (b) the combination of decision variables x_{ij} that the ith constraint may involve; and (c) the subset of decision matrices that the ith constraint may involve.

Further, each decision variable combination in Figure 3.7(b) constitutes a subset of the decision matrix X, as X_{i1}^Z to X_{i6}^Z in Figure 3.7(c). On this basis, a subset of these decision matrices can be expressed as follows:

$$Z_i(X) = \left\{ X_{ij}^Z \mid X_{ij}^Z \subset X, j = 1, 2, 3, \ldots \right\}, i \le \left| F^H \right|$$

(3.40)

where F^H is the constraint set X_{ij}^Z is the jth subset of decision matrix X that the ith constraint may involve. $Z_i(X)$ is set of subsets of decision matrix X that the ith constraint may involve, $Z_i(X) \supseteq X_{ij}^Z$.

As a note, each subset X_{ij}^Z in any decision matrix X is only related to the ith constraint, and does not change with the change of the values of the decision variables in the decision matrix X. In other words, given the constraints of the problem, screening and obtaining each decision subset X_{ij}^Z is a kind of pre-processing that provides an important basis for next frequent constraint value calculations.

3.4.3.2 Calculation of constraint values

Under the action of the constraint network, based on the current decision matrix X and each subset X_{ij}^Z, the constraint value $f_i^H(X)$ of the ith constraint can be calculated as follows:

$$f_i^H(X) = \sum_{j=1}^{|Z_i(X)|} f_i^H\left(X_{ij}^Z\right), \quad i \leq |F^H| \tag{3.41}$$

where $f_i^H(X)$, abbreviated as constraint value, is the value (≤ 0) calculated based on decision matrix X or its subset that violates the ith constraint.

It can be seen that by calculating the constraint values of each subset X_{ij}^Z in the decision matrix X, the originally complex and multiple constraint value calculation process of satellite task scheduling problems is divided into several independent and single constraint value calculation. The advantages of this calculation method are as follows:

(1) **It objectively reflects the action mechanism of constraints.** A constraint usually needs to be calculated several times in a cycle because the constraint essentially acts on a combination of many different decision variables x_{ij}, that is multiple subsets of decision matrices X_{ij}^Z. eq. (3.41) objectively presents the mechanism of this cyclic calculation.

(2) **It reduces the complexity of single constraint value calculation.** Whenever a decision variable x_{ij} is changed and the constraint value needs to be recalculated, it only needs to recalculate the constraint value of the subset X_{ij}^Z of the decision matrix containing the decision variable x_{ij} by eq. (3.41); There is no need to calculate most of the remaining subsets X_{ij}^Z since their constraint values are not affected. Therefore, this calculation method significantly reduces the complexity of single constraint value calculation, thus greatly shortening the constraint value calculation time, improving the calculation efficiency, and providing an important guarantee for the subsequent design of efficient algorithms.

3.4.4 Advantage analysis

Based on the above, compared with conventional studies on satellite task scheduling problems, the constraint model and constraint network built in this section have the following two advantages:

(1) **The complex constraints are described in a more general way and all kinds of satellite task scheduling constraints are integrated into a unified system.** There are many complex constraints in satellite task scheduling problems, but for the purpose of simplification or theoretical study, the constraint models built in conventional studies are often not comprehensive and general. For example, the constraint models in related studies of remote-sensing satellite task scheduling cannot be

applied to more than 100 constraints proposed by the management agency of Super-View-1 satellite, and it is also difficult to apply the models to other types of satellite task scheduling problems. Based on the reality-based and application-oriented principle of modeling and general decision model, this section analyzes and classifies the main constraints of satellite task scheduling problems, designs a general constraint template to uniformly describe the constraints of problems of remote-sensing satellite scheduling, relay satellite scheduling, navigation satellite scheduling and satellite task scheduling, and successfully integrates them into a unified system. It provides a general method for constraint modeling of various satellite task scheduling problems, as well as a default and convenient constraint modeling means for management agency.

(2) **The constraint values can be calculated more efficiently, which saves the calculation time for solving the satellite task scheduling problems.** Although there is usually no specific explanation of constraint calculation method in related studies, it is known that many studies and active satellite task scheduling systems pay insufficient attention to constraint calculation method, and the calculation time complexity is high. Considering the complex constraint environment of the actual satellite task scheduling problems, the time cost of constraint calculation will be further enlarged by the iterative process of the algorithms, which seriously affects the solving efficiency of the satellite task scheduling problems. As mentioned in Section 3.3, this section builds the constraint network that objectively reflects the mechanism of constraints and reduces the complexity of single constraint calculation. Especially in the case of complex constraint environment or a large number of iterative algorithms, this method can greatly save constraint calculation time, which is particularly important for the design of efficient algorithms.

To sum up, on the basis of constraint analysis and classification, this section designs a general constraint template and an efficient constraint calculation method for various satellite task scheduling problems, which provides important support for practical application and flexible expansion of constraint models, as well as the design of subsequent satellite task scheduling algorithms.

Besides, since the soft constraints and their corresponding constraint templates, networks and calculation methods of constraint values in satellite task scheduling problems are similar with those in this section, they will not be introduced separately.

3.5 Objective function model for satellite task scheduling

For the purpose of optimizing actual satellite operation, based on the 0–1 mixed integer decision model mentioned above, the objective function models of problems of remote-sensing satellite scheduling, relay satellites scheduling, navigation satellite scheduling and satellite range scheduling are established respectively, and the calculation methods of the objectives of agile remote-sensing satellites regarding the imaging quality as important and the time delay of navigation satellites are especially introduced, providing evaluation basis for satellite task scheduling schemes to guide the optimization direction for other algorithms.

3.5.1 Objective function of remote-sensing satellite task scheduling

3.5.1.1 General objective function
In most studies and practical application scenarios of remote-sensing satellite task scheduling, the objective function of remote-sensing satellite task scheduling can be expressed by eq. (3.42), that is, maximizing the total objective value of remote-sensing satellite task execution:

$$\max \; f(X) = \sum_{i=1}^{|T|} f_i \cdot x_{i1} \tag{3.42}$$

where T is the task set. X is the decision matrix. f_i is the objective value of remote-sensing satellite task t_i, which is given in advance by the satellite control agency, $t_i \in T$. x_{i1} are the decision variables of imaging event e_{i1} of task t_i of remote-sensing satellite, $0 \leq x_{i1} \leq 1$, $x_{i1} \in X$. $\lceil \; \rceil$ indicates upward rounding.

Common task objective value f_i are as follows: 1) When it is constant to one, the above equation indicates the maximum total number of tasks executed by the satellite; 2) When it is for the imaging time of task, the equation indicates the maximum imaging time of the satellite; 3) When it is for the priority or importance index of task, the above equation indicates the maximum priority or the sum of important indexes of the task; 4) When it is for the weighted value of the above indexes, the equation indicates the comprehensive index obtained by weighting the maximum imaging time and the priority of the task.

3.5.1.2 Objective function considering imaging quality
In recent years, with the rapid development of remote-sensing satellite and payload technology, the ability of on-orbit imaging has been continuously improved. Some scholars emphasize that the influence of satellite imaging quality should be further considered in the process of remote-sensing satellite task scheduling.

Remote-sensing satellites can be divided into agile remote-sensing satellites and non-agile remote-sensing satellites depending on whether they have pitch imaging capability. Based on the description of problems in this chapter, the main differences between the two satellites are as follows: the imaging time window of non-agile remote-sensing satellite only contains one event EO, and the satellite pitch angle is always 0, as shown in Figure 3.8(a). The imaging time window of agile remote-sensing satellite contains multiple event executable opportunities, and each opportunity corresponds to a satellite pitch angle respectively, as shown in Figure 3.8(b). In other words, when the imaging events of agile remote-sensing satellites are executed at different executable opportunities, the satellite pitch angles are different and the imaging quality is different.

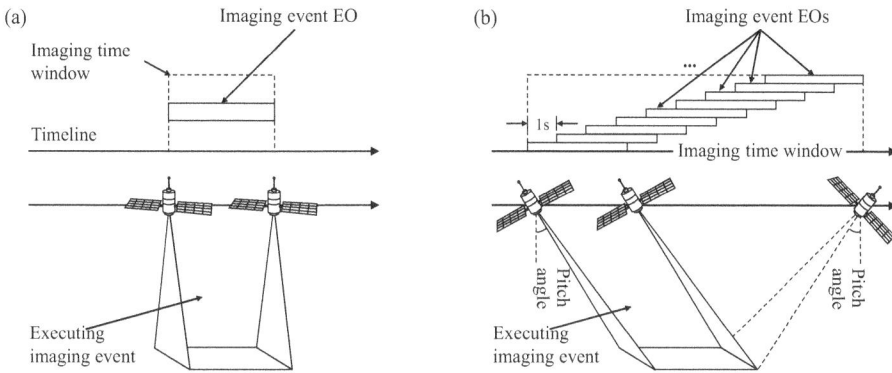

Figure 3.8: Examples of non-agile remote-sensing satellite and agile remote-sensing satellite: (a) non-agile remote-sensing satellite and (b) agile remote-sensing satellite.

In view of this phenomenon, there are usually two treatment methods in related studies and application at present:

(1) The imaging quality is considered in the constraints. For example, in the task scheduling problems of SuperView-1 satellite, the management agency requires that the results of satellite imaging events should achieve certain quality, that is, the events can only be executed at the executable opportunities that meet the imaging quality requirements. Therefore, such constraints belong to a kind of resource availability constraint, and the management agency has "cut" the time windows of imaging events in advance, eliminating the event executable opportunities that cannot meet the constraints to ensure that all imaging events can meet the imaging quality requirements when they are executed at all executable opportunities. In this case, the objective function of remote-sensing satellite task scheduling is not affected, and eq. (3.42) is still applicable. This method is suitable for the task scheduling problems of agile remote-sensing satellites with definite constraints on imaging quality. It is worth noting that because a certain number of event executable opportunities are excluded in advance, this processing method reduces the decision space of the original problems and improves the problem solving efficiency to a certain extent.

(2) The imaging quality is considered in the objective function. Peng et al. [174] directly considered the influence of imaging quality in the objective function, that is, according to the pitch angle of satellite imaging, the task objective value is penalized in a certain proportion. Therefore, on the basis of eq. (3.42), the objective function of remote-sensing satellite task scheduling is rewritten as

$$\max \ f(X) = \sum_{i=1}^{|T|} f_i \cdot \lceil x_{i1} \rceil \cdot \left[1 - \frac{|\theta(x_{i1})|}{0.5\pi} \right] \tag{3.43}$$

where $\theta(x_{i1})$ is the pitch angle (radian) of the satellite when imaging event e_{i1} is executed, based on eqs. (3.22) and (3.13c): $\theta(x_{i1}) = A_\theta(e_{i1}^k), k = \lceil x_{i1} \cdot | EO_{i1} | \rceil$.

It can be seen that the above $\frac{|\theta(x_{i1})|}{0.5\pi}$ is the penalty coefficient, which is proportional to the pitch angle $\theta(x_{i1})$. When the pitch angle is zero, the highest imaging quality can be obtained, and the penalty coefficient is zero, that is, there is no penalty; With the increase of pitch angle, the penalty coefficient increases and the task objective value decreases; When the pitch angle reaches 0.5 π (90°), the penalty coefficient reaches one and the objective value decreases to 0. This processing method directly quantifies the impact of imaging quality on task objectives, and is suitable for the task scheduling problems of agile remote-sensing satellites that require high imaging quality. Peng et al. [174] called the objective function a time-dependent objective function, that is, the task objective value depends on the time when the task is executed.

In view of the above two different processing methods, in the actual remote-sensing satellite task scheduling problems, it is necessary to selectively build an appropriate objective function model according to the specific requirements of the management agency.

3.5.2 Objective function of relay satellite task scheduling

Compared with remote-sensing satellite tasks, relay satellite tasks contain relatively few events, usually including downlink events and memory-erasing events. At the same time, based on the actual situation, the time dependence of task execution can be ignored for the time being. Therefore, the objective function in the task scheduling problems of relay satellites can be expressed as follows:

$$\max \ f(X) = \sum_{i=1}^{|T|} f_i \cdot x_{i1} \tag{3.44}$$

where f_i is the objective value of relay satellite task t_i, which is given in advance by satellite control agency, $t_i \in T. x_{i1}$ is the decision variable of downlink event e_{i1} of task t_i of relay satellite, $0 \le x_{i1} \le 1$, $x_{i1} \in X$.

Similar to the objective value of remote-sensing satellite tasks, the objective value f_i of relay satellite tasks is also given in advance by the management agency. Common objective values include 1, downlink time, and priority or weighted values of various indicators.

3.5.3 Objective function of navigation satellite task scheduling

In the navigation satellite system, since the satellites are scattered and many satellites are outside the tracking, telemetry and command range for a long time or periodically, the navigation data of these satellites can only be downlinked to the stations through inter-satellite links. Therefore, satellites in the navigation system can be divided into anchor satellites and non-anchor satellites, as defined below:

Definition 3.5 (Anchor satellite): A navigation satellite in the navigation satellite system, which establishes satellite-ground link with domestic stations and can downlink on-board data in real time.

Definition 3.6 (Non-anchor satellite): other navigation satellites except anchor satellites in the navigation satellite system. Navigation data of non-anchor satellite can only be downlinked to anchor satellites through inter-satellite links (including the transit of other non-anchor satellites), and then downlinked to the stations by anchor satellites.

Definition 3.7 (Time delay of navigation data downlink): abbreviated as time delay, in the navigation satellite system, the time (unit: seconds) or the number of timeslots that the navigation data generated by the satellites in a certain timeslot is downlinked to the anchor satellites.

According to the above definitions, the time delay of anchor satellites in the navigation system is 0, and the time delay of non-anchor satellites is at least one timeslot. Taking 30 satellites of BDS-3 navigation system as an example, the number of non-anchor satellites exceeds 15 most of the time. Therefore, the purpose of such navigation satellite task scheduling is to establish effective inter-satellite links for the non-anchor satellites to return navigation data, thus reducing time delay to improve the overall accuracy of the navigation satellite system.

Therefore, with the goal of minimizing the average delay of non-anchor satellites in the navigation system in each timeslot, the objective function of the navigation satellite task scheduling problems can be expressed by eq. (3.45) that corresponds to the optimization requirements of the management agency for reducing the delay of navigation data to improve the navigation accuracy:

$$\max f(X) = - \frac{\sum_{i=1}^{|T|} \sum_{j=1}^{|S|} f(x_{ij})}{\sum_{i=1}^{|T|} \sum_{j=1}^{|S|} [1 - a(x_{ij})]} \tag{3.45}$$

where $f(x_{ij})$ is the time delay of satellite s_j in the ith timeslot, calculated by the following eq. (3.46). $a(x_{ij})$ is a variable between 0 and 1, indicating whether the satellite s_j in the ith timeslot is an anchor satellite.

It is worth noting that since this book has stipulated that satellite task scheduling problems are all maximization problems, the average delay in the above equation is set to a negative value.

To show the calculation method of non-anchor satellite delay $f(x_{ij})$ in eq. (3.45) more clearly, this section illustrates the navigation satellite task scheduling problems of three timeslots for four satellites, as shown in Figure 3.9. In the task scheduling problems of navigation satellite, there are three timeslots, that is, task set $T = t_1, t_2, t_3$; four mutually visible navigation satellites, that is, satellite set $S = s_1, s_2, s_3, s_4$. Based on the decision matrix X shown in Figure 3.9(a), the inter-satellite link status of the navigation satellite system in each timeslot can be obtained, as shown in Figure 3.9(b).

To calculate the time delay of the navigation satellites in each timeslot, Figure 3.9(c) shows the downlink path of navigation satellite data between different timeslots from another perspective. Taking the satellite s_1 in the first timeslot as an example (bolded in the figure), the data generated by the satellite will first be downlinked to the satellite s_2, then downlinked by the satellite s_2 to the satellite s_4 in the second timeslot, and finally downlinked by the satellite s_4 back to the satellite s_1 in the third timeslot. Now, assuming that satellite s_4 is an anchor satellite, the downlink path of the data to the anchor satellites is: $s_1 \rightarrow s_2 \rightarrow s_4$. Since there are two timeslots, the time delay is two timeslots. Thus, based on the decision matrix X, the time delay $f(x_{ij})$ of the jth navigation satellite in the ith timeslot can be expressed as follows:

$$f\left(x_{ij}\right) = \begin{cases} 0, & \text{if } a\left(x_{ij}\right) = 1 \\ f\left(x_{i(i+1)k}\right) + 1, & \text{otherwise, } \quad i \le |T|, \ j \le |S|, \ k = \lceil x_{ij} \cdot |EO_{ij}| \rceil \end{cases} \tag{3.46}$$

where $a(x_{ij})$ is a variable between 0 and 1, indicating whether satellite s_j in the ith timeslot is an anchor satellite $.f(x_{ij})$ is the time delay of satellite s_j in the ith timeslot.

Therefore, by substituting eq. (3.46) into eq. (3.45), the average time delay of the navigation system can be obtained. In summary, this section introduces the objective function of navigation satellite task scheduling problems and the specific method of delay calculation, which provides evaluation basis for navigation satellite task scheduling schemes, that is, the evolution schemes of inter-satellite link network.

3.5.4 Objective function of satellite range scheduling

Similar to the problems of remote-sensing satellite scheduling and relay satellite scheduling, the problems of satellite range scheduling aims at maximizing the total objective of range tasks. Therefore, the objective function of satellite range scheduling can be expressed as follows:

$$\max \ f(X) = \sum_{i=1}^{|T|} f_i \cdot x_{i1} \tag{3.47}$$

where x_{i1} is the decision variables of downlink event e_{i1} of satellite range task t_i.

To sum up, for the purpose of optimizing actual satellite task scheduling, in this section, the objective function models of problems of remote-sensing satellite schedul-

(a)

$$X = \begin{bmatrix} x_{11} = 2/4 & x_{21} = 3/4 & x_{31} = 4/4 \\ x_{12} = 1/4 & x_{22} = 4/4 & x_{32} = 3/4 \\ x_{13} = 4/4 & x_{23} = 1/4 & x_{33} = 2/4 \\ x_{14} = 3/4 & x_{24} = 2/4 & x_{34} = 1/4 \end{bmatrix}^T \quad \cdots \cdots$$

$$x_{ij} = k/|S|$$
$$(x_{ik} = j/|S|)$$

(b)

① 1st satellite
② 2nd satellite
③ 3rd satellite
④ 4th satellite

(c)

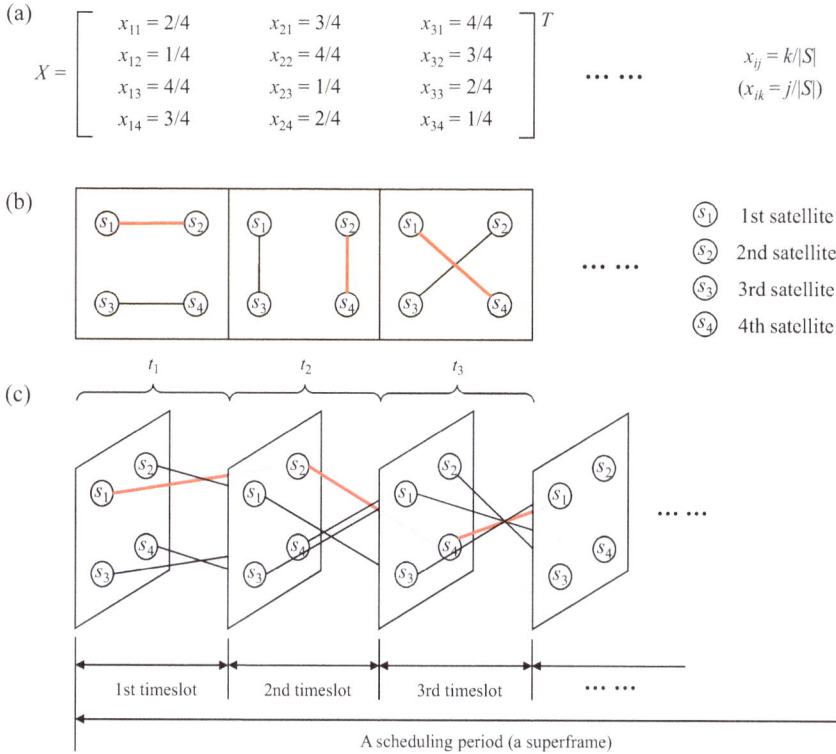

Figure 3.9: Navigation satellite task scheduling problems of three timeslots for four satellites: (a) the decision matrix X; (b) diagram of the inter-satellite link in each time; and (c) the downlink path of navigation satellite data between different timeslots..

ing, relay satellite scheduling, navigation satellite scheduling and satellite range scheduling are established respectively, and the calculation methods of the objectives of agile remote-sensing satellites regarding the imaging quality as important and the time delay of navigation satellites are especially introduced, providing evaluation basis for satellite task scheduling schemes to guide the optimization direction for other algorithms.

3.6 Chapter summary

To address the issues of insufficient generality of satellite task scheduling models and the current management situation of "one satellite, one system," following the reality-based and application-oriented principle, this chapter puts forward a general-purpose modeling method of satellite task scheduling that loosely decouples the decision, constraints, and objectives, specifically:

(1) The satellite task scheduling problems are described systematically and hierarchically. The satellite task scheduling problems are described as a quadruple including a task set, a resource set, a score set and a decision matrix. It is clarified that "event EO" is the underlying resource in the problems, which provides an objective basis for the modeling in this chapter.

(2) The general-purpose 0–1 mixed integer decision model for satellite task scheduling is built. The decision-relation between the task set and the resource set, specifically between the satellite event and its EO, is creatively explained to build a new and general decision model for various satellite task scheduling problems, providing a new and general decision model for various satellite task scheduling problems, laying an important foundation for subsequent general-purpose modeling of constraints and objectives and algorithm design, and playing the key coupling point in the loosely coupled Engine framework in this book.

(3) The constraint model and constraint network of satellite task scheduling are built. The constraints of satellite task scheduling are uniformly described, and a constraint calculation method based on the constraint network is designed, which provides a general constraint evaluation basis and an efficient constraint value calculation method for satellite task scheduling schemes, and a practical method for application and flexible expansion of related constraint models.

(4) The objective model of satellite task scheduling is built. The calculation methods of task objectives of agile remote-sensing satellite and time delay of navigation satellite considering imaging quality are especially introduced to guide the optimization direction for follow-up algorithms.

In the above modeling process, this chapter gives examples of problems including remote-sensing satellite scheduling, relay satellite scheduling, navigation satellite scheduling and satellite range scheduling. Through the above general modeling method that loosely decouples the decision, constraints, and objectives, this chapter properly model problems of the aforementioned four satellite task scheduling in a general-purpose manner. This manner opens up a new idea of general-purpose and refined modeling of satellite task scheduling problems, supporting the required general-purpose models for the satellite task scheduling engine in this book.

Chapter 4
General-purpose solving method for satellite conventional task scheduling

Facing the daily and weekly conventional task scheduling requirements of the satellite control agency, this chapter proposes a general-purpose adaptive parallel memetic algorithm (APMA). As the core algorithm for the satellite task scheduling engine in this book, APMA provides a general-purpose and efficient solution to the conventional task scheduling problems of remote-sensing satellite scheduling, relay satellite scheduling, navigation satellite scheduling, and satellite range scheduling.

Firstly, after analyzing the requirements of the satellite conventional task scheduling algorithm in terms of initial solution quality, local optimization, global optimization, adaptability, generality, and time complexity, this chapter designs a general-purpose APMA framework. To meet the requirements, the framework integrates the following four strategies: 1) heuristic-based fast initial solution construction strategy; 2) parallel search-based general-purpose local optimization strategy; 3) competition-based algorithm and operator adaptive selection strategy; 4) population evolution-based global optimization strategy. This chapter elaborates the implementation of the above four strategies, presenting algorithmic characteristics of synergistic and complementary strategies. On this basis, APMA is tested by benchmark such as orientation problems and simplified remote-sensing satellite task scheduling problems, proving its generality and effectiveness in solving standard problems and objectively verifying the algorithm performance experimentally.

4.1 General-purpose APMA framework

4.1.1 Requirements analysis

How to design efficient optimization algorithm is a constant theme in the study of combinational optimization problems. A comprehensive overview of exact solving algorithms, heuristic algorithms, and meta-heuristic algorithms commonly used in satellite task scheduling studies has been presented in the introduction of this book. Among them, Xiao et al. [19], Liu et al. [41], Bensana et al. [56], and Peng et al. [174] experimentally proved that the efficiency of exact algorithms in solving satellite task scheduling is very limited. For example, only 20 remote-sensing satellite tasks can be scheduled in 7,200 s of computation time, which is far from meeting the optimization requirements of actual satellite task scheduling [19]. Also, the satellite task scheduling model in the above literature is usually simplified to a large extent, so the efficiency of the algorithms is greatly reduced in practical application. Moreover, although heu-

https://doi.org/10.1515/9783111537191-004

ristic algorithms can quickly provide feasible solutions their optimization ability is very limited due to the lack of iteration and search mechanism for optimization. They are usually only used as algorithm support strategies such as the construction strategy of initial solution. It can be seen that exact solving algorithms and heuristic algorithms cannot meet the current daily and weekly conventional task scheduling requirements of the satellite control agency, so this chapter designs a satellite conventional task scheduling algorithm based on meta-heuristic algorithm.

In recent years, meta-heuristic algorithms represented by local search and evolutionary algorithms have become an important means to address combinational optimization problems and have been widely used in industry, logistics, aviation, aerospace, and other fields. Studies and practice have shown that the performance of meta-heuristic algorithms is usually affected by the following five aspects:

(1) Initial solution quality: The initial solution is the starting point of the search of meta-heuristic algorithms. A high-quality initial solution is helpful to guide meta-heuristic algorithms to enter the feasible domain space quickly and improve their early search efficiency. The role of initial solution is more prominent when solving satellite task scheduling problems with short computing time and complex constraints. In related studies, heuristic algorithms such as task allocation algorithms and conflict-avoidance algorithms are common algorithms for constructing initial solutions. Therefore, it is necessary to design a fast and effective heuristic initial solution construction algorithm for satellite task scheduling algorithm.

(2) Local optimization ability: Local optimization ability, also known as exploitation ability, refers to the ability of algorithms to search for optimization in a small neighborhood space. Since there are many complex constraints in satellite task scheduling problems, the feasible domain space is small, and the local optimization ability plays a very important role. In view of the characteristics of neighborhood search, local search algorithms such as hill climbing (HC) algorithm, Tabu search algorithm, simulated annealing algorithm, and late acceptance (LA) algorithm tend to have better local optimization ability and faster convergence rate, but they are also prone to fall into local optima in the neighborhood quickly, which is commonly known as "premature convergence." In this regard, the abovementioned algorithms such as Tabu search algorithm, simulated annealing algorithm, and LA algorithm, respectively, adopt meta-heuristic strategies such as Tabu table, probabilistic acceptance of inferior solutions, and backtracking, which provide the algorithms with the means to skip from local optima.

(3) Global optimization ability: Global optimization ability, also known as exploration ability, refers to the search optimization ability of algorithms in a large solution space, which is an important factor in determining whether the algorithm can skip from local optima. The Tabu table, probabilistic acceptance of inferior solutions, and backtracking strategies described above are all improvement measures for the local search algorithms to enhance their global optimization ability. In addition, in recent years, large neighborhood search algorithms [41, 53] have been studied and applied frequently, and the purpose of large neighborhood search is to improve the global optimization ability of the algorithms. Contrary to local search algorithms, in view of the characteristics of evolution and diversity of population, EAs such as genetic algorithms and particle swarm optimization algorithms usually have good global optimization ability, but they also show problems such as slow early convergence rate, high time complexity, and low search efficiency under complex constraints. In this regard, many scholars have designed a memetic algorithm (MA) that combines local search and evolutionary algorithm (EA), also known as cultural genetic algorithm, which provides a hybrid algorithm design idea that takes into account both local and global optimization abilities.

(4) **Adaptability and generality.** In the process of solving practical problems, the problem scale, the difficulty of solving, and even the constraints may change, so the adaptive ability of an algorithm can affect its practical application performance. The adaptive ability of an algorithm usually includes two aspects: 1) adaptive ability at the level of algorithm and operators, that is, the algorithm, with a certain degree of self-organization, operators, and related parameters can be adjusted appropriately along with the optimization process. For example, the adaptive large neighborhood search algorithms can dynamically adjust the probability of the use of operators according to the optimization results of the operators; 2) adaptive ability at the problem level, that is, with a certain degree of generality, the algorithm can be applied to address different problems. Since this study covers different satellite task scheduling problems such as remote-sensing satellite scheduling, relay satellite scheduling, navigation satellite scheduling and satellite range scheduling, which have huge differences and high complexity, the adaptive ability and generality of the algorithm are particularly important.

(5) **Time complexity.** Although the abovementioned points can enhance the optimization ability of meta-heuristic algorithms, they also increase the time complexity of the algorithm to varying degrees. For example, MA, which takes EA as the main loop and local search algorithm as the individual improvement strategy, takes into account both local and global optimization abilities, but it also leads to a sharp increase in the time of each evolutionary iteration. In addition, since the time complexity of the algorithm also includes the time complexity of computing constraints and profits during the iteration process, the constraint computation time cannot be ignored. In the context of this book, the solving time for satellite task scheduling is extremely limited. For example, the solving time for a navigation satellite task scheduling scenario is usually less than 1 min, and the solving time for a remote-sensing satellite task scheduling scenario is only 3–10 min. Therefore, the time complexity of an algorithm is also an important factor that cannot be ignored in the process of studying and designing the algorithm. Moreover, computer technology has undergone profound changes in recent years, and the popularity of high-performance computing and parallel computing has created a good hardware environment for algorithm design and application. In this regard, many scholars have carried out studies on the

design and application of parallel optimization algorithms, which provide a simple and feasible technical way to reduce the time complexity of algorithms and accelerate the algorithm optimization process.

In view of the above aspects of initial solution quality, local optimization ability, global optimization ability, adaptability, generality, and time complexity, this section designs an algorithm according to the listed countermeasures in Table 4.1 and proposes a general-purpose adaptive parallel memetic algorithm (APMA), which provides a general-purpose and efficient means of solving satellite conventional task scheduling problems.

4.1.2 Algorithm framework

Based on the above analysis, this section designs a general-purpose APMA, and the algorithm framework is shown in Figure 4.1. APMA mainly integrates the following four general strategies:

(1) **Heuristic-based fast initial solution construction strategy, referred to as "construction" strategy.** This strategy is the initialization strategy of APMA. Based on the abovementioned general-purpose 0–1 mixed integer decision model for satellite task scheduling, this strategy "arranges" the earliest EO that meets the constraints for each task event sequentially according to the heuristic algorithm with "pre-allocation and first-in-first-service," helping the algorithm quickly enter high-quality feasible domain space and providing an important basis for APMA to optimize iteratively.

(2) **Parallel search-based general-purpose local optimization strategy, referred to as "parallel" strategy.** This strategy is the first step of the APMA inner loop. According to the current optimal solution set (or initial solution set) and the probability of the use of algorithms and operators, this strategy runs different local search algorithms in parallel, calls different neighborhood operators, and carries out fast and efficient search optimization through different search trajectories to obtain diversified local optimization results. At the same time, the algorithms and operators that produce historical optimal solutions are recorded, which provides an important basis for APMA's subsequent algorithms, operator competition, and global optimization strategy.

(3) **Competition-based algorithm and operator adaptive selection strategy, referred to as "competition" strategy.** This strategy is the second step of APMA inner loop. It collects the historical optimal solution set obtained by each algorithm in the parallel strategy and selects the high-quality ones to construct the current optimal solution set. Then, it quantitatively evaluates the contribution of algorithms and operators and updates their use probability for the purpose of "survival of the fittest" so that APMA has the ability of self-organization and self-adaptation, which guarantees performance of APMA.

Table 4.1: Design considerations and countermeasures of satellite conventional task scheduling algorithm.

No.	Design considerations	Design ideas and countermeasures
1	Initial solution quality	To design a fast construction strategy for high-quality initial solutions.
2	Local optimization ability	To use local search algorithms such as hill climbing algorithm, Tabu search algorithm, simulated annealing algorithm, late acceptance algorithm, and iterative local search algorithm to realize fast search in small neighborhoods.
3	Global optimization ability	1. To let multiple local search algorithms work together for optimization and enhance the diversity of solutions. 2. To introduce evolutionary strategy to promote large-neighborhood optimization and avoid falling into local optima.
4	Adaptability and generality	1. To quantitatively assess the performance of algorithms and operators, realize the "survival of the fittest" of algorithms and operators, and enhance the adaptability of algorithms and operators. 2. To design general-purpose basic algorithms and operators to enhance their generality based on the general-purpose 0–1 decision matrix.
5	Time complexity	1. To run local search algorithm in parallel to save search time. 2. To take local search algorithm as the main loop and evolutionary strategy as the global optimization strategy to avoid the superposition of time complexity of local search and evolutionary algorithms in traditional memetic algorithms. 3. To use the abovementioned constraint computing methodology based on constraint network to improve the computing efficiency.

(4) Population evolution-based global optimization strategy, referred to as "evolution" strategy. This strategy is the final step of APMA inner loop. Besides the parallel and competition strategies, this strategy uses evolutionary operators such as crossover and repair operators to perform global optimization on the basis of the current optimal solution set so that APMA has the ability to skip from local optima for exploring a larger solution space, which further guarantees performance of APMA. At this point, if the termination condition is satisfied, the algorithm will output the current optimal solution; otherwise, it will return to run a new round of inner loop following the "parallel" strategy.

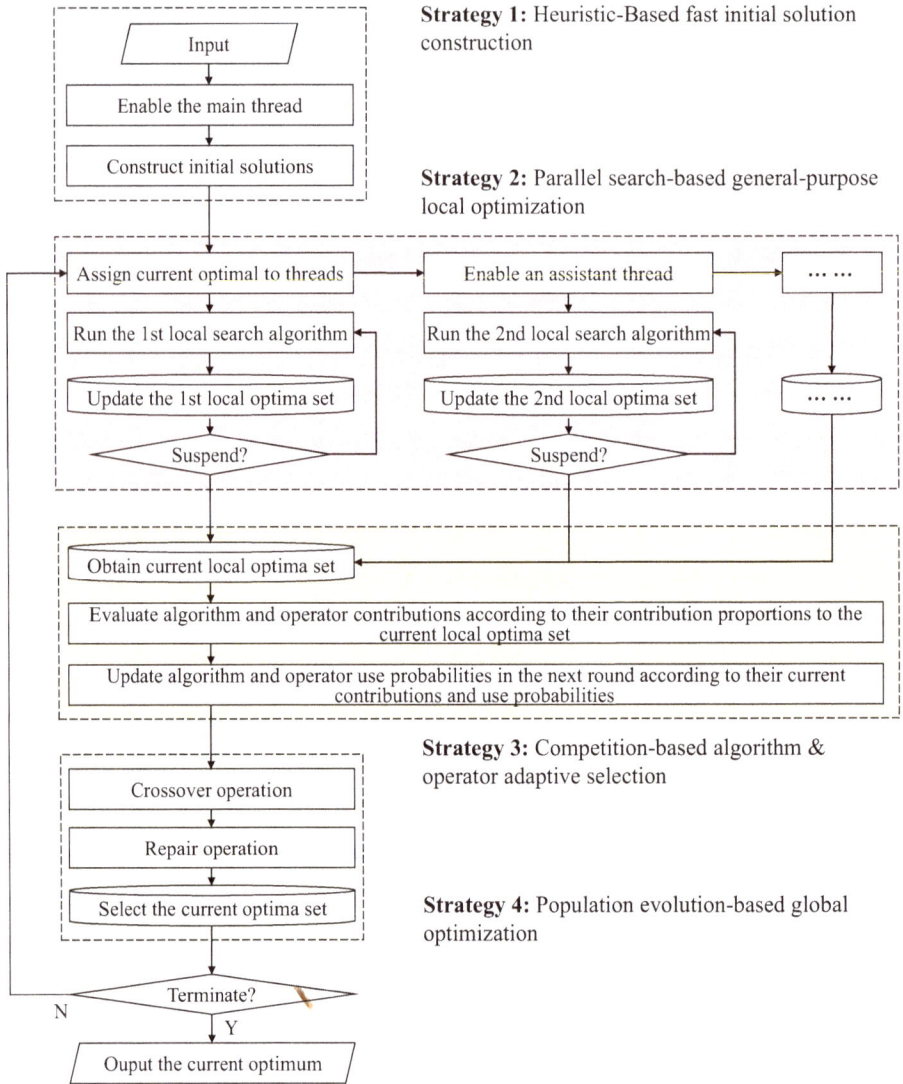

Figure 4.1: APMA framework.

4.1.3 Advantage analysis

In the above APMA framework, the four strategies work together and complement each other, which can meet the needs of satellite conventional task scheduling algorithms in terms of initial solution quality, local optimization ability, global optimization ability, adaptability, generality, and time complexity, as manifested in the following:

(1) **The "construction" strategy provides fast and high-quality initial solutions to meet the needs of the algorithm initial solution quality.** As a class of fast heuristic methods, the construction strategy can produce a feasible scheme that satisfies the constraints in a short time. Meanwhile, as a commonly used heuristic strategy for industrial sectors to address complex task scheduling problems, the heuristic algorithm with "first-in-first-service" involved in this strategy is widely used in many in-service satellite task scheduling systems, which plays a crucial role in improving the construction efficiency and quality of initial solutions, and provides an important basis for the iterative optimization of subsequent strategies.

(2) **The "parallel" strategy helps the algorithm to improve the search efficiency and makes the algorithm have good local optimization ability, diversity, and expandability to meet the needs of the algorithm's local optimization ability and time complexity.** Firstly, the competing local search algorithms run in parallel, which greatly reduces the time consumption of the algorithm and fully utilizes the local optimization ability of the local search algorithms. Secondly, different local search algorithms work together to search for more diverse solutions through different search trajectories, thereby meeting the needs of subsequent evolution strategy in terms of population diversity. In addition, a new algorithm or operator can easily join the competition and expand the algorithm by adding new threads, which makes the algorithm more flexible and easier to use. In other words, the parallel strategy is an important strategy to fully utilize the computing power of computers and enhance the advantages of the algorithm through hardware.

(3) **The "competition" strategy makes the algorithm have good adaptability to meet the needs of the algorithm adaptability.** The update of the probability of using algorithms and operators in this strategy helps to retain and use suitable algorithms and operators more frequently and eliminate unsuitable ones for the purpose of "survival of the fittest," which guarantees the adaptability and comprehensive performance of the algorithm.

(4) **The "evolution" strategy helps the algorithm skip from local optima to explore a larger solution space to meet the needs of the algorithm global optimization ability.** Given the hybrid framework of local search and EAs, APMA belongs to a class of MA, but in the earliest MA proposed by Moscato [220] and most of the MAs with EAs as the main loop and local search as the improved operators [182, 191, 192, 202], since the time complexity of the local search and EAs is superimposed, the running time of the algorithm surges in each round. Meanwhile, since EAs are not as good as local search algorithms for constraint optimization, the MAs with EAs as the main loop are not suitable for solving satellite task scheduling problems with complex constraints. On the contrary, with local search algorithms as the main loop and evolution strategy as the improved operators, APMA not only inherits both local and global optimization abilities of MA and meets the needs of the algorithm global optimization ability, but also avoids the superposition of time complexity. Therefore, APMA is bet-

ter suited for solving satellite task scheduling problems in this book, which further guarantees the performance of the algorithm.

To sum up, in the APMA framework, the four strategies of construction, parallel, competition, and evolution work together and complement each other to meet the design needs of satellite conventional task scheduling algorithms, providing a general-purpose and efficient solution to complex and diverse satellite task scheduling problems in this book.

4.2 Heuristic-based fast initial solution construction strategy

The heuristic-based fast initial solution construction strategy (referred to as construction strategy) provides the initial solution set for APMA. Based on the heuristic algorithm with "pre-allocation and first-in-first-service," this strategy helps the algorithm quickly enter the feasible domain space, constructs a fast and high-quality initial solution, and provides an important foundation for further iterative optimization of APMA.

4.2.1 General process

The general process of this strategy consists of the following four main steps as shown in Figure 4.2:

Step 1 (Task sequencing). The task set is resequenced according to the principle of descending order of task profits.

Step 2 (Pre-allocation). A task event is selected sequentially from the task set, the event EO set is traversed, the number of executable opportunities provided by each satellite and ground station for the event is recorded, and the event is "pre-allocated" to one of the satellites and ground stations according to the "roulette" mode.

Step 3 (First-in-first-service). On the basis of "pre-allocation" and according to the heuristic principle of "first-in-first-service," the earliest event EO that meets the constraints is "scheduled" for the event. If there is any event not traversed, return to step 2; otherwise proceed to step 4.

Step 4 (Output of initial solution set). Then, to enhance the randomness of the algorithm, n_{T-1} (where n_T is the number of threads enabled in the next stage) initial solutions are randomly generated to form the initial solution set.

It is thus clear that this strategy has two important characteristics: 1) fast constructivity: From the above steps, it can be seen that this strategy belongs to a fast heuristic algorithm, and a feasible scheme that satisfies the constraints can be produced in a short time; and 2) compactness: In step 3, the earliest event EO that meets the constraints is "scheduled" for each task and event one by one, which makes the task

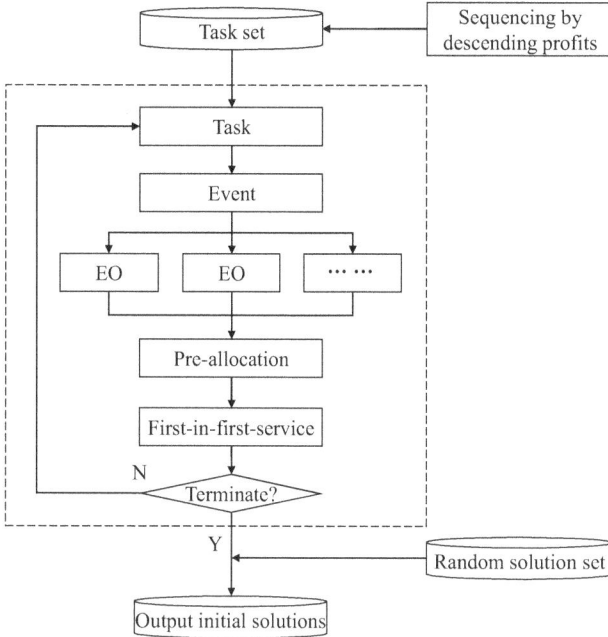

Figure 4.2: Flow of heuristic-based fast initial solution construction strategy.

scheduling scheme more compact, helps to improve the quality of the initial solution, and provides an important basis for iterative optimization of subsequent algorithms.

In the following section, the heuristic algorithm with "pre-allocation and first-in-first-service" involved in steps 2 and 3 is specifically described, and the time complexity of this strategy is further analyzed to illustrate its fast construction of the initial solution.

4.2.2 Heuristic algorithm

According to the general-purpose 0–1 mixed integer decision model and constraint model constructed above for satellite task scheduling, the heuristic algorithm with "pre-allocation and first-in-first-service" involved in steps 2 and 3 of this strategy is shown in Algorithm 4.1.

Algorithm 4.1: Heuristic algorithm with "pre-allocation and first-in-first-service."

Input: Task Set T, event EO set EO, constraint function $F^H(X)$
Output: Initial solution X_0

1	$X_0 \leftarrow \{x_{ij} \mid x_{ij} = 0, 0 \leq i \leq |T|, 0 \leq j \leq |E_i|\}$	//Initialize decision matrix, and decision variables are all 0
2	**for** $i = 1 : |T|$ **do**	//Traverse task set T
3	**for** $j = 1 : |E_i|$ **do**	//Traverse event set E_i of task t_i
4	$P \leftarrow \{p_m \mid p_m = 0, 0 \leq m \leq |P|\}$	//Empty satellite or ground station set, and record the number of event executable opportunities
5	**for** $k = 1 : |EO_{ij}|$ **do**	//Traverse event EO set EO_{ij} of event e_{ij}
6	$m \leftarrow \text{Id}(p(eo_{ij}^k))$	//Get the satellite or station id providing the EO
7	$p_m + +$	//Number of event executable opportunities provided by the satellite or station + 1
8	**end**	
9	$m \leftarrow \text{Roulette}(P)$	//Get the "pre-allocated" satellite or station id in a "roulette" manner
10	**for** $k = 1 : |EO_{ij}|$ **do**	//Traverse event EO set EO_{ij} of event e_{ij} again
11	$x_{ij} \leftarrow / \dfrac{k}{|EO_{ij}|}$	//Assign decision variable
12	**if** $\text{Id}(p(eo_{ij}^k)) = m \,\&\, F^H(X) = 0$ **then**	
13	**break**	//If the EO belongs to a "pre-allocated" satellite or ground station and meets the constraints, the "scheduling" is successful, and then return to line 3
14	**else**	
15	$x_{ij} \leftarrow 0$	//Otherwise, the decision variable is reset to 0, and return to line 10
16	**end**	
17	**end**	
18	**end**	
19	**end**	
20	**return** X_0	

In this algorithm, lines 4–9 are the "pre-allocation" steps: lines 4–8 sequentially count the total number of event executable opportunities provided by each satellite or ground station for the current event; line 9 selects a satellite or station "pre-allocated" for the event in a "roulette" manner. Here, the total number of event executable opportunities provided by the satellites and ground stations reflects their ability to perform the task event. The higher the total number is, the greater the ability and the higher the probability of being selected by the algorithm. This "pre-allocation" mechanism is in line with the subjective experience of satellite control and at the same time narrows the scope of event executable opportunities in the next "first-in-first-service," reduces computation complexity, and improves the construction efficiency of the initial solution.

Then, lines 10–17 are the steps for "first-in-first-service," which sequentially traverse the EO set of the current event to assign value to the decision variable. Among them, lines 12–14: If the EO belongs to a "pre-allocated" satellite or ground station and meets the constraints, the "scheduling" is successful, and the algorithm returns to line 3 to continue scheduling other task events; otherwise, lines 14–16: the decision variable is reset to 0, and the algorithm returns to line 10 to continue traversing other executable opportunities of the event. It is worth noting that, as a commonly used heuristic strategy for industrial sectors to address complex task scheduling problems, "first-in-first-service" is widely used in many in-service satellite task scheduling systems. The initial solution constructed by this strategy fits well with the actual situation of satellite control and the general-purpose 0–1 mixed integer decision model for satellite task scheduling in this book and meets the needs of APMA in terms of solving speed and initial solution quality.

4.2.3 Time complexity

To visualize the fast construction of the initial solution of this strategy, its time complexity is analyzed in this section. As can be seen from the above steps and flowchart, the time complexity of this strategy mainly arises from the following three aspects:

(1) Complexity of task sequencing: The strategy uses traditional sequencing algorithm, and the time complexity is at most $O(n^2)$.

(2) Complexity of "pre-allocation": The task set T contains $|T|$ tasks, where each task t_i contains $|E_i|$ (constant) events, the "pre-allocation" steps in lines 4–9 of the algorithm traverse EO set EO_{ij} of each event, so the time complexity of "pre-allocation" is $O(n)$.

(3) Complexity of "first-in-first-service": The "first-in-first-service" steps in lines 10–17 of the algorithm also traverse EO set EO_{ij} of each event, and the time complexity is also $O(n)$ regardless of the constraint computation complexity in this process.

The cumulative time complexity of the above three aspects is $O(n^2)$, indicating that the main loop of this strategy is polynomial time. Meanwhile, since the constraint computation complexity also affects the specific time complexity of the strategy, it is particularly important to design efficient and fast constraint computation methods. In this regard, based on the constraint network and constraint computation method in Section 3.4.3, an incremental constraint computation method is given in the next section, which provides an important guarantee for the initial solution construction efficiency of this strategy.

To sum up, this section elaborates on the construction strategy and the heuristic algorithm with "pre-allocation and first-in-first-service." The strategy helps APMA quickly enter the feasible domain space, constructs a high-quality initial solution that satisfies the constraints, and provides an important foundation for further iterative optimization of APMA.

4.3 Parallel search-based general-purpose local optimization strategy

Referred to as parallel strategy, parallel search-based general-purpose local optimization strategy is the first step of APMA inner loop. The strategy runs different local search algorithms in parallel, carries out fast and efficient search optimization through different search trajectories to obtain diversified local optimization results, which meets the needs of APMA in terms of local optimization ability and time complexity, and provides an important basis for the competition of algorithms and operators and global optimization in subsequent strategies. In the following, this section elaborates on the strategy flow, general algorithms and operators as well as the efficient and incremental constraint computation algorithm in the iteration process of the algorithm.

4.3.1 General process

Let's assume that n_T computation threads are enabled in APMA. The general process of the strategy consists of the following four steps as shown in Figure 4.3:

Step 1 (Initial solution allocation). The current optimal solution set is given, thread 1 selects n_T solutions from the set according to their scores in a "roulette" manner, and the solutions are allocated to n_T threads one by one as the initial solutions of the local search algorithm in each thread.

Step 2 (Algorithm selection (parallel)). A pool of algorithms is given, and each thread picks a local search algorithm from the pool according to the use probability of each algorithm.

Step 3 (Algorithm running (parallel)). A pool of operators is given, and each thread runs the selected local search algorithm in parallel. During each iteration, each algorithm picks a neighborhood operator from the operator pool and constructs a neighborhood solution according to the use probability of the operator. Meanwhile, whenever a new solution is accepted, the new solution is stored in the historical optimal solution set, and the neighborhood operator that produced the solution is recorded. If the size of the historical optimal solution set exceeds a preset value, the historical optimal solution set is updated based on the first-in-first-out (FIFO) rule.

Step 4 (Output of historical optimal solution set (parallel)). After a limited time or number of iterations, each thread outputs the historical optimal solution set.

It can be seen that the parallel search is the key to the strategy, and it has the following two functions: 1) Fast iterative optimization: By enabling multiple computer threads, more optimization results can be obtained at the same time, increasing the possibility of obtaining high-quality solutions; and 2) enhancement of solution diversity: In view of the characteristics of neighborhood search, it is usually difficult for local search algo-

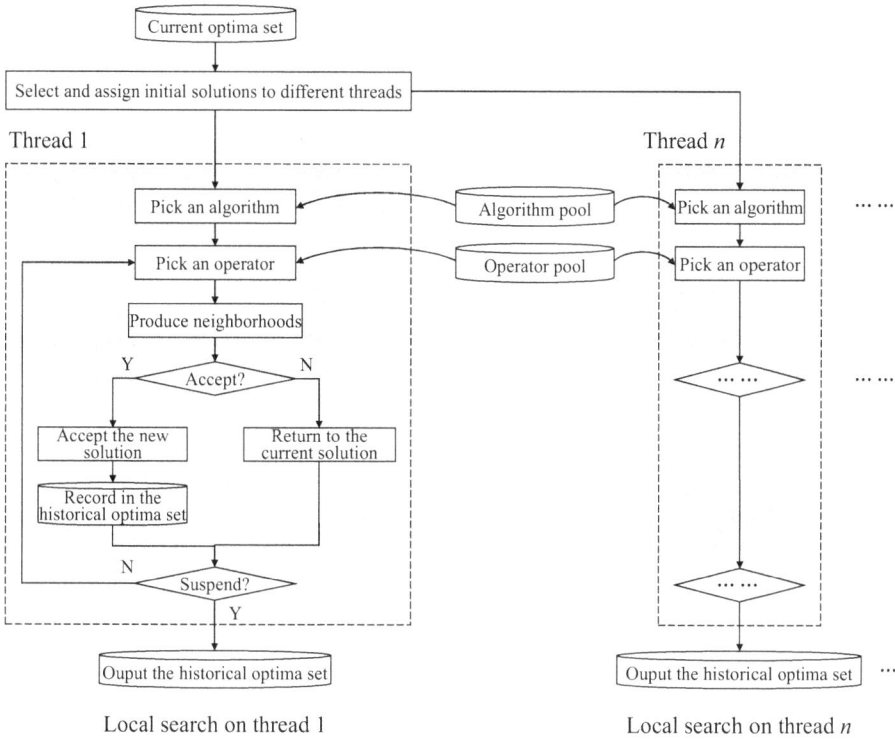

Figure 4.3: Flow of parallel search-based general-purpose local optimization strategy.

rithms to obtain diversified solution sets, but different algorithms can produce different search trajectories and solution sets, thus enhancing the diversity of solutions and providing a diversified population base for the subsequent "evolution" strategy. At the same time, all algorithms and operators that produced the historical optimal solution are recorded, providing evaluation basis for the subsequent "competition" strategy.

4.3.2 Algorithm pool

In this strategy, the algorithm pool stores a variety of local search algorithms for each thread to choose from. Each algorithm in the pool is assigned to use probability. Initially, the probability of each algorithm is the same, and in the subsequent "competition" strategy, the probability is updated iteration by iteration. Therefore, this section adds the following five commonly used local search algorithms to the algorithm pool. In particular, since the iterative optimization process of the algorithms can be regarded as the continuous optimization of the decision matrix, that is, that of the general-purpose 0–1 mixed integer matrix X in the general-purpose modeling of satellite task scheduling, this book uses the decision matrix X to represent a general-purpose satellite task scheduling scheme (solution).

4.3.2.1 Hill climbing algorithm

As the prototype of all local search algorithms, HC algorithm is the simplest and most universal local search algorithm. The specific steps of the algorithm are shown in Algorithm 4.2.

Algorithm 4.2: Hill climbing algorithm

Input: initial solution X_0, score function $F(X)$, operator pool O, empty historical optimal solution set U_X, and termination condition
Output: current solution X, and historical optimal solution set U_X

1 $X \leftarrow X_0$
2 **while** the termination condition is not met **do**
3   According to the use probability of operator, an operator $o \in O$ is selected from the operator pool
4   Construct neighborhood solution $X' \leftarrow o(X)$
5   **if** $F(X') \geq F(X)$ **then**   //If $F(X')$ is greater than or equal to $F(X)$
6        $X \leftarrow X'$ //Then the new solution X' is accepted
7   According to FIFO rule, X is stored   //And the historical optimal in the historical optimal solution   solution set is updated set U_X accordingly
8   **end**
9 **end**
10 **return** X, U_X

In particular, lines 3 and 4 of the algorithm: the algorithm picks an operator in a roulette manner according to the use probability of the operators in the operator pool O (see the description of the pool in the next section) and then constructs a neighborhood solution. All subsequent algorithms pick operator in the same way. The score function $F(X)$, introduced in Section 3.3.2.4, contains constraint values, profits, and soft constraint values at the same time, allowing for comprehensive and complete evaluation of the satellite task scheduling solutions.

Though HC is simple in principle and easy to implement, it has an obvious shortcoming, that is, the lack of a mechanism to skip from local optima. The HC given in this section accepts non-inferior solutions, that is, the acceptance condition is $F(X') \geq F(X)$, which can help the algorithm to transition to the gradient smoothing region and avoid falling into local optima prematurely. Nevertheless, the possibility of HC skipping from local optima is still low. At this point, in this section, four meta-heuristic local search algorithms are added to the algorithm pool to provide diverse and effective algorithms for the "parallel" strategy.

4.3.2.2 Tabu search algorithm

Tabu search algorithm (TS) is a class of local search algorithms with a memory strategy proposed by Glover [203, 204] in 1986. TS performs optimization based on HC and records local optima or the operations that generate local optima during the optimization process through a Tabu table to avoid repeated searches in the local optimal space and achieve the effect of skipping from local optima to explore other high-quality solution spaces. As one of the earliest algorithms used for solving satellite task scheduling problems, TS is simple and practical.

Though TS has been developed in many different versions so far, the Tabu idea of recording local optima to avoid repeated search is consistent among them. In this regard, based on the general-purpose model of satellite task scheduling in this book, this section designs a TS by taking local optima as the Tabu object and replacing the Tabu table with the Tabu solution set. Its specific steps are shown in Algorithm 4.3.

In particular, lines 6–8 of the algorithm: If the neighborhood solution obtained is tabued, TS directly discards the solution and re-do the search to avoid repeated searches of the local optimal space. All subsequent algorithms also use this Tabu strategy. Lines 12–14 of the algorithm: If the neighborhood solution is not as good as the current optimal solution, but better than or equal to the non-taboo optimal solution X^* in the historical optimal solution set U_X, TS accepts the solution and records it in the Tabu solution set, so as to skip from local optima for exploring a larger solution space.

For TS, the Tabu length, that is, the size of the Tabu solution set, is a major factor affecting the performance of the algorithm. To enhance the adaptability of the algorithm to address satellite task scheduling problems of different scales, this section sets the Tabu length to be a certain proportion of the size of the task set T in the problems, where α_T is the proportionality coefficient, which is 0.1–0.3 in the following experiments:

$$|U_T| \leq \alpha_T \cdot |T| \tag{4.1}$$

Algorithm 4.3: Tabu search algorithm

Input: initial solution X_0, score function $F(X)$, operator pool O, empty Tabu solution set U_T, historical optimal solution set U_X, and termination condition

Output: current solution X, and historical optimal solution set U_X

1   $X \leftarrow X_0$

2   $X^* \leftarrow X_0$ // X^* are the optimal solutions that are not tabued
                                                    in the historical optimal solution set U_X.

3   **while** the termination condition is not met **do**

4 | According to the use probability of operator, an operator $o \in O$ is selected from the operator pool.

5 | Construct neighborhood solution $X' \leftarrow o(X)$

6 | **if** $X' \in U_T$ **then**

7 | | continue   //Neighborhood solution is tabued, and search
                                                 again.

8 | **end if**

9 | **if** $F(X') \geq F(X)$ **then**

10 | | $X \leftarrow X'$

11 | | According to FIFO rule, X is stored in Tabu solution set U_T and historical optimal solution set U_X respectively

12 | **else if** $F(X') \geq F(X^*)$ **then**   //The optimal solutions that are not tabued in the
                                                       historical optimal solution set U_X are accepted.

13 | | $X \leftarrow X'$

14 | | According to FIFO rule, X is stored in Tabu solution set U_T

15 | **end**

16 **end**

17 **return** X, U_X

4.3.2.3 Simulated annealing algorithm

Simulated annealing algorithm (SA) was first proposed by Metropolis et al. [210] in 1953, and applied to combinational optimization by Kirkpatrick et al. [211] in 1983. The algorithm has been widely used in solving satellite task scheduling. SA is a class of local search algorithm originated from the principle of solid annealing, which probabilistically accepts inferior solutions in the process of simulated annealing to skip from local optima. Its specific steps are shown in Algorithm 4.4.

Because the principle of SA skipping from local optima is different from that of TS, lines 6–8 of SA also use the same Tabu strategy as that of TS to enhance the ability of SA to skip from local optima. The setting of Tabu length is the same as eq. (4.1). Line 9 of SA is annealing, and the current temperature gradually decreases through annealing. To improve generality of SA, the following three annealing functions are given in this book:

Algorithm 4.4: Simulated annealing algorithm (including Tabu strategy)

Input: initial solution X_0, score function $F(X)$, operator pool O, initial temperature t_0, empty Tabu solution set U_T, historical optimal solution set U_X, and termination condition
Output: current solution X, and historical optimal solution set U_X

1	$X \leftarrow X_0$	
2	$t \leftarrow t_0$	
3	**while** the termination condition is not met **do**	
4	According to the use probability of operator, an operator $o \in O$ is selected from the operator pool	
5	Constructing neighborhood solution $X' \leftarrow o(X)$	
6	**if** $X' \in U_T$ **then**	
7	continue	//Neighborhood solution is tabued, and search again
8	**end**	
9	$t \leftarrow$ annealing(t)	//Annealing, and calculate the current temperature
10	**if** $F(X') \geq F(X)$ **then**	
11	$X \leftarrow X'$	
12	According to FIFO rule, X is stored in Tabu solution set U_T and historical optimal solution set U_X respectively	
13	**else if** random$(0, 1) <$ exp $\left[\dfrac{F(X') - F(X)}{t}\right]$ **then**	//Inferior solutions are accepted probabilistically
14	$X \leftarrow X'$	
15	According to FIFO rule, X is stored in Tabu solution set U_T	
16	**end**	
17	**end**	
18	**return** X, U_X	

---

(1) Arithmetic annealing function, also known as linear annealing function, as shown in eq. (4.2a), where Δt is annealing temperature difference, so that the current temperature t is exactly 0 at the end of k_{max} iteration of the algorithm.

(2) Geometric annealing function, as shown in eq. (4.2b), where ξ is the annealing proportionality coefficient, so that the current temperature drops by 1,000 times at the end of k_{max} iteration of the algorithm.

(3) Logarithmic annealing function, as shown in eq. (4.2c), where k is the number of current iterations, so that the current temperature t is exactly 0 at the end of k_{max} iteration of the algorithm.

$$\text{annealing}_1(t) = t - \Delta t, \Delta t = \frac{t_0}{k_{max}} \tag{4.2a}$$

$$\text{annealing}_2(t) = \xi \cdot t, \xi = 1{,}000^{-\frac{1}{k_{max}}} \tag{4.2b}$$

$$\text{annealing}_3(t) = t \cdot \frac{\log_{k_{max}} k}{\log_{k_{max}} k - 1}, 2 \le k \le k_{max} \tag{4.2c}$$

According to the current temperature t, line 13 of the algorithm calculates the probability of accepting the inferior solution and decides whether to accept the inferior solution by random (0,1). It can be seen that the higher the temperature, the higher the probability that the algorithm accepts the inferior solution, which is helpful for the algorithm to skip from local optima. On the contrary, as the algorithm converges iteratively, the temperature gradually decreases, and the probability of accepting the inferior solution decreases accordingly, which is helpful for the algorithm to converge and locally search for the optima.

In addition, the initial annealing temperature t_0 of SA also affects its optimization results. In this section, the common initial temperature setting method is adopted, and the average value of 30 random initial solutions is used as the initial annealing temperature t_0.

4.3.2.4 Late acceptance hill climbing algorithm

The LAHC algorithm, also known as late acceptance algorithm, is a class of meta-heuristic algorithm proposed by Burke and Bykov [244, 245] in 2012. The LA means that the algorithm can accept a solution before a certain time (step) and use it as a way to escape from local optima. The algorithm is simple in principle. It not only retains the asymptotic convergence of HC but also has the intelligence to escape from local optima and has shown outperformance in solving many classic benchmark problems in operations research. The specific steps of LA are shown in Algorithm 4.5.

In particular, lines 5–7 of LA use the same Tabu strategy as that of TS to enhance the ability of LA to skip from local optima. LA has some similarities with TS in that they both achieve the purpose of skipping from local optima by recording historical optimal solutions, but there are three differences between them: (1) Tabu objects in TS are usually not repeated, while those in late solution set of LA can be repeated. For example, in line 15 of LA, even if the current solution X is not improved, it is repeatedly stored in the late solution set. (2) In TS, all elements in Tabu solution set participate in the judgment of Tabu, while in LA, only X_L, the first element in late solution set, participates in the judgment. (3) The basis of TS in this book for judging Tabu is whether the solutions are the same or not, while the basis of LA for judging whether to accept the late solution X_L is its score function $F(X_L)$. It is precisely because of the above differences that LA can work together with Tabu strategy, providing more opportunities for the algorithm to skip from local optima.

Algorithm 4.5: Late acceptance algorithm (including Tabu strategy)

Input: initial solution X_0, score function $F(X)$, operator pool O, empty late solution set U_L, Tabu solution set U_T, historical optimal solution set U_X, and termination condition
Output: current solution X, and historical optimal solution set U_X

1   $X \leftarrow X_0$
2   **while** the termination condition is not met **do**
3   According to the use probability of operator, an operator $o \in O$ is selected from the operator pool
4   Construct neighborhood solution $X' \leftarrow o(X)$
5   **if** $X' \in U_T$ **then**
6   **continue**   //Neighborhood solution is tabued, and search again
7   **end**
8   **if** $F(X') \geq F(X)$ **then**
9       $X \leftarrow X'$
10   According to FIFO rule, X is stored in Tabu solution set U_T and historical optimal solution set U_X respectively
11   **else if** $F(X') \geq F(X_L)$ **then**   // X_L is the first element in late solution set U_L
12      $X \leftarrow X'$
13   According to FIFO rule, X is stored in Tabu solution set U_T
14   **end**
15   According to FIFO rule, X is stored in   //The current solution of each iteration is late solution set U_L stored in late solution set

16 **end**
17 **return** X, U_X

Similar to TS, to enhance adaptability of LA, this section sets the late length of LA to be a certain proportion of the size of task set T, as shown in eq. (4.3), where a_L is the proportionality coefficient, which is 1.0–2.0 in the following experiments:

$$|U_L| = a_L \cdot |T| \qquad\qquad (4.3)$$

4.3.2.5 Iterated local search algorithm

Iterated local search (ILS) algorithm is a common and fundamental local search algorithm that has shown good results in solving many classic benchmark problems in operations research and simplified remote-sensing satellite task scheduling problems [174]. ILS achieves the purpose of searching large neighborhoods and skipping from local optima by periodically disturbing and repairing the current solution. Since the constraints are complex for the satellite task scheduling problems in this book, each process of constructing neighborhood solutions has a greater possibility of violating the constraints and obtaining a neighborhood solution that violates the constraints or has low profit, that is, such process can be regarded as a disturbance. In this context, the disturbance and repair mechanism in ILS can be regarded as a meta-heuristic mechanism that accepts inferior solutions and repairs them, as shown in Algorithm 4.6.

Algorithm. 4.6" Iterated local search algorithm (including Tabu strategy)

Input: initial solution X_θ, score function $F(X)$, operator pool O, Tabu solution set U_T, historical optimal solution set U_X, number of iterative search inner loop, repair condition, repair () as repair algorithm, and termination condition

Output: current solution X, and historical optimal solution set U_X

1 $X \leftarrow X_\theta$
2 **while** the termination condition is not met **do**
3 | According to the use probability of operator, an operator $o \in O$ is selected from the operator pool
4 | Construct neighborhood solution $X' \leftarrow o(X)$
5 | **if** $X' \in U_T$ **then**
6 | | **continue**   //Neighborhood solution is tabued, and search again
7 | **end**
8 | **if** $F(X') \geq F(X)$ **then**   //If the neighborhood solution is better, the solution is accepted.
9 | | $X \leftarrow X'$
10 | | According to FIFO rule, X is stored in Tabu solution set U_T and historical optimal solution set U_X respectively
11 | **else if** repair condition is satisfied **then**   //If the neighborhood solution is not better, the repair condition is satisfied
12 | | $X \leftarrow$ repair(X')   //The neighborhood solution is repaired and accepted
13 | | According to FIFO rule, X is stored in Tabu solution set U_T
14 | **end**
15 **end**
16 **return** X, U_X

In particular, lines 5–7 of ILS use the same Tabu strategy as that of TS to enhance the ability of ILS to skip from local optima. ILS has some similarities with SA in the mechanism of accepting and repairing the inferior solutions, but there are two differences between them: (1) SA has a certain probability of accepting the inferior solution every time it constructs the neighborhood, while ILS only accepts the inferior solution after a certain number of iterations (when the repair condition is satisfied, for example, line 11 of the algorithm). (2) Under the action of the annealing strategy in this book, the inferior solutions accepted by SA meet the constraints, and it is not necessary to repair them, while the inferior solutions accepted by ILS violate the constraints in most cases, and it is necessary to repair them, such as the repair () operator in line 12 of the algorithm. Here, the repair () operator uses the repair operator designed in Section 4.5, that is, a heuristic constraint resolution operator. In this process, ILS skips the current neighborhood space and implement a larger neighborhood transformation to achieve the effect of skipping from local optima for exploring other high-quality solution spaces.

To sum up, this section gives five local search algorithms including HC, TS, SA, LA, and ILS and their setting methods of related parameters. These five algorithms are added to the algorithm pool for each thread to choose from in the "parallel" strategy, and their probabilities of use are updated iteration by iteration in the subsequent "competition" strategy to achieve "survival of the fittest algorithm."

4.3.3 Operator pool

In each process of constructing neighborhood solution using the above local search algorithms, the operator pool provides available local search operators. In the operator pool, each operator is assigned a use probability, based on which the algorithms make their selection of operators. Similar to the algorithm pool, the use probability of each operator is initially the same and is updated iteration by iteration in the subsequent "competition" strategy. To visually show how operators change the decision variables in a general-purpose 0–1 decision matrix, this section takes a 3×3 decision matrix as an example to introduce the following local search operators.

4.3.3.1 Move operator
This operator randomly changes the value of one of the decision variables in the decision matrix, for example, changing the value of the decision variable x_{12} in the decision matrix X in Figure 4.4 from 0.1 to 0.5, denoted as Move(x_{12}). In the general-purpose 0–1 mixed integer decision model for satellite task scheduling in this book, the actual meaning of the Move operator is to change the EO of the second event of the first satellite task. Meanwhile, since the value range of each decision variable in the decision matrix X is [0, 1], the Move operator doesn't cause the decision variables to exceed the value range. Here, in the running process of local search algorithms, the Move operator is the most basic local search operator, and any complex local search operator can be described as a combination of multiple Move operators.

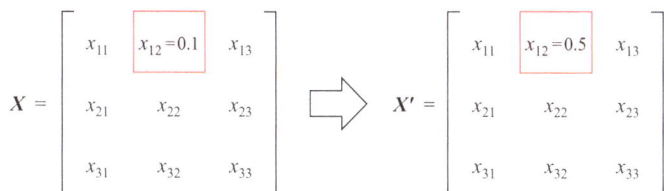

$$X = \begin{bmatrix} x_{11} & x_{12}=0.1 & x_{13} \\ x_{21} & x_{22} & x_{23} \\ x_{31} & x_{32} & x_{33} \end{bmatrix} \Rightarrow X' = \begin{bmatrix} x_{11} & x_{12}=0.5 & x_{13} \\ x_{21} & x_{22} & x_{23} \\ x_{31} & x_{32} & x_{33} \end{bmatrix}$$

Figure 4.4: Schematic diagram of Move operator in "parallel" strategy.

4.3.3.2 Swap operator

This operator randomly swaps the values of one or a series of decision variables in the decision matrix, as shown in Figure 4.5, and can be further categorized into the following three types:

(1) Random swap operator, that is, this operator randomly swaps the values of two decision variables in the decision matrix X. In Figure 4.5(a), the operator swaps the values of the decision variables x_{11} and x_{33}. This operator can also be described as a combination of two Move operators, namely Move(x_{11}) + Move(x_{33}). In the satellite task scheduling decision model of this book, the operator actually swaps the decision variables of an event of two satellite tasks.

(2) Colocated swap operator, that is, this operator randomly swaps the values of two decision variables located in the same column of the decision matrix. In Figure 4.5(b), the operator swaps the values of decision variables x_{12} and x_{32}, which can be described as a combination of Move(x_{12}) + Move(x_{32}). In the decision model of this book, the operator actually swaps the decision variables of the same event of two satellite tasks.

(3) Row swap operator, that is, this operator randomly swaps the values of decision variables in two rows of the decision matrix. In Figure 4.5(c), the operator swaps the values of the decision variables in rows 1 and 3, which can be regarded as a combination of six Move operators, namely, Move(x_{11}) + Move(x_{12}) + Move(x_{13}) + Move(x_{31}) + Move(x_{32}) + Move(x_{33}). In the decision model of this book, the operator actually swaps the decision variables of all events of two satellite tasks.

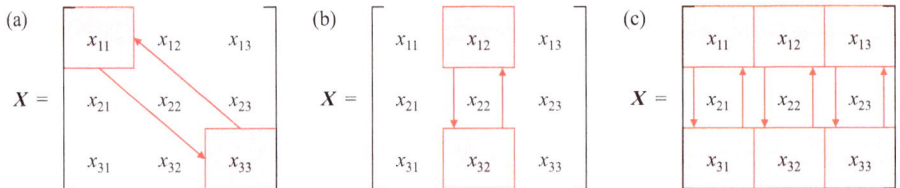

Figure 4.5: Schematic diagram of swap operators in "parallel" strategy: (a) random swap operator; (b) colocated swap operator; and (c) row swap operator.

It is worth noting that the above local search operators are reversible. Take the colocated Swap operator in Figure 4.5(b) as an example, the forward operation (constructing neighborhood solution) is Move(x_{12}) + Move(x_{32}), and the reverse operation (returning to the original solution) is Move(x_{32}) + Move(x_{12}). During the algorithm's search for neighborhood solutions, the reverse operation of returning to the original solution due to the discovery of inferior solutions needs to be performed frequently. Therefore, describing the Swap operator as a combination of multiple Move operators is helpful to record a series of neighborhood operations performed through the Move operator during the search process of the algorithm and to perform reverse operations in a fast and orderly manner when it is necessary to return to the original solution.

4.3.3.3 Replace operator

In the process of constructing the satellite task scheduling constraint model and constraint network in Section 3.4.3, a series of decision subsets $X_{Z_{ij}}$ (the jth subset of the decision matrix X that may be involved in the ith constraint) have been obtained in advance. In other words, the decision variables (corresponding events) in each decision subset $X_{Z_{ij}}$ are potentially violating constraints with each other. In view of this, the Replace operator randomly resets a decision variable in a decision subset $X_{Z_{ij}}$ to 0 (i.e., canceling the corresponding event) and randomly changes the value of another decision variable in the same decision subset (i.e., changing the EO of the corresponding event). Compared with the Move and Swap operators mentioned above, the Replace operator focuses on the constraint network existing in the current satellite task scheduling scheme, which is more targeted and embodies the idea of conflict avoidance. Similar to Swap operator, this operator can also be described as a combination of multiple Move operators.

In addition, operators can be created by combining multiple Move operators. All the above operators are added to the operator pool for each thread to choose from in the process of running the local search algorithms. The use probability of each operator in the operator pool is updated iteration by iteration in the subsequent "competition" strategy to achieve the "survival of the fittest operator."

4.3.4 Incremental constraint computation algorithm

Based on the design of algorithms and operators mentioned above, to guarantee the iteration efficiency of the local search algorithms in this strategy within satellite task scheduling environment with complex constraints and to save the time of repeated constraint computation during the iteration process, in this section, an incremental constraint computation algorithm is given according to the constraint network-based constraint computation method designed in Section 3.4.3 using the neighborhood operator Move(x) as an example. The specific steps of the algorithm are shown in Algorithm 4.7.

Algorithm 4.7: Incremental constraint computation algorithm

Input: original solution X, new solution X', constraint set F^H, original solution constraint value $F^H(X)$, neighborhood operator Move(x);

Output: new solution constraint value $F^H(X')$.

1    $F^H(X') \leftarrow F^H(X)$

2   **for** $i = 1: |F^H|$ **do**   //Traverse constraint set F^H

3   **for** $j = 1: |Z_i(X)|$ **do**   // Traverse decision subset that may be involved in the ith constraint

4   **if** $x \in X_{Z_{ij}}$ **then**   //If the changed decision variable x belongs *to* the subset $X_{Z_{ij}}$

5           $F^H(X') \leftarrow F^H(X') - f_i^H(X_{Z_{ij}})$ //First subtract the constraint value of the subset $X_{Z_{ij}}$ of the original solution X.

6           $F^H(X') \leftarrow F^H(X') + f_i^H(X'_{Z_{ij}})$ //Plus the constraint value of the corresponding subset $X'_{Z_{ij}}$ of the new solution X'

7   **end**

8   **end**

9   **end**

10 **return** $F^H(X')$

In particular, $X_{Z_{ij}}$ is the jth subset of the decision matrix X that may be involved in the ith constraint, and $Z_i(X)$ is the set of $Z_i(X)$, both of which are obtained in advance from the constraint network described in Section 3.4.3. The constraint computation algorithm is characterized in lines 4–7: since the neighborhood operator Move(x) only changes the value of one decision variable x in the decision matrix X, only the constraint values of the decision subset $X_{Z_{ij}}(x \in X_{Z_{ij}})$ associated with it need to be recomputed; on the contrary, since other decision subsets do not contain the changed decision variable x and their constraint values do not change, the constraint values do not need to be recomputed. In this way, the incremental constraint computation algorithm in this section can avoid repeated constraint computation during the neighborhood construction process, saving computation time and dramatically improving the computation efficiency of each neighborhood construction, which in turn improves the iteration and optimization efficiency of the local search algorithms in constraint environments, especially in the complex satellite task scheduling constraint environments in this book.

To sum up, this section elaborates the flow of "parallel" strategy and the general algorithms and operators and designs an incremental constraint computation algorithm. The strategy runs different local search algorithms in parallel to obtain diversified local optimization results, which meets the needs of APMA in terms of local optimization ability and time complexity, and provides an important basis for the competition of algorithms and operators and global optimization in subsequent strategies of APMA.

4.4 Competition-based algorithm and operator adaptive selection strategy

Referred to as competition strategy, competition-based algorithm and operator adaptive selection strategy is the second step of APMA inner loop. Based on the optimization results of the parallel strategy, this strategy constructs the current optimal solution set, quantitatively evaluates the contribution degree of algorithms and operators, updates their use probability iteration by iteration, so as to realize the competition and "survival of the fittest" at the level of algorithms and operators, to satisfy the needs of APMA in terms of adaptability, and to provide an important guarantee for the performance of the algorithms. In the following section, we elaborate on the flow of this strategy and introduce the method of updating the use probability of algorithms and operators as well as the elimination mechanism.

4.4.1 General process

Assuming that the population size required by the subsequent "evolution" strategy of the algorithms is n_p, then the "competition" strategy needs to provide a solution set of size n_p. The general process of this strategy consists of the following four main steps as shown in Figure 4.6.

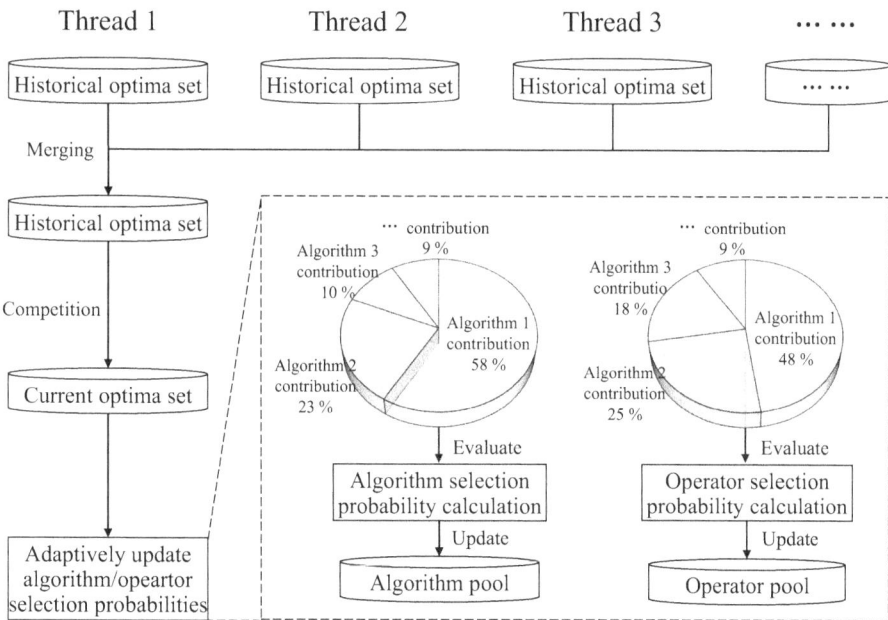

Figure 4.6: Flow of competition-based algorithm and operator adaptive selection strategy.

Step 1 (Merging historical optimal solution set). After the completion of the "parallel" strategy, each thread outputs a historical optimal solution set. First, thread 1 (the main thread) merges these historical optimal solution sets to obtain the historical optimal solution set of "parallel" strategy.

Step 2 (Selecting the current optimal solution set). According to the descending order of scoring, the historical optimal solution set is sorted, and the first n_P solutions are selected to construct a current optimal solution set. If the number of solutions in the historical optimal solution set is less than n_P, the historical optimal solution set is recognized as the current optimal solution set.

Step 3 (Competition of algorithms and operators, and update of use probability). The algorithms and operators that produce the solutions in the current optimal solution set are queried one by one, and the proportion of the contribution of each algorithm and operator to the current optimal solution set is calculated, which is denoted as the contribution degree. Based on the contribution degree and use probability of the current algorithms and operators, the use probability of each algorithm and operator in the next "parallel" strategy is updated.

Step 4 (Eliminating algorithms and operators). Individual algorithms and operators with consistently low contribution degree are eliminated, and they are removed from the algorithm pool and operator pool to achieve "survival of the fittest algorithms and operators."

It can be seen that the competition strategy is a connecting link between the preceding strategy and the following strategy, and it has the following two functions: 1) Constructing population: Because the subsequent evolution strategy is based on population with fixed size, the competition strategy constructs a population solution set that meets the needs of evolution by merging the solution sets obtained from parallel strategy and selecting the high-quality ones; and 2) enhancing self-adaptability: Different algorithms and operators perform differently. This strategy quantitatively evaluates the contribution degree of algorithms and operators in the "parallel" strategy, adjusts their use probability in the next iteration accordingly, and adaptively guides the algorithm to use more effective algorithms and operators to achieve the "survival of the fittest algorithms and operators." The following section introduces the specific steps of this strategy in detail.

4.4.2 Current optimal solution set

The current optimal solution set is the concentrated reflection of the optimization results of the parallel strategy, the basis for the competition strategy to evaluate the contribution degree of algorithms and operators and update their use probability, and the population basis of the subsequent evolution strategy. In this regard, this section adopts the following method to obtain the current optimal solution set.

As mentioned in steps 1 and 2 of Section 4.4.1, firstly, the historical optimal solution sets obtained by n_T threads in the parallel strategy are denoted as U_{1_X}, U_{2_X}, U_{3_X}, ..., $U_{n_{TX}}$, respectively, and then the set U_X of these historical optimal solution sets can be expressed as

$$U_X = \bigcup_{i=1}^{n_T} U_X^i \tag{4.4}$$

where n_T is the number of threads, U_X^i is the historical optimal solution set obtained by the ith thread, and U_X is the historical optimal solution set.

Secondly, according to the descending order of scoring, the solutions in the historical optimal solution set U_X are resequenced. Because this book uses the score function $F(X)$ to evaluate the solution quality and defines the criteria for judging the grade of the score values, the sequenced historical optimal solution set U_X satisfies

$$U_X = \left\{ X_i \mid F(X_{i_1}) \geq F(X_{i_2}), \forall i_1 \leq i_2 \leq |U_X| \right\} \tag{4.5}$$

Finally, the first n_P solutions in the historical optimal solution set U_X are selected to construct the current optimal solution set P_X. If the number of solutions in the historical optimal solution set U_X is less than n_P, the historical optimal solution set U_X is recognized as the current optimal solution set P_X, that is

$$P_X = \begin{cases} \{X_i \mid X_i \in U_X, \ i \leq n_P\}, \ |U_X| \geq n_P \\ U_X, |U_X| < n_P \end{cases} \tag{4.6}$$

where n_P is the preset current optimal solution set size, that is, the population size of the subsequent evolution strategy, U_X is the historical optimal solution set, P_X is the current optimal solution set, and X_i is the ith solution in U_X and P_X.

4.4.3 Competition of algorithms and operators and update of use probability

Based on the current optimal solution set P_X, this section quantitatively evaluates the contribution degree of each algorithm and operator and updates their use probability in the next parallel strategy. As described in step 3 of Section 4.4.1, the ith algorithm in algorithm pool A has contributed n_A^i solutions to the current optimal solution set P_X, and the ith operator in operator pool O has contributed n_O^i solutions to P_X, then the contribution degrees g_A^i and g_O^i of the ith algorithm and operator can be expressed by the following equations, respectively:

$$g_A^i = \frac{n_A^i}{|P_X|}, \forall i \leq |A| \tag{4.7a}$$

$$g_O^i = \frac{n_O^i}{|P_X|}, \forall i \le |O| \tag{4.7b}$$

where P_X is the current optimal solution set, A and O are the algorithm pool and operator pool, respectively, n_A^i and n_O^i are, respectively, the number of solutions contributed to P_X by the ith algorithm and operator in the algorithm pool and operator pool, and g_A^i and g_O^i are, respectively, the contribution degrees (proportions) of the ith algorithm and operator in the algorithm pool and operator pool to P_X.

Then the probabilities of use of the algorithms and operators are updated according to their contribution degrees. Considering the randomness of the optimization process, it may not be possible to objectively evaluate the algorithm and operator based on the optimization result of the parallel strategy only once. Therefore, for the ith algorithm and operator, this section updates the probabilities of use $p_A^i(k+1)$ and $p_O^i(k+1)$ of the $k+1$th-iteration algorithm and operator by averaging their current probabilities of use $p_A^i(k)$ and $p_O^i(k)$ and contribution degrees ($g_A^i(k)$ and $g_O^i(k)$), respectively, that is

$$p_A^i(k+1) = \frac{1}{2}\left[p_A^i(k) + g_A^i(k)\right] \tag{4.8a}$$

$$p_O^i(k+1) = \frac{1}{2}\left[p_O^i(k) + g_O^i(k)\right] \tag{4.8b}$$

where $g_A^i(k)$ and $g_O^i(k)$ are, respectively, the contribution degrees of the ith algorithm and operator in the kth-iteration algorithm pool and operator pool, and $p_A^i(k)$ and $p_O^i(k)$ are, respectively, the probabilities of use of the ith algorithm and operator in the kth-iteration algorithm pool and operator pool.

In this way, the unusually low probabilities of use of the algorithm and operator due to the poor results of a particular iteration of the parallel strategy can be avoided to some extent. Moreover, if the ith algorithm or operator is not used in the kth iteration, its use probability is not updated. Here, it is agreed that the initial probabilities of use of each algorithm and operator are equal, that is, $p_A^i(0) = \frac{1}{|A|}, p_O^i(0) = \frac{1}{|O|}$.

Finally, the probabilities of use for all algorithms and operators are processed through 0–1 normalization:

$$p_A^i = \frac{p_A^i}{\sum_{i \in A} p_A^i}, \forall i \le |A| \tag{4.9a}$$

$$p_O^i = \frac{p_O^i}{\sum_{i \in O} p_O^i}, \forall i \le |O| \tag{4.9b}$$

In summary, this section quantitatively evaluates the contribution degrees of algorithms and operators and updates their probabilities of use generation by generation, which provides important support for enhancing the self-organization and self-adaptability of the algorithms.

4.4.4 Elimination mechanism

On the basis of the abovementioned competition of algorithms and operators and the evaluation of their contribution degrees, it is necessary to eliminate the algorithms and operators that have been making little contribution and showing poor performance in the iterative process, so as to avoid the waste of computational resources and further promote the "survival of the fittest algorithms and operators."

This section specifies that for the ith algorithm and operator in the algorithm pool and operator pool, if at their kth generations, the contribution degrees of their previous n_E successive generations are lower than the threshold g_{min}, that is, eqs. (4.10a) and (4.10b) are satisfied, they are regarded as "eliminated" and removed from the algorithm pool and operator pool:

$$g_A^i(k-n) \le g_{min}, \forall i \le |A|, n \le n_E \le k \tag{4.10a}$$

$$g_O^i(k-n) \le g_{min}, \forall i \le |O|, n \le n_E \le k \tag{4.10b}$$

where $g_A^i(k)$ and $g_O^i(k)$ are, respectively, the contribution degree of the kth generations of the ith algorithm and operator in the algorithm pool and operator pool, g_{min} is the contribution degree threshold of the algorithm and operator, and n_E is the maximum number of consecutive iterations allowed when the contribution degrees of the algorithm and operator are below the threshold value.

To sum up, this section details the flow of the competition strategy. As the second step of the APMA inner loop, this strategy collects the historical optimal solution set obtained by the parallel strategy, the first step of the APMA inner loop. Then, this strategy quantitatively evaluates the contribution degrees of the algorithms and operators and updates their probabilities of use generation by generation, thus realizing the "survival of the fittest algorithms and operators," meeting the needs of APMA in terms of adaptability and generality, and providing an important guarantee for the performance of APMA.

4.5 Population evolution-based global optimization strategy

Referred to as evolution strategy, the population evolution-based global optimization strategy is the last step of APMA inner loop. Besides the parallel and competition strategies, the evolution strategy uses evolutionary operators such as crossover operators and repair operators to perform a global optimization on the basis of the current optimal solution set so that APMA has the ability to skip from local optima for exploring a larger solution space, which further guarantees comprehensive performance of APMA. In the following, this section elaborates on the specific process of this strategy and introduces the specific implementation methods of operators such as crossover operators and repair operators.

4.5.1 General process

Similar to the process of EA, as shown in Figure 4.7, the general process of this strategy consists of the following four main steps:

Step 1 (Constructing the parent population). The current optimal solution set obtained by the competition strategy is used as the parent population, the population size is n_P, and the individuals (solutions) in the parent population are randomly paired.

Step 2 (Crossover and repair). According to the crossover probability, the crossover operator is applied to the paired individuals in the parent population to obtain the offspring individuals. If any offspring individual violates the constraints, the repair operator is used to repair it.

Step 3 (Constructing the offspring population). The offspring individuals obtained in step 2 are stored in the offspring population one by one.

Step 4 (Updating the current optimal solution set). After merging the parent and offspring populations, n_P individuals with the highest scores in the merged population are selected to construct a new current optimal solution set.

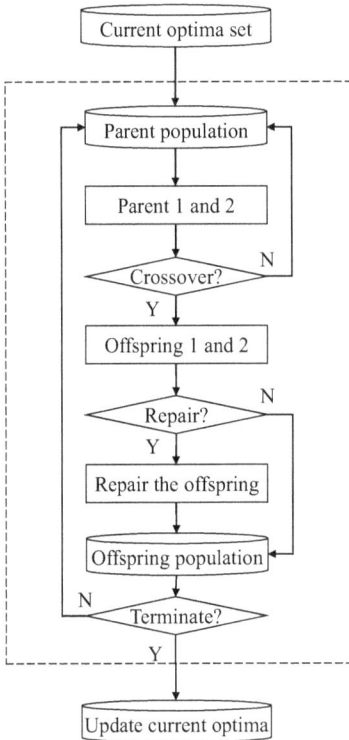

Figure 4.7: Flow of population evolution-based global optimization strategy.

As the last step of APMA inner loop, the evolution strategy has following two main functions: 1) Global optimization: Through the crossover and repair operators, a larger neighborhood is constructed, and a larger solution space is explored, which helps the algorithm to skip from local optima for achieving global optimization; and 2) enhancement of solution diversity: The current optimal solution set is optimized by the local search algorithms in the parallel strategy, and the diversity provided by the strategy is limited, while this strategy disturbs the current optimal solution set through the probabilistic crossover operators, and once again enhances the solution diversity, and promotes the performance of the next generation of the parallel and evolution strategies, forming a virtuous circle.

In other studies, local search algorithms, such as adaptive large neighborhood search, have also been used to achieve the goal of skipping from local optimization for achieving global optimization through operators such as disturbance operators, destruction operators, and repair operators. This chapter adopts the evolution strategy to achieve this goal, mainly because the authors intend to further enhance the global performance of APMA by utilizing the rich optimization results obtained by multithread parallel algorithm and by leveraging the diversity and adaptability of the populations.

4.5.2 Crossover operator

In the design of EA and evolution strategy, the crossover operator should meet the following two conditions: (1) It does not affect the value of decision variables, that is, the decision variables will not exceed the range of the original values and affect the expression of solutions after crossover; and (2) the constraints are violated as little as possible. Any operator may cause the original feasible solution to violate the constraints, while the crossover operator usually changes the values of many decision variables at one time, which is more likely to violate the constraints. In this regard, on the basis of the general-purpose 0–1 mixed integer decision model for satellite task scheduling in this book, this section gives two crossover operators, as shown in Figure 4.8.

(1) Single-point crossover operator, that is, the operator randomly selects an identical position in the decision matrices X_1 and X_2 and swaps the values of all decision variables above that position, as shown in Figure 4.8(a). Based on the general-purpose 0–1 mixed integer decision model for satellite task scheduling in this book, the actual meaning of this operator is to swap the executable opportunities of all events of some satellite tasks.

(2) Two-point crossover operator, that is, the operator randomly selects two identical positions in the decision matrices X_1 and X_2 and swaps the values of all decision variables in the middle of two positions, as shown in Figure 4.8(b). Similar to the single-point crossover operator, the actual meaning of the double-point crossover operator is to swap the executable opportunities of all events of some satellite tasks.

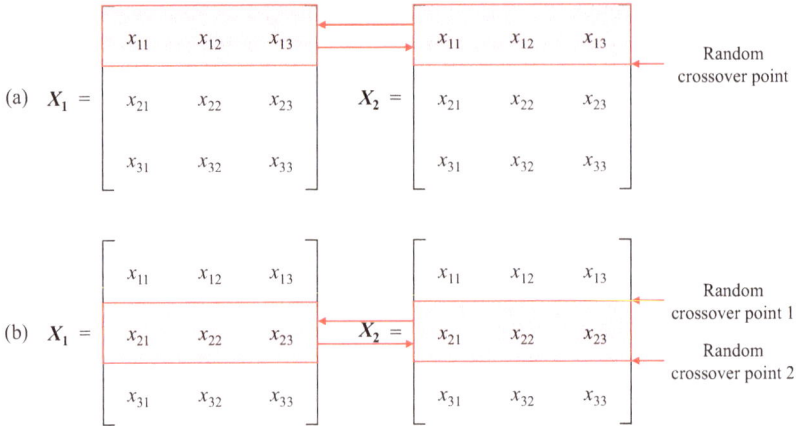

Figure 4.8: Schematic diagram of crossover operators in evolution strategy: (a) single-point crossover operator and (b) double-point crossover operator.

It is worth noting that the crossover operator can also be described as a combination of multiple Swap or Move operators, so the crossover operator will not cause the decision variables to exceed the range of the original values. In the case that the decision matrices X_1 and X_2 both satisfy the constraints, the crossover operator can preserve the execution of some of the satellite events and satisfy some of the constraints, but there is still a possibility that other constraints are violated. In this regard, it is necessary to design a repair operator to quickly and appropriately repair the solutions that violate the constraints, so as to guarantee the feasibility of the global optimization results of the evolution strategy in this section.

4.5.3 Repair operator

In view of the complexity of the constraints of satellite task scheduling problems, the process of using the crossover operator to generate offspring individuals may produce some constraint-violating solutions, which affects the crossover operator performance and population diversity. For these constraint-violating solutions, based on the incremental constraint computation algorithm in Section 4.3.4, a repair operator is designed in this section to improve the crossover operator performance and maintain the diversity of the population with less time cost, so as to ensure the feasibility of the global optimization results of the evolution strategy in this section. The specific steps of the repair operator are shown in Algorithm 4.8.

Algorithm 4.8: Repair operator based on incremental constraint computation

Input: decision matrix X, decision variable set X_C ($X_C \subset X$) involved in crossover operator, constraint set F^H

Output: decision matrix X satisfying constraints

```
1   for x∈X_C  do              //Traverse decision variables involved in the crossover operator
2      for i = 1: |F^H|  do    //Traverse the constraints and the decision matrix subset that may
                                be involved, and perform incremental constraint checking
3         for j = 1: |Z_i(X)|  do
4            if x∈X^Z_{ij} & f_i^H(X^Z_{ij}) ≠ 0 then
5               x ← 0           //If any decision variable leads to a constraint violation, reset it
                                to 0
6            end
7         end
8      end
9   end
10  return F^H(X′)
```

In summary, this section details the evolution strategy and the specific crossover and repair operators. As the last step of APMA inner loop, this strategy performs a global optimization of the current optimal solution set obtained by the parallel and competition strategies through crossover operators, repair operators, and other evolutionary operators, which helps APMA skip from local optima for exploring a larger solution space, meets the needs of APMA in terms of global optimization ability, and provides another important guarantee for the comprehensive performance of APMA.

4.6 Performance test of APMA in benchmark problems

Due to the high degree of simplification of the relevant problems in traditional satellite task scheduling studies, the algorithms designed in the studies are insufficiently practical and general to address actual problems. Therefore, it is not possible to objectively compare the performance of such algorithms with APMA in later experiments. In this regard, this section tries to examine the performance of APMA in benchmark problems to give objective experimental verification. Here, considering that Peng et al. [174] have pointed out that the simplified conventional task scheduling problems of remote-sensing satellites are very similar to the orienteering problem (OP), this section selects benchmark problems such as OP and its variants, and the simplified conventional task scheduling problems of remote-sensing satellites to carry out algorithm tests. The test results show that compared with the latest algorithms, APMA exhibits good optimization performance and generality and can provide a general-purpose and efficient means of solving conventional satellite task scheduling problems.

4.6.1 Orienteering problem

Proposed by Golden et al. [246] in 1987 and named after "orientation movement," OP is a classical combinational optimization problem in operations study. OP can be simply described as selecting a subset from a given vertex set and sorting it to maximize the sum of vertex profits in the subset, subject to time constraints. As a result, OP is often regarded as a combination of knapsack problem and traveling salesman problem (TSP), and selective TSP. This section takes OP benchmark [247] provided by the KU Leuven in Belgium as the test set. In 1996, Chao et al. [248] solved each example in the test set through a fast heuristic algorithm, and the results obtained were recognized as the optimal solutions for the test set.

Based on this, an algorithm testing experiment is conducted in this section. In this book, five local search algorithms, including HC, TS, SA, LA, and ILS, are used in APMA parallel and competition strategies. Each parallel iteration is 1,000 generations, in which the Tabu length coefficient of TS is set to 0.1, the annealing method of SA is set to geometric annealing, the late length coefficient of LA is set to 2.0, and the disturbance and repair frequency of ILS is set to 10 times. In the evolution strategy, the population size is set to 100, and the crossover and mutation probability is set to 1.0. To quantitatively compare the performance of the algorithms, the algorithm running time in each example is consistent with that in the literature. The average results of 10 independent computations are taken, and the results are listed in Table 4.2.

Table 4.2: Algorithm test and comparison results in solving OP benchmark.

Example	Optima	APMA	Example	Optima	APMA	Example	Optima	APMA
1.1	10	10	1.18	285	285	3.6	430	430
1.2	15	15	2.1	120	120	3.7	470	470
1.3	45	45	2.2	200	200	3.8	520	520
1.4	65	65	2.3	210	210	3.9	550	550
1.5	90	90	2.4	230	230	3.10	580	580
1.6	110	110	2.5	230	230	3.11	610	610
1.7	135	135	2.6	265	265	3.12	640	640
1.8	155	155	2.7	300	300	3.13	670	670
1.9	175	175	2.8	320	320	3.14	710	710
1.10	190	190	2.9	360	360	3.15	740	740
1.11	205	205	2.1	395	395	3.16	770	770
1.12	225	225	2.11	450	450	3.17	790	790
1.13	240	240	3.1	170	170	3.18	800	800
1.14	260	260	3.2	200	200	3.19	800	800
1.15	265	265	3.3	260	260	3.20	800	800
1.16	270	270	3.4	320	320			
1.17	280	280	3.5	390	390			

As given in Table 4.2, APMA can obtain the optimal solutions in equal running time, and the results of 10 independent operations are exactly the same as the optimal solutions, which demonstrates the effectiveness and robustness of APMA in solving the test set. In the following, the performance of APMA is further tested based on the OP variant problem.

4.6.2 OP with time windows

Proposed by Kantor et al. [249] in 1992, OP with time windows (OPTW) is a variant of OP, which is mainly characterized by setting a time window constraint for the access time of each vertex in OP. In this section, the OPTW benchmark [247] provided by KU Leuven is used as the test set, for which the latest algorithm is ant colony system algorithm (ACS) proposed by Verbeeck et al. [247] in 2017. Based on this, this section carries out the algorithm testing and comparative experiments, and the algorithm parameters and termination conditions remain unchanged. The experimental results are listed in Table 4.3. The experimental results in the table are the average results obtained by several independent operations of APMA.

Table 4.3: Algorithm test and comparison results in solving OPTW benchmark.

Example	ACS	APMA	Outperformance	Example	ACS	APMA	Outperformance
20.1.1	177	177	0.00%	50.3.1	380	380	0.00%
20.1.2	193	193	0.00%	50.3.2	444	444	0.00%
20.1.3	201	201	0.00%	50.3.3	403	403	0.00%
20.2.1	213	213	0.00%	50.4.1	498	498	0.00%
20.2.2	219	219	0.00%	50.4.2	463	463	0.00%
20.2.3	211	211	0.00%	50.4.3	479	**493**	**2.90%**
20.3.1	306	306	0.00%	100.1.1	297	297	0.00%
20.3.2	262	262	0.00%	100.1.2	320	320	0.00%
20.3.3	286	286	0.00%	100.1.3	373	373	0.00%
20.4.1	293	293	0.00%	100.2.1	397	397	0.000%
20.4.2	299	299	0.00%	100.2.2	393	**396.8**	**1.00%**
20.4.3	283	283	0.00%	100.2.3	394	394	0.00%
50.1.1	314	314	0.00%	100.3.1	490	490	0.00%
50.1.2	290	290	0.00%	100.3.2	511	511	0.00%
50.1.3	316	316	0.00%	100.3.3	523.4	**525**	**0.30%**
50.2.1	322	322	0.00%	100.4.1	489.3	**522.3**	**6.70%**
50.2.2	347	347	0.00%	100.4.2	529.4	**552**	**4.30%**
50.2.3	346	346	0.00%	100.4.3	581.7	**588.3**	**1.10%**
Number of results better than/equal to/ inferior to the literature result			6/30/0	Average outperformance			0.50%

As given in the table, the average results obtained from 10 independent computations of APMA are better than or equal to the literature results in the same computing time. Six of the examples achieved better optimization results than the literature results, and the average outperformance of the 36 examples is 0.5%. The experimental results illustrate the effectiveness of APMA in solving the OPTW test set.

4.6.3 Time-dependent OPTW

Proposed by Fomin et al. [251] in 2002, time-dependent OPTW (TD-OPTW) is a variant of OPTW. Its main feature is that time-dependent transfer time is introduced into OPTW, that is, the transfer time between two vertices is affected by access time. In view of this feature, Peng et al. [174] pointed out that the simplified conventional task scheduling problems of agile remote-sensing satellites (only considering imaging events) are quite similar to TD-OPTW.

In this section, TD-OPTW benchmark [247] provided by KU Leuven is used as the test set, for which the latest algorithm is the ACS proposed by Verbeeck et al. [250]. Based on this, this section carries out the algorithm testing and comparative experi-

Table 4.4: Algorithm test and comparison results in solving TD-OPTW benchmark.

Example	ACS	APMA	Outperformance	Example	ACS	APMA	Outperformance
20.1.1	159	159	0.00%	50.3.1	339	**346**	**2.10%**
20.1.2	173	173	0.00%	50.3.2	404	**415**	**2.70%**
20.1.3	183	**184**	**0.50%**	50.3.3	366	366	0.00%
20.2.1	188	188	0.00%	50.4.1	476.6	**478**	**0.30%**
20.2.2	201	201	0.00%	50.4.2	439.8	**441**	**0.30%**
20.2.3	195	195	0.00%	50.4.3	450	450	0.00%
20.3.1	277	277	0.00%	100.1.1	275	275	0.00%
20.3.2	245	**246**	**0.40%**	100.1.2	278	278	0.00%
20.3.3	259	259	0.00%	100.1.3	343	<u>342</u>	−0.30%
20.4.1	274	274	0.00%	100.2.1	351.2	<u>351.1</u>	0.00%
20.4.2	275	275	0.00%	100.2.2	366.6	**367**	**0.10%**
20.4.3	268	268	0.00%	100.2.3	370	370	0.00%
50.1.1	288	288	0.00%	100.3.1	436	**437**	**0.20%**
50.1.2	274	274	0.00%	100.3.2	446.6	**454**	**1.70%**
50.1.3	289	289	0.00%	100.3.3	467	**470**	**0.60%**
50.2.1	298	298	0.00%	100.4.1	480	480	0.00%
50.2.2	310	310	0.00%	100.4.2	494.6	**497**	**0.50%**
50.2.3	340	340	0.00%	100.4.3	526.8	**532**	**1.00%**
Number of results better than/equal to/ inferior to the literature result			12/23/1	Average outperformance			0.30%

Note: Bold numbers indicate better results, and underlined numbers indicate worse results.

ments, and the algorithm parameters and termination conditions remain unchanged. The experimental results are listed in Table 4.4.

As can be seen from Table 4.4, except for Example 100.1.3, the average results achieved by APMA for 10 independent computations are better than or equal to the literature results in the same computing time, 12 examples achieved better optimization results than those in the literature, and the average outperformance of the 36 examples is 0.3%. The experimental results illustrate the effectiveness of APMA in solving the TD-OPTW test set.

4.6.4 Simplified remote-sensing satellite conventional task scheduling problems

Peng et al. [174] pointed out that the simplified conventional task scheduling problems of agile remote-sensing satellites are similar to TD-OPTW. This section simulates the task scheduling problems of an agile remote-sensing satellite AS-01 imaging domestic and global random targets, considering the imaging events of the remote-sensing satellite and their time-dependent transition time constraints, but not considering downlink events, memory-erasing events and many other related constraints. To address these simplified problems, Peng et al. [174] proposed a dynamic programming based ILS algorithm (BDP-ILS). The main feature of this algorithm is that the subjective decoding strategy of heuristic rules such as "imaging quality first" and "imaging time first" is replaced by decoding the start time of each imaging events in its time window through dynamic programming. However, since the decoding strategy and operators are specially designed based on the simplified problems, the algorithm can't be applied to the actual satellite task scheduling problems. Nevertheless, to fully examine the performance of APMA, this section carries out algorithm testing with this benchmark as the test set and compares it with BDP-ILS. The algorithm parameters and termination conditions remain unchanged, and the experimental results are listed in Table 4.5.

From the results of the 24 examples in Table 4.5, it can be seen that APMA has achieved better optimization results in 14 examples, the optimization efficiency is consistent with the literature in 4 examples, but the optimization results are not good in 6 examples, and the average outperformance of 24 examples is 1.8%. The experimental results indicate that APMA is competitive compared with the latest algorithm. Moreover, APMA is designed for real satellite task scheduling problems, such as remote-sensing satellite scheduling, relay satellite scheduling, navigation satellite scheduling and satellite range scheduling. Since APMA follows the reality-based and application-oriented study principle in this book, and its strategies and operators are loosely coupled with the problems, it has higher generality, practicability, and flexibility and can better meet the objective needs of the satellite control agency in terms of task scheduling algorithms.

To sum up, to objectively test APMA performance, this section carries out algorithm testing and comparison in benchmark problems such as OP and simplified re-

Table 4.5: Algorithm test and comparison results in solving AS-01 benchmark.

Example	ACS	APMA	Outperformance	Example	ACS	APMA	Outperformance
100_A	568.4	**569.2**	**0.10%**	100_W	550	550	0.00%
200_A	894	**910.8**	**1.90%**	200_W	1,004	997.4	−0.70%
300_A	998.6	**1,037.6**	**3.90%**	300_W	1,622	1,619.7	−0.10%
400_A	1,162.8	**1,224.7**	**5.30%**	400_W	2,263	2,255.4	−0.30%
500_A	1,297.6	**1,345.8**	**3.70%**	500_W	2,686	**2,691.3**	**0.20%**
600_A	1,399.8	**1,510.2**	**7.90%**	600_W	3,122	3,122.2	0.00%
100_ATD	491.1	**492.3**	**0.20%**	100_WTD	528	528	0.00%
200_ATD	747.6	**763**	**2.10%**	200_WTD	973.9	971.3	−0.30%
300_ATD	854.9	**889.2**	**4.00%**	300_WTD	1,557.1	1,550.2	−0.40%
400_ATD	1,053.2	**1,102.1**	**4.60%**	400_WTD	2,149.9	2,141.2	−0.40%
500_ATD	1,153.7	**1,197.8**	**3.80%**	500_WTD	2,558.4	**2,561**	**0.10%**
600_ATD	1,260.1	**1,360.1**	**7.90%**	600_WTD	2,847.4	2,847.8	0.00%
Number of results better than/equal to/inferior to the literature result			2014/4/6	Average outperformance			1.80%

Note: Bold numbers indicate better results, and underlined numbers indicate worse results.

mote-sensing satellite conventional task scheduling problems. The test results show that compared with the latest algorithms, APMA exhibits good generality and optimization performance and can provide a general-purpose and efficient means of solving satellite conventional task scheduling problems.

4.7 Chapter summary

Facing the daily and weekly conventional task scheduling requirements of the satellite control agency, this chapter proposes a general-purpose APMA, which mainly includes

(1) Heuristic-based fast initial solution construction strategy is designed. The strategy helps APMA to enter the high-quality feasible domain space quickly through the heuristic algorithm with "pre-allocation and first-in-first-service," which provides an important foundation for further iterative optimization of APMA.

(2) Parallel search-based general-purpose local optimization strategy is designed. The strategy runs different general-purpose local search algorithms and operators in parallel to obtain diversified local optimization results efficiently, which meets the needs of APMA in terms of local optimization ability and time complexity, and provides an important basis for the competition of algorithms and operators and global optimization in subsequent strategies of APMA.

(3) Competition-based algorithm and operator adaptive selection strategy is designed. The strategy quantitatively evaluates the contribution degree of algorithms

and operators and updates their probabilities of use generation by generation, so as to realize the competition and "survival of the fittest" at the level of algorithms and operators, to satisfy the needs of APMA in terms of adaptability, and to provide an important guarantee for the performance of APMA.

(4) Population evolution-based global optimization strategy is designed. This strategy performs a global optimization of the current optimal solution set through crossover operators, repair operators and other evolutionary operators, which helps APMA skip from local optima to explore a larger solution space, meets the needs of APMA in terms of global optimization ability, and provides another important guarantee for the comprehensive performance of APMA.

The above four strategies work together and complement each other, which not only retains the advantages of the local and global optimization abilities of traditional MA, but also avoids the disadvantages of time complexity superposition and low efficiency of constrained optimization, and performs well in a series of benchmark experiments. Therefore, APMA in this chapter can meet the practical optimization needs of satellite conventional task scheduling, provide a general-purpose and efficient means of solving satellite conventional task scheduling problems in this book, and provide core algorithm for the satellite task scheduling engine in this book.

Chapter 5
General-purpose solving method for satellite emergency task scheduling

In the long-term process of satellite control, since dynamic events such as adding or deleting tasks and satellite failures occur frequently and the timeliness of conventional task scheduling algorithms is insufficient, satellite emergency task scheduling has become the new normal. In this regard, for the needs of the satellite control agency in terms of emergency task scheduling, this chapter designs a general-purpose distributed dynamic rolling optimization algorithm (DDRO) to perform real-time and effective dynamic optimization for the current satellite task scheduling scheme to meet the practical needs of satellite emergency task scheduling.

Firstly, this chapter analyzes the requirements of the satellite emergency task scheduling algorithm and builds a general-purpose algorithm framework of DDRO. To implement the framework and meet the application requirements, the following four strategies are sequentially designed: 1) Task negotiation and allocation strategy based on dynamic contract net; 2) single-platform task rescheduling strategy based on rolling time domain; 3) task rapid insertion strategy based on schedulable prediction; and 4) real-time conflict-avoidance strategy based on constraint network. This chapter elaborates on the implementation methods of strategies of DDRO, which provides a general-purpose and flexible means of solving satellite emergency task scheduling problems and another algorithm for the satellite task scheduling engine.

5.1 General-purpose distributed dynamic scrolling algorithm framework

5.1.1 Request analysis

In the long-term process of satellite control, some dynamic events inevitably occur, affecting the original satellite task scheduling schemes and requiring the control agency or satellite to spontaneously make effective emergency responses. From the perspective of the satellite control agency, these dynamic events include subjective dynamic events and objective dynamic events, that is, dynamic events triggered by the control agency and dynamic events triggered by the satellite itself. Table 5.1 lists some common dynamic events and their impact on satellite task scheduling problems.

As given in Table 5.1, the impact of the abovementioned common dynamic events on satellite task scheduling problems mainly includes resource increase, resource decrease, task increase, task decrease, and violation of constraints. Among them, the increase in resources and the decrease in tasks lead to increase in available resources,

https://doi.org/10.1515/9783111537191-005

Table 5.1: Common dynamic events in satellite control process and their impacts on task scheduling problems.

Type	Dynamic event	Impact on task scheduling problem
Subjective dynamic events triggered by the control agency	Temporary addition of satellites, orbits, and windows	Increase in resources, which is in favor of optimization
	Temporary decrease of satellites, orbits, and windows	Decrease in resources, which is not conducive to optimization
	Sudden and temporary new tasks	Increase of tasks, which is not conducive to optimization
	Temporary cancelation of tasks	Decrease of tasks, which is in favor of optimization
	Manual and customized adjustment of schemes	Possible violation of constraints, which is not conducive to optimization
	Rescheduling of tasks that have to be canceled owing to reduced resources	Increase of tasks, which is not conducive to optimization
	Rescheduling of tasks that have to be canceled owing to violation of constraints	Increase of tasks, which is not conducive to optimization
Objective dynamic events triggered by the satellite itself	Window shortening or failure due to cloud cover	Decrease of resources, which is not conducive to optimization
	Satellite was unable to carry out its tasks owing to its malfunction	Decrease of resources, which is not conducive to optimization
	New tasks are triggered in autonomous perception mode	Increase of tasks, which is not conducive to optimization
	New tasks allocated by other satellites in autonomous coordination mode	Increase of tasks, which is not conducive to optimization

which reduce the degree of constraints and the scale of the task scheduling, and is conducive to solving the task scheduling problems. On the contrary, the decrease in resources and the increase in tasks lead to a decrease in available resources, which increases the degree of constraints and the scale of the task scheduling, and is not conducive to solving the task scheduling problems. It is worth noting that when some tasks are forced to be canceled or need to be rescheduled due to the decrease in resources and violation of constraints, these tasks can also be regarded as "new tasks." In view of this, based on the actual situation of satellite control, the requirements for the design of satellite emergency task scheduling algorithm can be summarized as follows:

(1) Requirement for real-time and rapid scheduling of "new tasks" under the unfavorable impact of decreased resources, increased tasks, and constraint violations. In Table 5.1, most of the emergencies produce adverse impacts such as resource decrease, task increase, and constraint violation, so mitigating such adverse impacts is the most common and important requirement of the satellite control agency in the process of emergency task scheduling. Especially in recent years, the maritime security situation has been grim, with frequent disputes and controversies, and the demand for emergency tasks for maritime targets has increased, placing new requirements on the emergency response capability of the control agency. Meanwhile, with the continuous improvement of onboard hardware and computing power, more and more satellites have autonomous and independent scheduling capabilities, commonly known as "autonomous satellites." Autonomous satellites can generate and execute tasks spontaneously with little intervention from the ground control center, and this process also requires the support of real-time and effective emergency scheduling algorithms.

(2) Requirement for continuous optimization of the original scheme under the favorable impact of increased resources and increased tasks. From the perspective of the task scheduling scheme, the feasibility and benefits of the current scheme have not changed under the impact of increased resources and decreased tasks. Compared to other unfavorable impacts, the requirement for rescheduling the current scheme is not urgent. Meanwhile, such impacts lead to an increase in currently available resources, and there is a certain room for optimization of the current program. Under such impacts, the main requirement of the satellite control agency is further optimization of the original scheme, and the profits of satellite task scheduling are further enhanced by using the favorable conditions of increased available resources.

(3) Operational requirement to minimize changes to the original scheme during the rescheduling process. Changes to satellite task scheduling scheme involve not only the recompilation and annotation of task instructions but also adjustments to the work plan and shifts of ground-based controllers. Therefore, the satellite control agency does not want drastic changes in the task scheduling scheme, no matter it is real-time and rapid scheduling for unfavorable impacts or continuous optimization for favorable impacts. Consequently, the original task scheduling scheme should be changed as little as possible to meet the actual operational requirements in the real-time and rapid rescheduling or continuous optimization of the scheme.

(4) Requirement for intelligent control of multi-satellite and multi-station task negotiation and allocation under the growing trend of autonomous satellite systems. With the continuous increase in the number of satellites in orbit and the continuous improvement of onboard hardware performance and computing power, there is a trend for more and more satellite systems to have the capability of autonomous management. Under this trend, the traditional ground-centered control mode will gradually shift to the autonomous and intelligent control mode of satellite-ground task negotiation and allocation. Therefore, a task scheduling framework that meets the dual requirements of rapid response and intelligent control is indispensable. It should be noted that although the requirements are based on the background of the future development trend of satellite systems, the design of the task scheduling framework should still follow the reality-based and application-oriented principle to meet the basic application requirements of the current satellite systems.

In response to the above requirements, this section designs an emergency task scheduling algorithm according to the countermeasures listed in Table 5.2, as well as a DDRO algorithm framework based on multi-agent system (MAS), which provides a real-time and effective means of solving satellite emergency task scheduling problems.

Table 5.2: Design ideas and countermeasures of satellite emergency task scheduling algorithm.

No.	Design requirements	Design ideas and countermeasures
1	Real-time and rapid scheduling of new tasks	1. Designing a task rescheduling strategy based on rolling time domain to respond to new tasks in a timely manner 2. Designing a real-time conflict avoidance strategy to meet the application requirements of rescheduling strategy
2	Continuation optimization of the original scheme	Using the simplified version of APMA designed in Chapter 4
3	Minor changes to the original scheme	1. Rescheduling local tasks following the rescheduling strategy based on the rolling time domain 2. Using local optimization as much as possible for continuous optimization to reduce the use of global optimization strategy
4	Multi-satellite task negotiation and allocation	1. Building the MAS framework by considering each satellite as an agent 2. Designing the task negotiation and allocation strategy based on the MAS framework

5.1.2 Algorithm framework

Based on the above analysis, a general-purpose DDRO algorithm (DDRO) is designed in this section. As shown in Figure 5.1 (taking the task scheduling problems of remote-sensing satellites as an example), the algorithm framework mainly contains the following four layers:

(1) Requirement layer
The requirement layer is to obtain the specific requirements of dynamic task scheduling, that is, under the unfavorable dynamic impacts of decreased resources, increased tasks, constraint violations, etc. as mentioned above, the control agency specifies the

"new tasks" that need to be scheduled and uploads them to the MAS negotiation layer. Compared to the traditional control mode, the control agency no longer directly schedules the "new tasks," but uploads them to the MAS negotiation layer, where MAS completes the subsequent task negotiation for scheduling.

(2) Negotiation layer

The negotiation layer is used to build MAS and execute the communication, negotiation, and task allocation among the agents in the MAS. In this regard, a task negotiation and allocation strategy based on a dynamic contract net is designed in this chapter, and the strategy is jointly maintained by the ground control center and the satellite system. It should be noted that the negotiation layer is built based on the communication blackboard, that is, on a hypothetical communication platform organized by the "satellite-satellite-ground" communication network technology. This type of MAS is often referred to as blackboard model-based MAS. Since the methods of establishing, maintaining, and updating the communication blackboard belong to the study field of communication technology, they are outside the scope of this book. Through the strategy of task negotiation and allocation based on the dynamic contract net, the communication, negotiation, and task allocation results of the negotiation layer are eventually synchronized to the satellite or station platform, guiding the platform to perform subsequent task scheduling

(3) Instance layer

The instance layer instantiates the agents in MAS to give the agents real physical meaning. In the MAS framework, each agent represents a station or a satellite (including both autonomous and nonautonomous satellites) in orbit. The autonomous satellite agent is operated by the onboard hardware, while the nonautonomous satellite agent is operated by the ground control center. In this way, the framework has solved the problem that traditional MAS is only applicable to a few autonomous satellites and enhanced the generality and practical value of MAS under the current situation that the satellite system is still dominated by nonautonomous satellites, which follows the reality-based and application-oriented study principle in this book.

(4) Single-platform layer

The single-platform layer is the terminal optimization layer of the framework, which is to implement task scheduling at the single-platform level. For the generality of the algorithm in this chapter, in different satellite task scheduling problems, the single-platform layer is represented as a satellite (in the task scheduling problems of remote-sensing satellites, relay satellites, and navigation satellites) or a station (in the satellite range scheduling problems). Since the single-platform layer only involves one satellite or one station performing tasks, the task scheduling problems of this layer can be regarded as a subproblem of the original problems. For the emergency task scheduling problems in the single-platform layer, this chapter designs a single-platform task rescheduling strategy based on a rolling time domain, as well as a task rapid insertion strategy based on schedulable prediction and a real-time conflict-avoidance strategy based on constraint network, etc., required by the rescheduling strategy. When the window is rolling or the task pool receives a new task, the single-platform layer schedules the tasks in the task pool, reschedules the tasks in the window in real time, and updates the task pool and the current satellite task scheduling scheme in time to satisfy the emergency task scheduling needs of the single platform in the dynamic MAS environment.

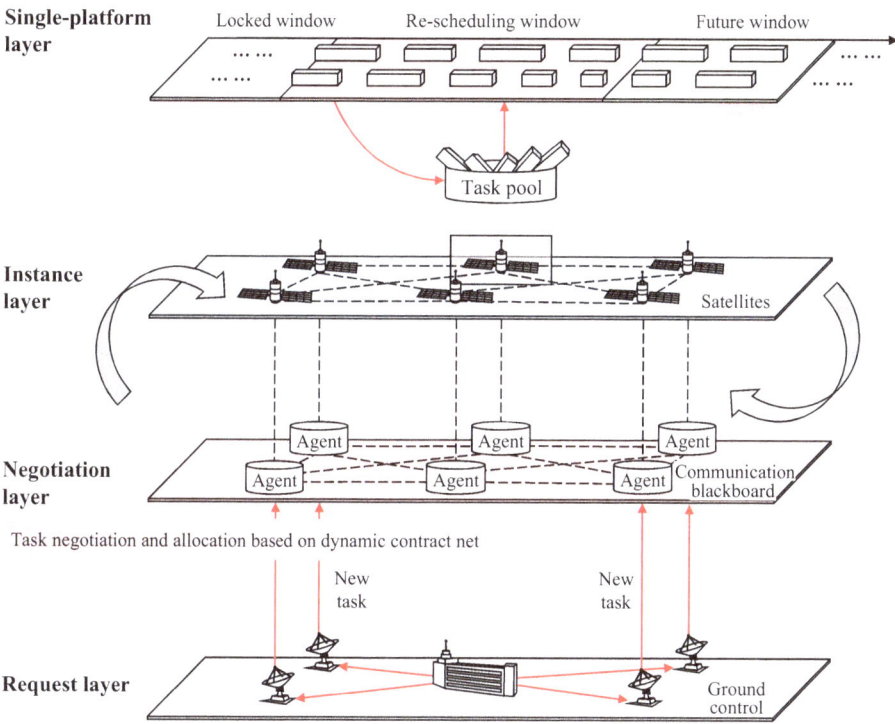

Figure 5.1: DDRO framework (taking remote-sensing satellite task scheduling problems as an example).

5.1.3 Advantage analysis

Based on the above introduction, oriented to the actual requirements of the satellite control agency, the advantages of DDRO optimization framework can be summarized as follows:

(1) **An emergency scheduling framework that is different from the traditional conventional scheduling is designed to meet the requirements for real-time emergency optimization and autonomous optimization of task scheduling scheme in the long-term process of satellite control.** Different from the conventional scheduling algorithm designed in the previous chapter, the algorithm designed in this chapter aims to address the real-time emergency task scheduling problems in dynamic environments. In other words, the goal of the framework in this chapter is no longer the original conventional optimization, but flexible and rapid emergency optimization. Therefore, in the long-term process of satellite control, the algorithms in this chapter and the previous chapter are designed to address satellite emergency and conventional task scheduling problems, respectively, which together provide the necessary technical support for the long-term control of satellite systems.

(2) **A distributed task negotiation and allocation environment is built to meet the requirements for multi-satellite and multi-station intelligent and collaborative control under the new trend.** For the purpose of emergency optimization and autonomous optimization, the framework builds a distributed task negotiation and allocation environment, which realizes the transformation of the traditional ground-centered control mode to the intelligent control mode of satellite-ground negotiation and collaboration, reduces the pressure and complexity of ground-based control, and meets the requirements of multi-satellite and multi-station intelligent and collaborative control under the growing trend of autonomous satellite system.

(3) **A rescheduling mechanism based on rolling time domain is designed to meet the requirements for minor changes and continuous optimization of task schemes in dynamic environment.** Through the rescheduling mechanism based on rolling time domain, the framework flexibly responds to the requirements of satellite emergency task scheduling and updates the task scheduling scheme in continuous time domain, which guarantees the consistency of the scheme and provides the necessary means of solving the satellite emergency task scheduling problems. Moreover, due to the existence of the rolling window, this mechanism only affects a part of the current satellite task scheduling scheme and does not cause substantial changes to the original scheme, which meets the relevant operation requirements of the satellite control agency.

(4) **A real-time task insertion and conflict avoidance strategy is integrated to meet the requirements of the control agency for real-time and rapid response.** Finally, the framework guarantees the real-time performance of the algorithm and the feasibility of the results through two strategies including rapid task insertion and real-time conflict avoidance to meet the requirements of the control agency for real-time and rapid response to emergencies, especially new tasks and satellite malfunctions, in the long-term process of satellite control.

To sum up, the DDRO optimization framework proposed in this section can meet all the practical requirements of the control agency. In particular, the framework presents comprehensive features of satellite-ground collaboration, rolling optimization, flexible response, and rapid response, which provides an important framework for solving satellite task scheduling problems in dynamic environments.

5.2 Task negotiation and allocation strategy based on dynamic contract net

On the basis of the abovementioned general-purpose DDRO framework, to meet the requirements for MAS communication and negotiation and task allocation in the negotiation layer, this section designs a task negotiation and allocation strategy based on a dynamic contract net. This strategy simulates the communication and negotiation mechanism of bid invitation, bidding, bid evaluation, and bid winning in the contract net protocol [252, 253]. At the same time, based on the actual situation of satellite emergency task scheduling, it introduces new improved strategies such as dynamic bid invitation, all-member bidding, and differentiated bid evaluation for flexible and self-organized task negotiation and allocation, which provides an important basis of task negotiation for the DDRO framework.

5.2.1 General process

This strategy consists of the following four steps, and the general process is shown in Figure 5.2:

Step 1 ((First round of communication) bid invitation). In MAS, the agent that receives the task to be allocated automatically acts as the bid-invitation agent, broadcasts the task bidding information to all agents (including itself) in MAS through the communication blackboard, and sets the bidding deadline.

Step 2 ((Second round of communication) bidding). Each agent that receives the task bidding information acts as the bidding agents, starts the bidding procedure, and makes an evaluation: that is, based on the information about its latest state such as task pool, profit ratio, constraint margin, and so on, each agent formulates its bid and feeds it back to the bid-invitation agent. If the bidding task must be executed, the bidding agent initiates a simulated rescheduling (without changing the task scheme it is executing), that is, the agent simulates the impact of inserting the task into the current task scheme in real time and adjusts the tender accordingly. The details of the bid and the rescheduling strategy are described below.

Step 3 ((Third round of communication) bid evaluation). After receiving all the bids by the bidding agents or the predetermined deadline has passed, the bid-invitation agent ends the bidding and starts the bid evaluation procedure; that is, according to the bids of the bidding agents, the bid-winning agent is determined

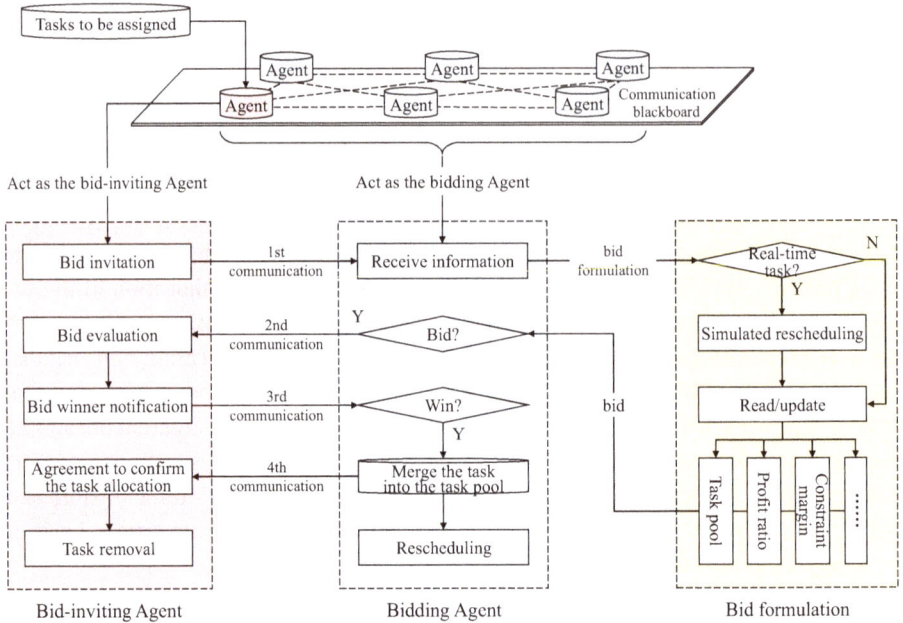

Figure 5.2: Task negotiation and allocation process based on dynamic contract net.

and notified through the bid evaluation algorithm before the task is assigned to the bid-winning agent.

Step 4 ((Fourth round of communication) contract signing). After receiving the bid-winning notice, the bid-winning agent informs the bid-invitation agent to remove the task, and both agents reach an agreement to confirm the task allocation. Meanwhile, the bid-winning agent merges the task into its task pool, starts a real rescheduling, and updates the task scheme it was executing.

In the above steps, the tasks to be allocated are the input of this strategy, and the main types are shown in Table 5.3.

According to the above steps, the task negotiation and allocation strategy designed in this section have the following characteristics:

(1) Dynamic bid invitation: Any agent that receives the task to be allocated automatically acts as a bid-invitation agent, which shows higher flexibility compared to the traditional contract net protocol with a fixed bid-invitation agent and is more suitable for satellite task scheduling with dynamic communication topology.

(2) All-member bidding: On the one hand, since the bid-invitation agent itself represents an autonomous satellite or station and has the resources to perform the task, all agents including the bid-invitation agent act as bidding agents and participate in the bidding, thus guaranteeing the bidding rights of the bid-invitation agent itself. On the other hand, on the basis of the communication blackboard, nonautonomous satel-

Table 5.3: Main types, sources, and characteristics of tasks to be allocated in satellite emergency task scheduling.

No.	Task type to be allocated	Source	Mandatory or not
1	Important emergency task	Notified by ground-based control center	Yes
2	General emergency task	Notified by ground-based control center	No
3	Onboard autonomous task	Triggered by the satellite autonomously	Yes
4	Unscheduled tasks in the onboard task pool	When there is no any other task, the autonomous satellites periodically select such kind of tasks from the task pool for execution.	No

lites also participate in the bidding through the agents operated by the ground-based control center (although there may be cases where communication and bidding cannot be completed within the specified time), which enhances the diversity of the contract net protocols to a certain extent and also has a high degree of feasibility in the current situation, where the satellite system is still dominated by nonautonomous satellites.

(3) Differentiated bidding: On the one hand, for the mandatory emergency tasks with high timeliness and scheduling urgency, to ensure the accurate and effective allocation of such tasks, the bidding agent evaluates the impact of the tasks through a simulated rescheduling (fast) to provide a detailed and accurate allocation basis for the bidding agents. On the other hand, for other nonmandatory emergency tasks with relatively low timeliness and scheduling urgency, the bidding agent does not start the simulated rescheduling, but directly formulates the bids according to the information of their latest states, thus saving their computing resources. For such kinds of tasks, whether they can be successfully scheduled in the end is determined by the task rescheduling strategy of the subsequent single-platform layer.

It can be seen that on the basis of the traditional contract net protocols, combined with the actual situation of satellite emergency task scheduling, this strategy introduces new strategies such as dynamic bid invitation, all-member bidding , and differentiated bid evaluation, which are helpful to improve the flexibility and generality of the contract net in a complex dynamic environment. In the following, the formulation of the bids in step 2 and the bid evaluation algorithm in step 3 of this strategy are specifically introduced.

5.2.2 Bid formulation

Each bid is a comprehensive index set formulated by one bidding agent according to the bidding task information and the agent's own state, and it is an important basis for the bid-invitation agent to determine the bid-winning agent. The agent set is denoted as A, in which the jth agent is a_j and its bid is B_j and the kth index in B_j is denoted as b_{jk}. Meanwhile, the task set of the satellite task scheduling subproblem involved in agent a_j is denoted as T_j ($T_j \subseteq T$), and the decision matrix is denoted as X_j ($X_j \subseteq X$). On this basis, the following indexes are considered in this section to formulate the bids.

5.2.2.1 Task pool size
Definition 5.1 (Task pool): This refers to a set of tasks that have not yet been scheduled or have not been successfully scheduled in a task scheduling problem of the single-platform layer, reflecting the degree of oversubscription in the current task scheduling problem. Here, the task pool of agent a_j is denoted as T_j^P ($T_j^P \subseteq T_j$); for the convenience of the subsequent computation, the negative value of the size of the task pool $-|T_{Pj}|$ is denoted as b_{j1}, that is

$$b_{j1} = -\left|T_j^P\right| = -\left|\{t_i | t_i \in T_j, x_{i1} = 0\}\right|, x_{i1} \in X_j, X_j \subseteq X, T_j \subseteq T \tag{5.1}$$

where T_{Pj} is the task pool of agent a_j, $a_j \in A$, $T_j^P \subseteq T_j . t_i$ is the ith task in task pool T_j^P, $t_i \in T_j . x_{i1}$ is the decision variable about the first event (e_{i1}) of task $t_i . X_j$ is a subset of the decision matrix involved in agent a_j, $X_j \subseteq X$.

5.2.2.2 Profit ratio
Definition 5.2 (Profit ratio): This refers to the ratio of the profits of successfully scheduled tasks to the total profit of all tasks in a task scheduling problem of the single-platform layer, reflecting the degree of completion of the current task scheduling problem and the degree of resource utilization. Denoted as b_{j2}, the profit ratio of agent a_j is calculated by the following equation:

$$b_{j2} = \frac{\sum_{t_i \in T_j} f_i \cdot \lceil x_{i1} \rceil}{\sum_{t_i \in T_j} f_i}, x_{i1} \in X_j, X_j \subseteq X, T_j \subseteq T \tag{5.2}$$

where f_i is the profit of task t_i.

5.2.2.3 Average constraint margin
Definition 5.3 (Average constraint margin): This refers to the average value of the difference between each constraint object within the constraints and the constraint threshold in the task scheduling subproblems of the single-platform layer, reflecting

the resource utilization degree and reserve capacity of the current task scheduling problem. Based on the constraint network and constraint value calculation method in Section 3.4.3, the average constraint margin of agent a_j that is denoted as b_{j3} can be calculated by the following equation:

$$b_{j3} = \frac{1}{|F^H|} \cdot \sum_{i=1}^{|F^H|} \sum_{k=1}^{|Z_i(X)|} |c_i^H(X_{ik}^Z) - y_i^H(X_{ik}^Z)|, \ X_{ik}^Z \subseteq X_j, \ X_j \subseteq X \tag{5.3}$$

where F^H is a set of constraints, c_i^H is the constraint object of the ith constraint obtained by eq. (3.26), y_i^H is the constraint threshold of the ith constraint obtained by eq. (3.26), X_{ik}^Z is the kth subset of decision matrices that may be involved in the ith constraint obtained by eq. (3.40), and $Z_i(X)$ is the set of subsets of decision matrices that may be involved in the ith constraint obtained by eq. (3.40).

It is worth mentioning that in traditional-related studies, the payload quantity and payload rate are usually used to assess the degree of utilization of satellite and station resources. However, in actual satellite task scheduling problems, the constraint complexity is high, and numerous complex constraints jointly limit the degree of resource utilization. For example, even a saturated satellite task scheduling scheme may exhibit a low payload rate under the combined effect of single-orbit imaging times, time constraints, and time windows of the imaging satellites. In this regard, this section proposes a new index, average constraint margin, based on the constraint network above. Compared to payload quantity and payload rate, the average constraint margin can intuitively reflect the degree to which the satellite task scheduling scheme is constrained under the current constraints: the smaller the value, the higher the degree to which the scheme is constrained (tight constraint), and the more saturated the task scheme, leading to the difficult addition of new tasks; on the contrary, the bigger the value, the lower the degree to which the scheme is constrained, and the more unsaturated the task scheme, leading to the easy addition of new tasks. Therefore, the average constraint margin reflects both the utilization degree of resources and their potential to perform other tasks, which is more suitable as a reference basis for task allocation in the context of complex satellite emergency task scheduling in this book.

In addition to the above indexes, the bidding agents also start a simulated rescheduling for the emergency tasks that must be executed in step 2 of the strategy in Section 5.2.1, that is, they simulate the impact of inserting the tasks into the current satellite task scheme in real time. Based on the results of the simulated rescheduling, the task pool, profit ratio, and average constraint margin of the bidding agents are updated and recorded as b_{j4}, b_{j5}, and b_{j6}, respectively, which are written into the bids as new indexes. Therefore, according to the task information published by the bid-invitation agent, the bids formulated by the bidding agents can be denoted as B_j:

$$B_j = \begin{cases} \{b_{j1}, b_{j2}, b_{j3}, b_{j4}, b_{j5}, b_{j6}\} & \text{if the bidding tasks must be scheduled in real time} \\ \{b_{j1}, b_{j2}, b_{j3}\}, & \text{other} \end{cases}$$

(5.4)

5.2.3 Bid evaluation algorithm

In step 3 of the strategy in Section 5.2.1, after receiving all the bids or the predetermined deadline has passed, the bid-invitation agent ends the bidding and starts the bid evaluation procedure. Considering that each bid contains multiple indexes, the bid evaluation procedure can also be regarded as a multiobjective evaluation and selection process. In most cases, there is no single bid that is optimal in all indexes. In this regard, the set of bids received by the bid-invitation agent is denoted as U_B, and this section draws on the multiobjective dominance relationship to define the dominance relationship of the bids as follows:

$$B_k \succ B_j \text{ if } b_{km} > b_{jm}, \ \forall m \leq |B_j|, B_j \in U_B, B_k \in U_B$$

(5.5)

where any index b_{km} in the bid B_k is higher than the similar index b_{jm} in the bid B_j, which is called B_k dominating B_j, and is denoted as $B_k \succ B_j$. Based on this dominance relationship, the specific steps of the bid evaluation algorithm are shown in Algorithm 5.1.

Algorithm 5.1: Bid evaluation algorithm

Input: bid set U_B;
Output: the bid-winning bid B_j with the highest comprehensive ranking of the indexes.
```
1  for j = 1: |U_B| do           //Traverse the bids and conduct dominance test in pairs.
2  |   for k = 1: |U_B| do
3  |   |   if B_k≻B_j then        //If there is a bid B_k dominating another bid B_j
4  |   |   |  U_B ← U_B - {B_j}    //Remove the dominated bid B_j
5  |   |   break
6  |   end
7  end
8  end
9  C ← {c_j | c_j = 0, 0≤j≤|U_B|} for j = 1: |U_B| do
10 |   for k = 1: |B_j| do
11 |   |  c_j ← c_j + Order(b_jk)   //Accumulate the ranking of index b_jk of bid B_j among similar
   |   |                             indexes (the same column)
12 |   end
13 end
14 j ← MinValueId(C)              //Find id of the bid with the highest comprehensive ranking of
   |                                the statistical indexes
15 returnB_j
```

Algorithm 5.1 firstly, based on the multiobjective domination relationship, removes the dominated bids from the bid set U_B through the steps of lines 1–8 and retains the bids that are not dominated by each other, that is, the set of the Pareto optimal bids generated through multiobjective evaluation. Secondly, to select a bid from the Pareto bid set as the final bid-winning bid, the algorithm further counts the comprehensive ranking of the indexes for each bid: the ranking value c_j of each index of each bid B_j is accumulated through the steps of lines 10–14, where Order(b_{jk}) is the ranking value of index b_{jk} of bid B_j in the same kind of indexes (in the same column of the matrix). Based on this ranking, the algorithm finally outputs the bid with the highest comprehensive ranking of the indexes (the minimum value) as the bid-winning bid.

It can be seen that under the background of multi-index evaluation, the algorithm, on the one hand, draws lessons from the multiobjective dominance relationship, eliminates the dominated bids, and scientifically gives the Pareto optimal bid set. To quickly find the bid-winning agent from the Pareto set, the algorithm, on the other hand, comprehensively considers the comprehensive ranking of the indicators of the bids, which provides a direct and reasonable basis for selecting the bid-wining agent.

To sum up, this section designs a task negotiation and allocation strategy that simulates the mechanism of comprehensively considered in the contract net protocols, introduces the specific methods of bid formulation and bid evaluation, and provides a flexible and self-organized task negotiation and allocation scheme, which offers an important negotiation basis for the DDRO optimization framework.

5.3 Single-platform task rescheduling strategy based on rolling horizon

Rolling horizon scheduling, also known as rolling/sliding window scheduling, is a dynamic task scheduling method that conditionally implements task scheduling based on predetermined windows and constantly advances and updates the windows over time. It is a common strategy for handling complex long-cycle task scheduling. To address the rescheduling requirements in the abovementioned task negotiation and allocation process, this section designs a single-platform real-time task rescheduling strategy based on rolling horizon, which is centered around the two steps of window rolling and rescheduling, and continuously iterates the response in a continuous horizon to dynamically update the satellite task scheduling scheme, which provides a general-purpose means of solving the satellite emergency task scheduling problems of the single-platform layer under the DDRO framework.

5.3.1 General process

This strategy consists of the following four steps, and the general process is shown in Figure 5.3:

Step 1 (Judgment of precondition of window rolling). The mixed rolling trigger conditions of "task trigger" and "periodic trigger" are adopted. The former applies to the case when an agent wins the bid and obtains a new task in the task negotiation and allocation steps as introduced above, and the latter applies the case when the control agency presets the rolling cycle.

Step 2 (Window rolling). A window rolling is performed to update the start and end time of each window and the tasks the window contains. Here, the continuous horizon of a single platform is divided into three continuous and nonoverlapping windows such as the locked window, real-time window, and future window. The specific definitions and scrolling modes of the windows are described in the next section.

Step 3 (Rescheduling). A real-time and fast rescheduling of tasks in the task pool and real-time window is performed to update the tasks in the task pool and real-time window. The specific rescheduling algorithm is also described later.

Step 4 (Execution of the tasks by the platform). Based on the latest task scheduling scheme, the satellite or station continues to execute the tasks. After completing the execution, the satellite or station returns to step 1 and waits for the trigger of the next window rolling.

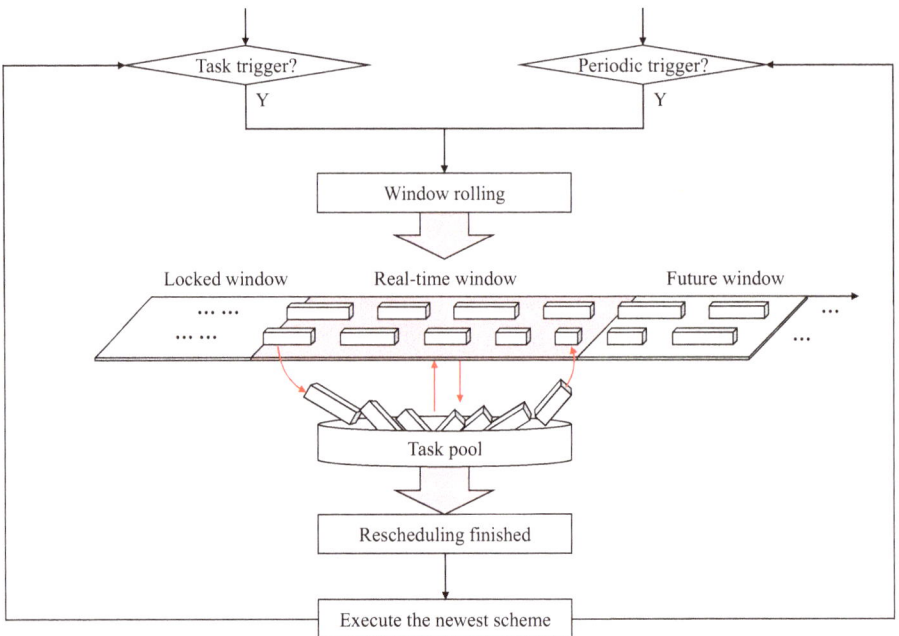

Figure 5.3: Flow of single-platform task rescheduling strategy based on rolling horizon.

It can be seen that this strategy focuses on window rolling and rescheduling and dynamically updates the satellite task scheduling scheme by continuously iterating the response in a continuous horizon. Meanwhile, this strategy only reschedules the tasks in the task pool and the real-time window, which reduces the scale of the problems, minimizes the changes to the original scheme, especially to the scheme that has been in existence for a long time, and can satisfy the dual requirements of the control agency in terms of minor changes and real-time optimization of the task scheduling scheme, which provides a feasible scheme for solving the satellite emergency task scheduling problems in this book.

5.3.2 Definitions of windows and rolling modes

Before defining the rolling windows, the definitions of time nodes are first given as follows:

Definition 5.4 (Scheduling starting point): This refers to the starting time of the satellite emergency task scheduling scenario, which is the time starting point of the satellite task scheduling in this book.

Definition 5.5 (Scheduling ending point): This refers to the ending time of the satellite emergency task scheduling scenario, which is the time ending point of the satellite task scheduling in this book.

Definition 5.6 (Rescheduling starting point): This refers to the earliest time node of the tasks involved in rescheduling in the current task scheduling scheme within the time frame of the satellite emergency task scheduling scenario, which is the starting point of the satellite task rescheduling in this section. In the context of this book, the rescheduling starting point is typically the current time point (or a buffer time delayed backward from the current time point) at which the rescheduling is triggered.

Definition 5.7 (Rescheduling ending point): This refers to the latest time node of the tasks involved in rescheduling in the current task scheduling scheme within the time frame of the satellite emergency task scheduling scenario, which is the time ending point of the satellite task rescheduling in this section. In the context of this book, the rescheduling ending point usually varies with the rescheduling starting point, and the two always maintain a fixed time interval.

As can be seen from the above definitions, the satellite emergency task scheduling problems studied in this book involve all the tasks between the scheduling starting point and the scheduling ending point, while the satellite task rescheduling discussed in this section involves only the tasks between the rescheduling starting point and the rescheduling ending point. The tasks before the rescheduling starting point and after the rescheduling ending point are not involved in rescheduling. Here, if a satellite task contains multiple events, where the execution time points of any of the events are between the rescheduling starting point and the rescheduling ending point, the task is regarded as a task between the rescheduling starting point and the rescheduling ending point.

According to the above definitions, the four time nodes divide the time axis into three continuous windows, as shown in Figure 5.4. In this regard, this section defines these three consecutive windows as follows:

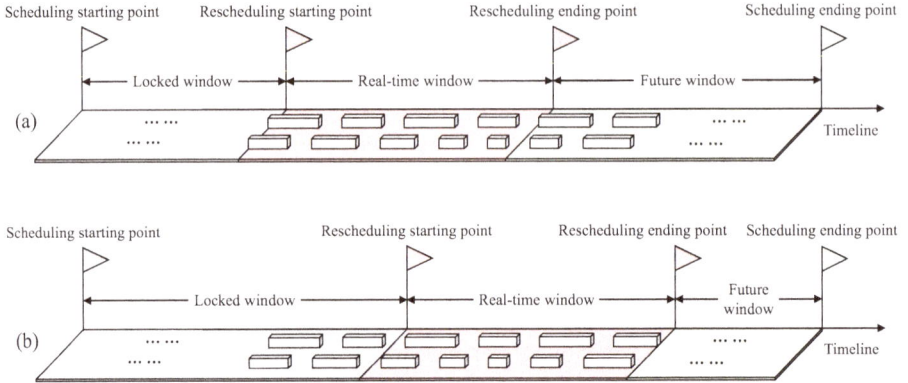

Figure 5.4: Schematic diagram of time nodes and rolling windows in continuous horizon: (a) time nodes and windows before window rolling and (b) time nodes and windows after window rolling.

Definition 5.8 (Locked window): This refers to a closed interval on the time axis with the scheduling starting point as the starting time and the rescheduling starting point as the ending time, which contains the tasks in the current task scheduling scheme that have been executed or are about to be executed by the satellite or ground station and cannot be adjusted.

Definition 5.9 (Real-time window): This refers to an open interval on the time axis with the rescheduling starting point as the starting time and the rescheduling ending point as the ending time, which contains the tasks in the current task scheduling scheme that will be executed by the satellite or ground station in the near future and can still be adjusted.

Definition 5.10 (Future window): This refers to a closed interval on the time axis with the rescheduling ending point as the starting time and the scheduling ending point as the ending time, which contains the tasks in the current task scheduling scheme that will be executed after the real-time window and can still be adjusted.

According to the above flow, this strategy only reschedules the tasks in the real-time window and task pool, and the tasks in the locked window and future window are not affected. Under the dual requirements of the control agency in terms of minor changes and real-time optimization of the task scheduling scheme, this strategy can significantly reduce the task scheduling problem scale, minimize the changes to the original scheme, and improve the timeliness and practicability of the rescheduling scheme.

5.3.3 Rescheduling algorithm

According to the abovementioned rescheduling flow and definitions of windows, this section designs a rescheduling algorithm as shown in Figure 5.5.

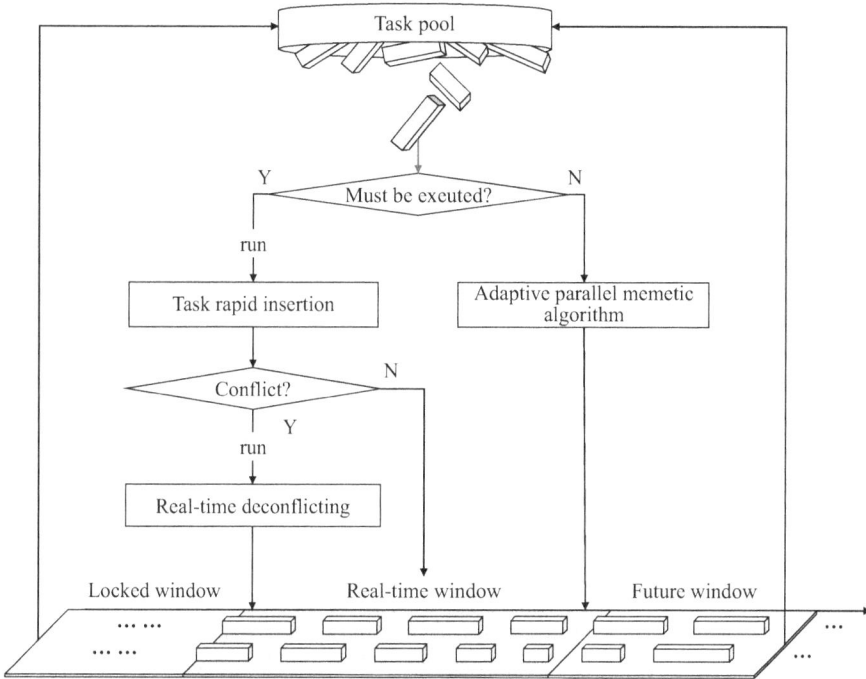

Figure 5.5: Rescheduling algorithm for satellite task scheduling subproblems.

First of all, the rescheduling problem of tasks in the task pool and real-time window can be regarded as a (conventional scheduling) subproblem with smaller task and resource sizes in the original satellite task scheduling problems. Based on this, combined with the actual situation of satellite control, this section further categorizes the rescheduling algorithm into the following two cases:

(1) The task pool contains emergency task that must be executed. In this case, the control agency puts forward the scheduling requirements of the emergency tasks that must be executed, and the scheduling time is extremely limited. In this regard, this algorithm calls the strategy of task rapid insertion and real-time conflict avoidance to ensure that the task scheduling is completed in a very short period of time to give a feasible scheme, thus meeting the requirements for executing related tasks. The strategy of task rapid insertion and real-time conflict avoidance is elaborated in Sections 5.4 and 5.5.

(2) The task pool does not contain emergency task that must be executed. In this case, the requirement for rescheduling is not urgent, and the scheduling time is relatively abundant. In view of this, this algorithm directly calls the simplified version of the abovementioned APMA, parallel memetic algorithm (PMA), to carry out iterative optimization to a certain extent, thus meeting the requirements for optimization of the task scheduling scheme. Here, to avoid drastic changes to the original scheme and meet the operation requirements of the control agency, only the "parallel" strategy in APMA is retained in PMA, and the "competition" and "evolution" strategies are no longer used; that is, the original scheme will be slightly adjusted mainly by local optimization, and the global optimization strategy will no longer be used to implement large-scale scheme changes.

To sum up, this section designs a single-platform task rescheduling strategy based on rolling horizon around window rolling and rescheduling, which provides a general-purpose means of solving satellite emergency task scheduling problems of the single-platform layer within the framework of DDRO. Based on this, Sections 5.4 and 5.5 design the strategies of task rapid insertion and real-time conflict avoidance, respectively, to provide the necessary technical support for the task rescheduling strategy in this section.

5.4 Task rapid insertion strategy based on schedulability prediction

Schedulability refers to the possibility that tasks can be successfully scheduled in task scheduling problems, which is usually expressed in the form of probability. For the emergency tasks that must be executed in the rescheduling strategy in the previous section, this section designs a task rapid insertion strategy based on schedulability prediction. Based on a pretrained schedulability prediction model, this strategy predicts the schedulability of task events at different executable opportunities and assigns the highest schedulability executable opportunities to them, thus realizing the rapid insertion of tasks. With a large amount of data training and learning, this strategy enhances the rationality and accuracy of task insertion, provides the necessary technical support for the rescheduling strategy in Section 5.3, and provides important support for DDRO to address the real-time satellite emergency task scheduling problems.

5.4.1 General process

This strategy consists of the following four steps, and the general process is shown in Figure 5.6.

Step 1 (Pretraining of the schedulability model). Based on a large amount of historical data, the schedulability prediction model is pretrained to provide the schedulability prediction basis for this strategy.

Step 2 (Establishment of the task set to be inserted). For the emergency tasks that must be executed in the rescheduling strategy described in Section 5.3, a "task set to be inserted" is established, and the time windows and executable opportunities of each task event are calculated and updated to provide necessary data input for the next step of the task rapid insertion algorithm.

Step 3 (Prediction and rapid insertion). Based on the schedulability prediction model, the schedulability of each task and event at different executable opportunities is predicted in turn, and the EO with the highest schedulability is assigned to the corresponding event, thus realizing the rapid insertion of tasks.

Step 4 (Output). If the task insertion is completed, it outputs the current task scheduling scheme; otherwise, it returns to step 3. The constraints of the current scheme are checked to judge whether to call further conflict avoidance strategies subsequently.

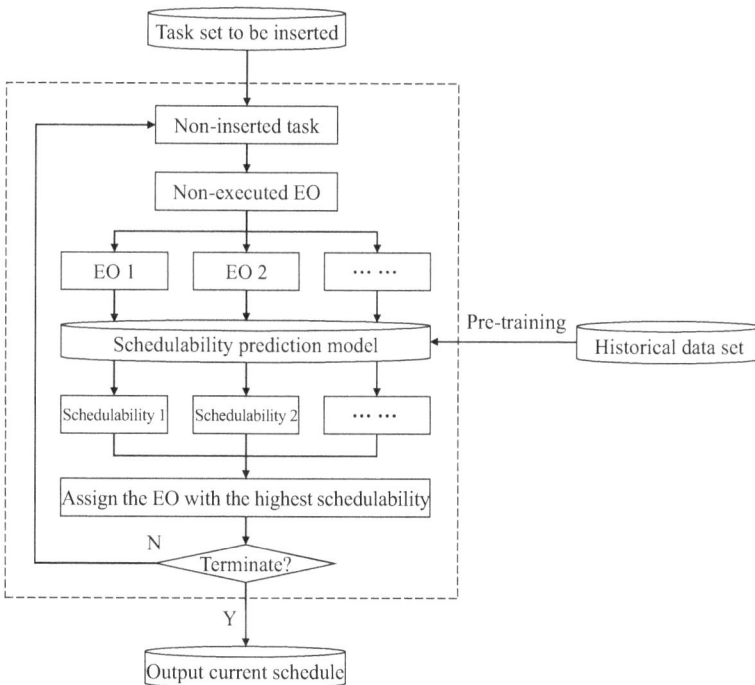

Figure 5.6: Flow of schedulability-based fast initial solution construction strategy.

It can be seen that this strategy, for the emergency tasks that must be executed in the rescheduling strategy mentioned in Section 5.3, satisfies the execution requirements of related tasks through schedulability prediction, rapid insertion, and other means. In the process of task insertion, it relies on the schedulability prediction model to guarantee the reasonableness of the insertion scheme, and the constraint conflicts that may be triggered are not dealt with temporarily, which significantly improves the efficiency of task insertion. After the task is inserted, considering the constraint conflicts that may occur, the strategy also checks the constraints once to judge whether to call further conflict avoidance strategies subsequently.

5.4.2 Schedulability prediction model

In the above process, the strategy adopts an artificial neural network as the schedulability prediction model, with 10 indexes such as tasks, events, and executable opportunities as the feature inputs and adopts neuroevolution of augmenting topologies (NEAT) [159, 254] to carry out model training.

5.4.2.1 Coding method

As shown in Figure 5.7(a), the neural network consists of nodes and links, where nodes 1, 2, and 3 are input nodes (i.e., feature input nodes), node 4 is a hidden node, and node 5 is an output node (i.e., schedulability output node). To facilitate model training and prediction, the nodes and links in the neural network shown in Figure 5.7(a) are coded by the coding method shown in Figure 5.7(b) and (c), respectively.

Compared to the traditional feedforward neural network, this coding method does not define the input layer, hidden layer, and output layer, can establish links between any two nodes to downlink data, for example, link 5(3→5) directly links the input and output nodes, and allows the links to be in an inactive state such as link 4 (2→5). Meanwhile, for the excitation functions, there are Log-sigmoid, Tan-sigmoid, Relu, and other types of functions. As a result, the coding method is not restricted to the traditional feedforward neural network, which makes the topological form of the neural network more diversified and more capable of characterization, and provides a prerequisite for the strategy to subsequently carry out the model training through neuroevolutionary algorithm.

5.4.2.2 Feature input

Based on the above neural network model, the strategy uses the 10 indexes listed in Table 5.4, such as tasks, events, and executable opportunities, as the feature inputs:

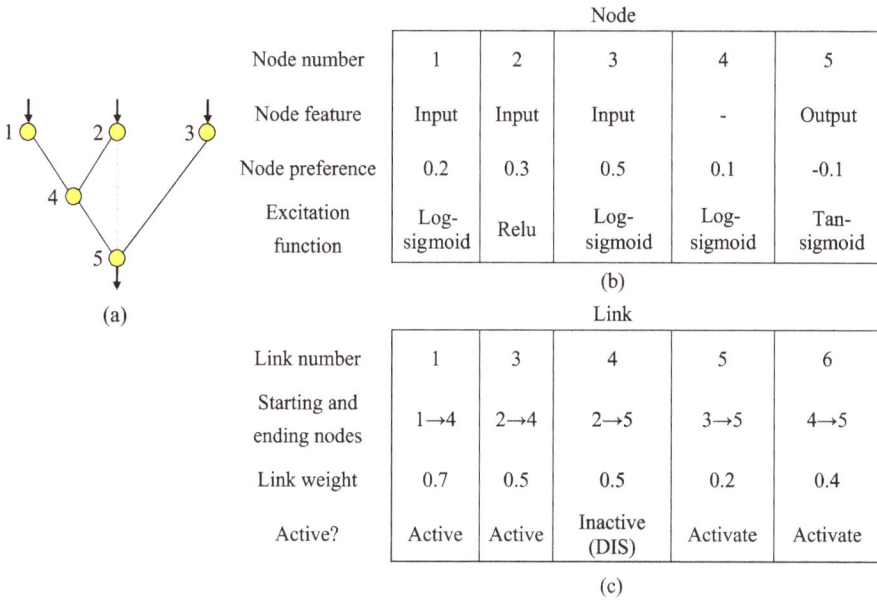

	Node				
Node number	1	2	3	4	5
Node feature	Input	Input	Input	-	Output
Node preference	0.2	0.3	0.5	0.1	-0.1
Excitation function	Log-sigmoid	Relu	Log-sigmoid	Log-sigmoid	Tan-sigmoid

(b)

	Link				
Link number	1	3	4	5	6
Starting and ending nodes	1→4	2→4	2→5	3→5	4→5
Link weight	0.7	0.5	0.5	0.2	0.4
Active?	Active	Active	Inactive (DIS)	Activate	Activate

(c)

(a)

Figure 5.7: Example of coding method of schedulability prediction model: (a) neural network; (b) node coding; and (c) link coding.

Table 5.4: Feature inputs of task schedulability prediction model.

No.	Features of tasks and events	No.	Features of event executable opportunities
1	Total number of tasks	6	Beginning time of EO
2	Task profit	7	Ending time of EO
3	Number of events contained in a task	8	Beginning time of the time window corresponding to an EO
4	Execution time allowed for an event	9	Ending time of the time window corresponding to an EO
5	Total number of executable opportunities of an event	10	Constraint value of the current task scheduling scheme at the time the event is executed at the EO

Among them, the features of tasks and events are mainly used to reflect the objective features of the tasks and events themselves, which embody their importance and scheduling difficulty to some extent. The features of event executable opportunities are mainly used to reflect the time features of the events as well as the conflicts that may be triggered, which embody the degree of reasonableness of the task execution. It is worth noting that since the task to be inserted in the strategy has been allocated to a certain platform (satellite or ground station) in advance, and the original satellite task scheduling problems have been converted into a subproblem of the single-platform layer, the table only lists the event executable opportunities of the single-platform layer.

5.4.2.3 Training algorithm

Based on the above neural network prediction model, the strategy carries out model training with the help of NEAT. The related algorithm flow is detailed in the literature [159, 254], and the algorithm is not introduced further in this section. This section mainly describes the mutation and crossover operators used in the evolution of the neural network.

Table 5.5: Mutation operators used in the evolution of the neural network.

Mutation type	Mutation operator	Typical example
Numerical mutation	Link weight mutation operator	0.1→0.5
	Node preference mutation operator	0.1→0.5
Functional mutation	Link activation state mutation operator	Active → Inactive
	Node excitation function mutation operator	Log-sigmoid→Tan-sigmoid
Structural mutation	Link mutation operator	Figure 5.8
	Node mutation operator	Figure 5.9

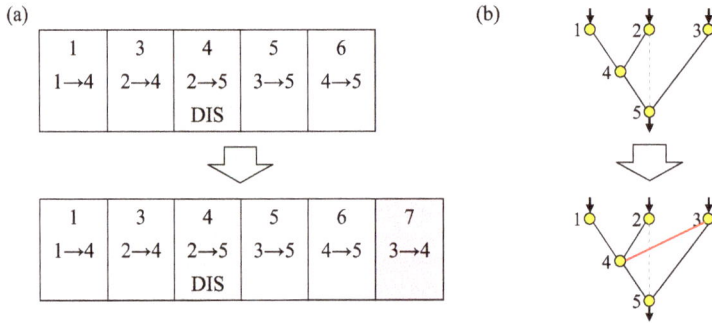

Figure 5.8: Example of link mutation operator of the neural network: (a) link coding before and after the mutation and (b) neural network before and after mutation.

(1) Mutation operator: The mutation operators used in the evolution of the neural network are shown in Table 5.5. Among them, structural mutation operators are divided into link mutation operators and node mutation operators. A link mutation operator is to add a new link or delete an existing link. For example, in Figure 5.8, a new link 7 (3→4) is added between nodes 3 and 4. Node mutation operator is to add a new node

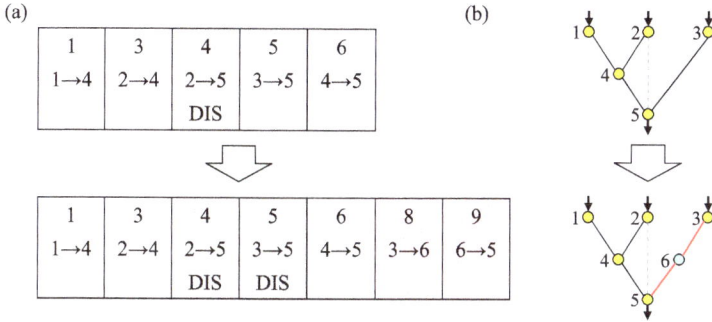

Figure 5.9: Example of node mutation operator of the neural network: (a) link coding before and after the mutation and (b) neural network before and after mutation.

to a link or delete an existing node. For example, in Figure 5.9, a new node 6 is added between the original link 5(3→5) between nodes 3 and 5, and the link is split into two new links 8(3→6) and 9(6→5). In other words, a node mutation operator is usually accompanied by multiple link mutation operators.

(2) Crossover operator: The crossover operators used in the evolution of the neural network are shown in Figure 5.10. In the crossover process, the colocated parent

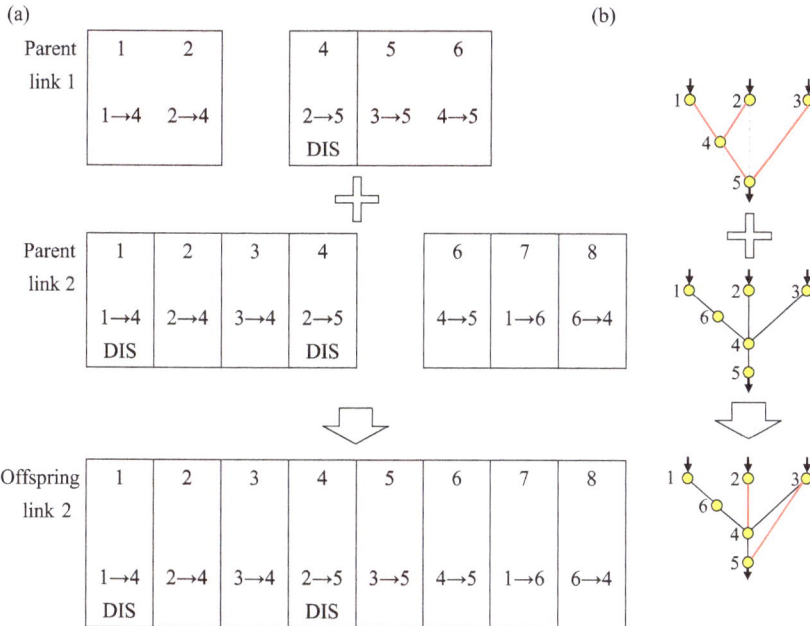

Figure 5.10: Example of crossover operators of the neural network: (a) link coding of parent links and offspring links before and after the crossover and (b) parent neural network and offspring neural network before and after the crossover.

links are paired according to the link numbers in the link coding, and a parent link at the same numbered position is randomly inherited to an offspring link. As a result, the crossover operators can preserve the main structure of the parent neural network, which is in line with the characteristics of EA.

5.4.2.4 Task rapid insertion algorithm

Based on the task schedulability prediction model described in the previous section, to meet the application requirements of task rapid insertion in rescheduling strategy, the following task rapid insertion algorithm is designed in this section, and its specific steps are shown in Algorithm 5.2.

Algorithm 5.2: Rapid task insertion algorithm based on schedulability

---

Input: current solution X, task set T, set of tasks to be inserted T_x, and schedulability prediction model Pre
Output: current solution X after the tasks have been inserted

1	while $T_x \neq \varnothing$ do	//Insert the tasks one by one until they are all inserted.
2	$t \leftarrow RandomFrom(T_x)$	//Randomly selects a task to be inserted
3	if $t \in T$ then	//If the task is an unscheduled task of the task set
4	$i \leftarrow Id(t)$	//Then get the task id
5	else	
6	$T \leftarrow T \cup \{t\}$	//Otherwise, merge the task into the task set
7	$X \leftarrow X \cup 0$	//Add a new row to the decision matrix (decision variables are all 0)
8	$i \leftarrow |T|$	//Assign the task the current maximum number as its id
9	end	
10	for $j = 1: |E_i|$ do	//Traverse all events of the task
11	$P \leftarrow \{p_k \mid p_k = 0, 0 \leq k \leq |EO_{ij}|\}$	//Initialize the schedulability prediction set (values are all 0 initially)
12	for $k = 1: |EO_{ij}|$ do	//Traverse all executable opportunities of the event
13	$p_k = Pre(X, t_i, eo_{ij}^k)$	//Predict the schedulability of the event at this EO
14	end	
15	$k \leftarrow MaxValueId(P)$	//Find the id of the event EO with the highest schedulability
16	$x_{ij} \leftarrow \dfrac{k}{|EOij|}$	//Assign values to the decision variables according to eq. (3.21)
17	end	
18	$T_x \leftarrow T_x - \{t\}$	//Remove the tasks after the insertion is completed
19	end	
20	return X	

In lines 3–8 of Algorithm 5.2, two cases are considered: 1) the task already exists in the task set T and has a task id but has not been successfully scheduled; and 2) the task to be inserted is a brand-new task and needs to be merged into the task set and given a new id. Accordingly, the task is denoted as t_i, the algorithm traverses the event set E_i of the task and the execution opportunity set EO_{ij} of the events in lines 9–16 and predicts and records the schedulability in turn. Then, the EO with the highest schedulability is assigned to each event, and value of the decision variable x_{ij} is assigned by eq. (3.21) to realize the task insertion. Thereafter, in line 17, the algorithm removes the inserted tasks from the set of tasks to be inserted T_X and returns to line 2 to cycle through the above steps until all the tasks in the task set T_X are inserted. It can be seen that though the algorithm is not only simple in operation and low in time complexity, by relying on the schedulability prediction model, it can satisfy the dual requirements of the control agency in terms of timeliness and reasonableness of task insertion.

It should be noted that although the feature inputs of the schedulability prediction model in the algorithm have taken into account the constraint violation that may be caused by task insertion, the task scheduling scheme output by the algorithm may still violate the constraints, for example, there may be a situation in which the insertion of a task at any position violates the constraints. In this regard, DDRO further carries out the necessary conflict avoidance after the tasks are rapidly inserted with the help of real-time conflict avoidance strategy to ensure the reasonableness and feasibility of the overall task scheduling scheme.

To sum up, this section designs a task rapid insertion strategy based on the schedulability prediction model. By predicting the schedulability of the task events at different executable opportunities, the strategy can quickly allocate an EO with the highest schedulability to each event, which provides a real-time and reasonable task insertion means for the rescheduling strategy in Section 5.3 and an important support for DDRO to address the real-time satellite emergency task scheduling problems.

5.5 Real-time conflict avoidance strategy based on constraint network

To address the possible violation of constraints by the task rapid insertion strategy in Section 5.4, a conflict avoidance strategy is designed in this section. The strategy quickly calculates the task conflict degree through the constraint network and preferentially removes the tasks with high conflict degree according to the heuristic principle, which has the advantages of simple operation and low time complexity, meets the requirement of real-time response, provides the necessary conflict avoidance means for the task rapid insertion strategy in the previous section, and provides an important guarantee for the feasibility of the optimization results of DDRO.

5.5.1 General process

This strategy consists of the following three steps, and the general process is shown in Figure 5.11.

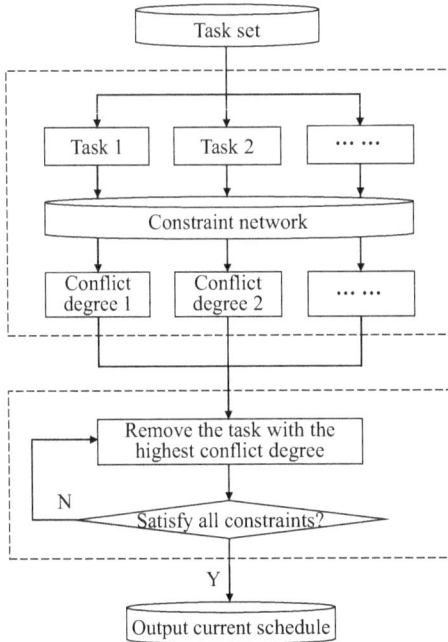

Figure 5.11: Real-time conflict avoidance strategy based on constraint network.

Step 1 (Task conflict degree calculation). Based on the satellite task scheduling constraint model and constraint network, the conflict degree of each task in the current task scheduling scheme is calculated.

Step 2 (Conflict avoidance). According to descending order of task conflict degree, the tasks that cause conflicts are removed in turn. The removed tasks are put into the task pool of the single-platform layer, which will not be considered in this strategy.

Step 3 (Constraint update and output). The constraint value of the current task scheduling scheme is recalculated. If the current scheme meets all constraints, the current scheme is output; otherwise, it returns to step 2.

It can be seen that this strategy can quickly calculate the conflict degree of each task in the current scheduling scheme through the constrain network and preferentially remove the tasks with a high conflict degree by heuristic principle, which has the advantages of simple operation and low time complexity, meets the requirement of real-time response, and provides the necessary conflict avoidance means for the task rapid insertion strategy in Section 5.4.

5.5.2 Task conflict degree calculation algorithm

The calculation of task conflict degree is the premise of conflict avoidance. To fulfill the requirement of real-time conflict avoidance, it is particularly important to quickly and accurately identify the tasks that cause conflicts from the current task scheduling scheme (with conflicts) and quantify the conflict degree caused by each task. In this regard, this section first provides the definition of task conflict degree:

Definition 5.11 (Task conflict degree): This refers to the sum of the constraint values involved in a task among the constraint values of the current task scheduling scheme.

It reflects the degree of conflict caused by the task in the current scheduling scheme.

Given that the satellite task scheduling constraint model and constraint network constructed in Section 3.4.3 present a fast and efficient method of calculating the constraint values, which provides favorable conditions for the calculation of the task conflict degree in this section, a task conflict degree calculation algorithm based on the constraint network is given in this section, whose specific steps are shown in Algorithm 5.3.

Algorithm 5.3: Task conflict degree calculation algorithm based on constraint network

Input: current solution X, task set T, constraint value calculation function $F^H(X)$, and constraint network;
Output: task conflict degree set C.

1	$C \leftarrow \{c_k \mid c_k = 0,\ 0 \le k \le |T|\}$	//Task conflict degree set, values are all 0 initially
2	**for** $i = 1: |F^H|$ **do**	//Traverse constraint set
3	**for** $j = 1: |Z_i(X)|$ **do**	//Traverse the decision subsets that may be involved in the ith constraint
4	**if** $f_i^H(X_{Z\,ij}) \ne 0$ **then**	//If the decision subset $X_{Z\,ij}$ violates the ith constraint
5	**for** $x \in X_{Z\,ij}$ **do**	//Traverse the decision variables of the decision subset $X_{Z\,ij}$
6	$k \leftarrow \text{Id}(x)$	//Get id of the task involved in the decision variable x
7	$c_k \leftarrow c_k + |\ f_i^H(X_{Z\,ij})|$	//Accumulate the conflict degree of task t_k
8	**end**	
9	**end**	
10	**end**	
11	**end**	
12	**return** C	//Output the task conflict degree set

As can be seen, on the basis of the satellite task scheduling constraint model and constraint network, Algorithm 5.3 traverses all constraints from line 2 and traverses the decision subset that may be involved in each constraint from line 3. Subsequently, Algorithm 5.3 filters out the decision subset that triggers conflicts in line 4, obtains the id of each task involved in the decision subset in lines 5–7, records and accumulates the conflict degree of the tasks, and finally outputs the task conflict degree set under the current satellite task scheduling scheme, which provides a basis for quantitatively evaluating the degree of conflicts triggered by each task.

It is not difficult to find that this task conflict degree calculation algorithm has some similarity with the incremental constraint calculation algorithm in the local search process described in Section 4.3.4, both of which traverse the decision subset that may be involved in the constraints based on the preconstructed constraint network, and then complete the related filtering and calculation work with low time complexity. The task conflict degree set obtained by the algorithm further guides the subsequent conflict avoidance and task scheduling scheme update.

5.5.3 Conflict avoidance heuristic algorithm

To meet the request for real-time conflict avoidance, based on the above conflict avoidance algorithm and the resulting task conflict degree set, a fast conflict avoidance heuristic algorithm is designed in this section. According to the descending order of task conflict degree, the algorithm removes the tasks that cause conflicts in turn until the current task scheduling scheme satisfies all the constraints. Its specific steps are shown in Algorithm 5.4.

Algorithm 5.4 Conflict avoidance heuristic algorithm

Input: current solution X, task set T, locked task set T_X, constraint value calculation function $F^H(X)$, and task conflict degree set C

Output: solution X after conflict avoidance

1	**while** true **do**	//Remove the tasks with high conflict degrees in turn until all constraints are satisfied
2	$i \leftarrow \text{MaxValueId}(C)$	//ID of the task with the highest conflict degree at present
3	**if** $t_i \in T_X$ **then**	
4	continue	//Skip locked tasks
5	**end**	
6	**for** $j = 1 : |E_i|$ **do**	//Traverse related decision variables of task t_i
7	$x_{ij} \leftarrow 0$	//Remove the task and clear the relevant decision variables
8	$c_i \leftarrow 0$	//Update the conflict degree set
9	**end**	
10	**if** $F^H(X) = 0$ **then**	//Recalculate the constraint value of the current task scheduling scheme
11	break	//If all constraints are satisfied, break the loop
12	**end**	
13	**end**	
14	**return** X	

It should be noted that to meet the actual operation requirements for satellite control, Algorithm 5.4 takes into account the locked task set T_X (i.e., the set of tasks that are not allowed to be modified by the control agency) and skips the nonadjustable tasks in lines 3–5 without removing them. Here, the locked task set T_X is given in advance by the control agency such as specialized tasks scheduled by the user agency, temporary tasks manually adjusted by the controllers, and newly inserted tasks based on schedulability prediction in Section 5.4. If this kind of situation is not considered, the conflict avoidance algorithm may remove tasks that are manually adjusted by the controllers and cause conflicts. It can be seen that the algorithm design in this section fully considers the operation requirements for satellite control and is close to the actual application.

To sum up, this section designs a real-time conflict avoidance strategy based on the constraint network, which meets the requirements of real-time response and satellite operation control, provides a fast and practical conflict avoidance means for the above task rapid insertion strategy, and provides an important guarantee for the feasibility of the optimization results of DDRO.

5.6 Chapter summary

To meet the emergency task scheduling needs of satellite control agency under dynamic influence such as task additions and deletions, and satellite failures, this chapter designs a general-purpose distributed DDRO, which mainly includes

(1) A task negotiation and allocation strategy based on a dynamic contract net is designed. By simulating the communication and negotiation mechanism in the contract net protocol, this strategy introduces new improved strategies such as dynamic bid invitation, all-member bidding, and differentiated bid evaluation, which provides an important basis of task negotiation in the general-purpose DDRO optimization framework.

(2) A single-platform task rescheduling strategy based on a rolling horizon is designed. The strategy continuously iterates the response in a continuous horizon to dynamically update the satellite task scheduling scheme through the mechanisms of window rolling and rescheduling, which provides a general-purpose means of solving the satellite emergency task scheduling problems of the single-platform layer under the DDRO optimization framework.

(3) A task rapid insertion strategy based on schedulability prediction is designed. By predicting the schedulability of the task events at different executable opportunities, this strategy realizes reasonable and rapid insertion of tasks, which provides the necessary task insertion means for the rescheduling strategy and important support for DDRO to address real-time satellite emergency task scheduling problems.

(4) A real-time conflict avoidance strategy based on constraint network is designed. This strategy quickly calculates the task conflict degree through the constraint network and completes real-time conflict avoidance, which provides the necessary

conflict avoidance means for the task rapid insertion strategy and also provides an important guarantee for the feasibility of DDRO optimization results.

Under the framework of general-purpose DDRO, the above four strategies implement real-time and dynamic optimization for the current satellite task scheduling scheme, which provides a general-purpose and flexible means of solving satellite emergency task scheduling problems. Consequently, DDRO can meet the needs of emergency optimization and autonomous optimization of the task scheduling scheme in the long-term process of satellite control, providing another algorithm for the satellite task scheduling engine in this book.

Chapter 6
Application experiments of satellite task scheduling engine

Based on the abovementioned general-purpose modeling method of satellite task scheduling as well as the general-purpose solving methods adaptive parallel memetic algorithm (APMA) and dynamic rolling optimization algorithm (DDRO) for conventional and emergency task scheduling, this chapter carries out application experiments of the satellite task scheduling engine by taking the four types of satellite task scheduling problems including remote-sensing satellite scheduling, relay satellite scheduling, navigation satellite scheduling, and satellite range scheduling as examples.

Firstly, this chapter carries out the experiment setup such as algorithm parameter setting. Secondly, the conventional and emergency scheduling experiments are carried out, respectively, for the SuperView-1 commercial remote-sensing satellite scheduling, TianLian-1 relay satellite scheduling, BDS-3 navigation satellite scheduling, and the satellite range scheduling from the U.S. Air Force Institute of Technology. Through the experiments in this chapter, the general-purpose modeling method of satellite task scheduling is practiced, the APMA for satellite conventional task scheduling is fully tested, and the distributed DDRO for satellite emergency task scheduling is experimented with. Finally, this chapter introduces some application systems supported by the satellite task scheduling engine to illustrate the feasibility and application prospect of the engine.

6.1 Experiment setup

6.1.1 Experimental purposes

Based on the general-purpose modeling method of satellite task scheduling in this book, as well as the general-purpose solving methods APMA and DDRO for conventional and emergency task scheduling, this chapter carries out the experiments. The experimental purposes of this chapter include:

(1) **To practice the general-purpose modeling method for satellite task scheduling.** Through the general-purpose modeling method of satellite task scheduling in Chapter 3, this book has completed the modeling and optimization experiments of four types of satellite task scheduling, such as SuperView-1 commercial remote-sensing satellite scheduling, TianLian-1 relay satellite scheduling, BDS-3 navigation satellite scheduling, and satellite range scheduling, to verify the feasibility and gener-

https://doi.org/10.1515/9783111537191-006

ality of the modeling method in the application process of actual complex satellite task scheduling problems.

(2) To comprehensively test the APMA for the satellite conventional task scheduling. For the four types of satellite conventional task scheduling problems, conventional scheduling experiments are carried out through APMA in Chapter 4, the average and standard deviation of the results of multiple experiments are calculated to quantitatively compare with other algorithms to illustrate the optimization performance and robustness of the algorithm, and the iteration curves of the algorithm are drawn to show the convergence rate of the algorithm. A large number of conventional scheduling experiments are compared to demonstrate the comprehensive convergence optimization performance and robustness of APMA as well as its generality in solving satellite task scheduling problems in various scenarios.

(3) To experiment the distributed dynamic rolling optimization algorithm DDRO for the satellite dynamic task scheduling problems. For the four types of satellite emergency task scheduling problems, on the basis of the conventional scheduling results, emergency scheduling experiments are carried out through the distributed DDRO in Chapter 5 to meet the emergency scheduling requirements in various scenarios, and the real-time optimization results of satellite emergency task scheduling are given. The profit change curves during the experiments are drawn to show the generality and effectiveness of DDRO in various satellite emergency task scheduling experiments in combination with the emergency scheduling requirements of each scenario.

(4) To illustrate the feasibility and application prospect of the satellite task scheduling engine. Under the framework of the top-level design of the satellite task scheduling engine in Chapter 2, the above experiments are successfully completed. The results support the development of satellite task scheduling operation or simulation systems, meet the scheduling needs of various satellite tasks such as related remote-sensing satellite scheduling, relay satellite scheduling, navigation satellite scheduling, and satellite range scheduling, provide the satellite control agency with practical and efficient task scheduling tools, and further demonstrate the application value and promotion prospects of the satellite task scheduling engine in this book.

6.1.2 Engine deployment

According to the above experimental purposes, the deployment and experimental flow of the satellite task scheduling engine in this chapter are shown in Figure 6.1. In the figure, the three modules in the satellite task scheduling engine framework are coupled with each other and operate together. The main workflow of each module is as follows:

(1) Based on the general-purpose modeling method for satellite task scheduling, the satellite task scheduling engine completes the modeling for four different types of

satellite task scheduling problems including remote-sensing satellite scheduling, relay satellite scheduling, navigation satellite scheduling, and satellite range scheduling. Here, the general-purpose modeling method for satellite task scheduling presents its advantage of generality. It incorporates various satellite task scheduling problems into unified modeling system and provides a general-purpose task scheduling model to support the solving of subsequent algorithm.

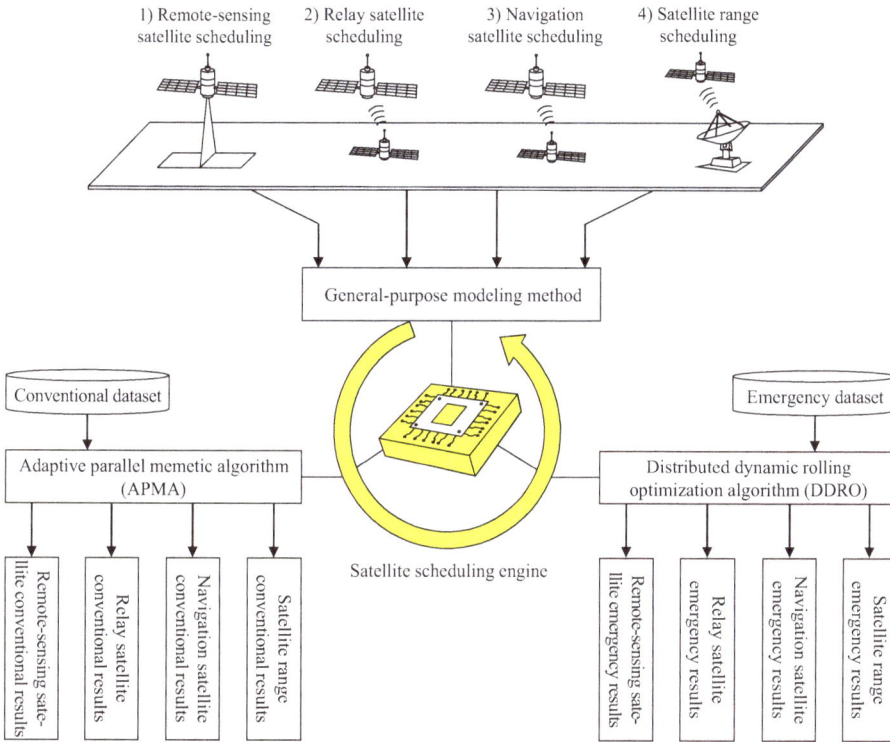

Figure 6.1: Schematic diagram of satellite task scheduling engine deployment for the experiments.

(2) For the requirements of satellite conventional task scheduling, based on the general-purpose APMA and related data sets, experiments of four types of satellite conventional task scheduling are completed, including remote-sensing satellite scheduling, relay satellite scheduling, navigation satellite scheduling and satellite range scheduling, and the results of the task scheduling of each scenario are output. Here, APMA presents its advantages in generality and performance. It optimizes the above satellite task scheduling models with a set of general-purpose algorithms, and provides a general-purpose and efficient means of solving various satellite conventional task scheduling problems.

(3) For the requirements of satellite emergency task scheduling, based on the general-purpose DDRO algorithm, conventional task scheduling results of each scenario and related data sets, experiments of four types of satellite emergency task scheduling are completed, including remote-sensing satellite scheduling, relay satellite scheduling, navigation satellite scheduling and satellite range scheduling, and the results of the satellite emergency task scheduling are output. Here, the DDRO algorithm also presents its advantages of generality and dynamic. It optimizes the satellite task scheduling models again with a set of general-purpose algorithms and provides a general-purpose and flexible means of solving various satellite emergency task scheduling problems.

Through the above flows, this chapter deploys the satellite task scheduling engine on a computer with a quad-core eight-thread Intel(R) Core(TM) i7-7700 CPU @ 3.60 GHz, 16 GB RAM and a compilation environment of Java 1.8.0. For each experimental scenario, conventional and emergency scheduling experiments are carried out, respectively. In the conventional scheduling experiments of each scenario, 10 independent operation experiments are carried out, and the average value and standard deviation of the experimental results are calculated to demonstrate the optimization performance and robustness of the algorithm. The iteration curves of the algorithm are drawn based on the medians of the experiments to show the convergence rate of the algorithm. Then, in the emergency scheduling experiments of each scenario, a dynamic operation experiment is carried out in this chapter and the dynamic change curve of the experimental results is drawn to show the emergency scheduling effect of the algorithm.

6.1.3 Algorithms and parameter settings

6.1.3.1 Algorithm parameter setting experiments

To unify the experimental parameters of the algorithms, create a fair experimental environment, and at the same time enhance the experimental effect of conventional satellite task scheduling as much as possible, and reflect the adaptive advantages of this book's APMA in solving conventional satellite task scheduling problems, this section carries out algorithm parameter setting experiments. Considering that the remote-sensing satellite task scheduling problems are often the largest, most difficult, and most urgent among satellite task scheduling problems, this section carries out algorithm parameter testing and setting based on the conventional task scheduling problems of the "SuperView-1" commercial remote-sensing satellite. The experimental scenes are elaborated in detail in Section 6.2. Based on the experimental results, this section analyzes the influence of the Tabu length coefficient of TS, the annealing method of SA, the late length coefficient of LA, the disturbance and repair frequencies of ISL and other related parameters on the performance of the algorithms, and sets the relevant parameters of each algorithm and APMA, which creates a basis for the experiments in this chapter.

APMA integrates five local search algorithms such as HC, TS, SA, LA, and ISL to implement parallel search and algorithm competition. Therefore, for the main parameters with the most typical characteristics of the meta-heuristic algorithms, such as TS, SA, LA, and ISL, this section sets three parameter levels, respectively, as listed in Table 6.1, and sets the total number of iterations to $|T| \times 500$ generations (T is the task set). Then, this section carries out algorithm parameter comparison experiments, and the experimental results are shown in Table 6.2 and Figure 6.2.

Table 6.1: Main parameters and parameter level settings of local search algorithms.

Parameter level	Tabu length coefficient of TS a_T	Annealing method of SA	Late length coefficient of LA a_L	Disturbance and repair frequencies of ISL
Level 1	0.1	Arithmetic annealing	1	10
Level 2	0.2	Geometric annealing	1.5	25
Level 3	0.3	Logarithmic annealing	2	50

As shown in the statistical results in Table 6.2 and Figure 6.2, for the three Tabu length coefficients a_T of TS, in the 11 examples: when $a_T = 0.1$, TS obtains the highest average value 9 times and the lowest standard deviation 4 times; when $a_T = 0.3$, TS obtains the highest average value twice and the lowest standard deviation 5 times; and when $a_T = 0.2$, TS obtains the highest average value 0 times and the lowest standard deviation only 1 time. Thus, when $a_T = 0.1$, TS has the best optimization performance and good robustness; when $a_T = 0.3$, TS has the best robustness but insufficient optimization performance; and when $a_T = 0.2$, TS has poor optimization performance and robustness, which shows that the Tabu strategy can help the algorithm skip from local optima, but over-long Tabu length reduces the optimization performance of the algorithm. In view of this, this chapter sets the Tabu length coefficient a_T of TS to 0.1.

For the three annealing methods of SA, in 11 examples: SA based on geometric annealing obtains the highest average value 9 times and the lowest standard deviation 6 times, SA based on logarithmic annealing obtains the highest average value twice and the lowest standard deviation 5 times, and SA based on arithmetic annealing only obtains the highest average value once and the lowest standard deviation 3 times. Thus, SA based on geometric annealing has the best optimization performance and robustness, while SA based on logarithmic or arithmetic annealing has far inferior optimization performance and robustness, which shows that the geometric temperature descent method of "fast and then slow" is more efficient in helping SA converge quickly in the early stage and skip from local optima in the later stage. In view of this, this chapter sets the annealing method of SA as geometric annealing.

Table 6.2: Experimental results of parameter comparison of local search algorithms.

Scenario	TS			SA			LA			ISL		
	Level 1 ($\alpha_T = 0.1$)	Level 2 ($\alpha_T = 0.2$)	Level 3 ($\alpha_T = 0.3$)	Level 1 (geometric annealing)	Level 2 (arithmetic annealing)	Level 3 (logarithmic annealing)	Level 1 ($\alpha_L = 1.0$)	Level 2 ($\alpha_L = 1.5$)	Level 3 ($\alpha_L = 2.0$)	Level 1 (10)	Level 2 (25)	Level 3 (50)
1	70.4 (4.2)	71.1 (3.3)	**71.3 (2.2)**	**78.3 (1.6)**	73.1 (1.6)	77.4 (1.6)	76.8 (1.6)	78.2 (1.9)	**78.6 (1.3)**	**68.7 (3.5)**	66.8 (4.7)	62.8 (4.8)
2	**101.4 (5.3)**	101.1 (5.0)	101.0 (7.6)	**98.0 (0.0)**	97.8 (0.6)	98.0 (0.0)	98.0 (0.0)	**100.3 (4.9)**	98.8 (2.9)	95.9 (5.1)	**97.3 (11.0)**	94.1 (5.1)
3	**236.0 (4.9)**	235.0 (5.6)	231.9 (5.8)	**260.9 (2.5)**	246.0 (2.8)	258.7 (3.6)	**261.4 (2.2)**	261.3 (3.0)	261.2 (3.0)	**234.2 (6.6)**	228.4 (8.6)	221.0 (11.5)
4	**464.5 (5.8)**	460.9 (5.7)	449.4 (2.4)	**451.5 (13.0)**	429.0 (7.7)	444.6 (17.7)	**478.3 (14.7)**	470.7 (12.1)	468.8 (14.5)	**461.6 (11.3)**	451.4 (15.4)	436.0 (15.3)
5	**222.3 (3.5)**	216.7 (3.9)	211.5 (2.9)	**235.1 (3.1)**	218.2 (3.0)	231.8 (1.5)	235.4 (2.5)	236.2 (1.9)	**237.0 (1.4)**	**219.1 (4.3)**	212.4 (8.2)	207.8 (12.1)
6	278.2 (5.8)	278.1 (7.2)	**279.0 (3.9)**	**300.9 (3.8)**	286.9 (3.2)	297.8 (3.4)	296.9 (3.7)	296.9 (4.4)	**297.4 (3.5)**	**277.8 (8.0)**	269.9 (8.0)	261.6 (7.7)
7	**525.0 (10.0)**	517.8 (9.7)	515.8 (9.4)	470.1 (27.6)	**474.4 (33.8)**	468.4 (19.0)	**538.2 (4.3)**	538.2 (6.7)	519.5 (20.4)	**512.7 (10.6)**	502.2 (13.6)	488.6 (13.8)
8	**310.0 (3.5)**	304.7 (5.9)	303.6 (5.6)	**335.4 (2.3)**	323.3 (2.6)	333.8 (4.2)	330.1 (4.6)	**334.4 (2.6)**	331.8 (3.7)	**298.2 (8.8)**	291.9 (16.3)	286.5 (9.8)
9	**77.1 (3.0)**	76.1 (3.5)	74.5 (3.7)	**87.5 (0.7)**	82.5 (1.8)	85.8 (1.5)	84.6 (2.3)	85.6 (2.4)	**86.8 (1.4)**	70.0 (10.5)	**70.0 (6.0)**	69.5 (7.8)
10	**51.5 (0.8)**	50.9 (1.1)	48.6 (1.6)	**52.4 (0.8)**	49.9 (1.7)	52.0 (1.3)	52.1 (0.7)	52.5 (0.7)	**52.7 (1.1)**	**49.8 (3.0)**	49.3 (1.9)	43.7 (5.1)
11	**78.4 (2.8)**	77.1 (2.1)	77.1 (2.7)	**83.6 (0.9)**	80.0 (1.8)	83.3 (0.8)	82.7 (1.9)	**84.1 (1.7)**	83.7 (1.7)	72.2 (6.5)	**74.4 (3.1)**	61.9 (5.6)
Statistics	**9 (4)/11**	0 (2)/11	2 (5)/11	**9 (6)/11**	1 (3)/11	2 (5)/11	3 (4)/11	4 (4)/11	5 (5)/11	**9 (7)/11**	3 (3)/11	0 (2)/11
ranking	1	3	2	1	3	2	3	2	1	1	2	3

Note: 1) Data format in the table: bold average values (standard deviations) represent the best results among those of the same algorithm; and 2) algorithm ranking principle: the more occurrences of the best average values and the higher the average values, the more prioritized the ranking.

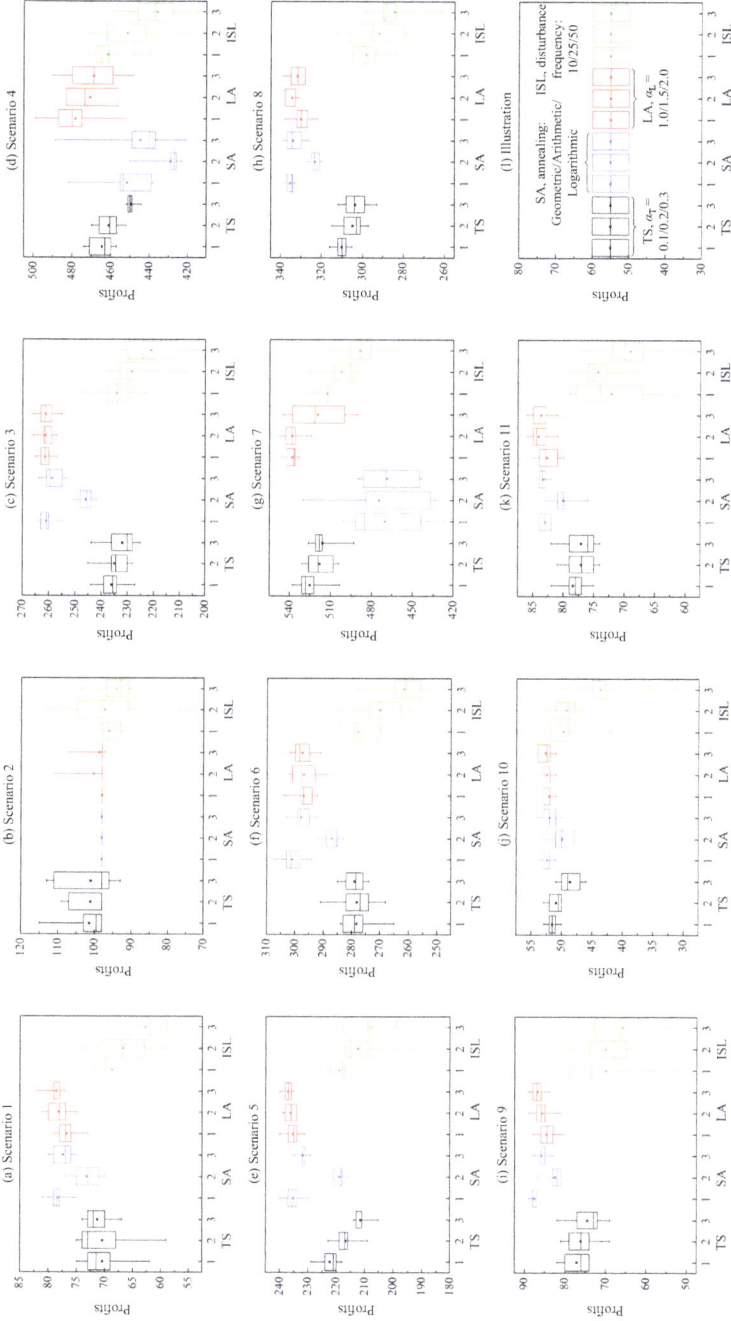

Figure 6.2: Experimental results of parameter comparison of local search algorithms: (a) scenario 1; (b) scenario 2; (c) scenario 3; (d) scenario 4; (e) scenario 5; (f) scenario 6; (g) scenario 7; (h) scenario 8; (i) scenario 9; (j) scenario 10; (k) scenario 11; and (l) illustrations.

For the three late length coefficients α_L of LA, in 11 examples: when $\alpha_L = 2.0$, LA obtains the highest average value 5 times and the lowest standard deviation 5 times; when $\alpha_L = 1.5$, LA obtains the highest average value 4 times and the lowest standard deviation 5 times; and when $\alpha_L = 1.0$, LA obtains the highest average value 3 times and the lowest standard deviation 4 times. Thus, when $\alpha_L = 2.0$, LA has the best optimization performance and robustness, and when $\alpha_L = 1.0$ or $\alpha_L = 1.5$, LA has slightly inferior optimization performance and robustness, which shows that the increase in late length can enhance the optimization performance and robustness of LA to some extent. In view of this, this chapter sets the late length coefficient α_L of LA to 2.0.

For the three kinds of disturbance and repair frequencies of ISL, in 11 examples: ISL with a frequency of 10 obtains the highest average value 9 times and the lowest standard deviation 7 times; ISL with a frequency of 25 obtains the highest average value 3 times, and the lowest standard deviation 3 times; and ISL with a frequency of 50 obtains the highest average value 0 times and the lowest standard deviation only 2 times. Thus, ISL with a frequency of 10 has the best optimization performance and robustness, while ISL with a frequency of 25 or 50 has far inferior optimization performance and robustness, which shows that the disturbance and repair strategy can help the algorithm skip from local optima, but over-frequent disturbance and repair reduce the optimization performance and robustness of the algorithm. In particular, this phenomenon is more obvious in this book because of the complexity and variety of satellite task scheduling constraints. In view of this, this chapter sets the disturbance and repair frequency of ISL to 10.

Through Table 6.2 and Figure 6.2, it is easy to find that algorithms of the same type and parameters may also exhibit significantly different optimization performance in different scenarios. For example, SA significantly outperforms TS in scenarios 1, 3, 6, 8, 9, and 11, but significantly underperforms TS in scenarios 2, 4, and 7. Therefore, to adapt to different types and scenarios of satellite task scheduling problems, self-organization and self-adaptation are particularly important for APMA in this book.

6.1.3.2 Comparison algorithm selection and parameter setting

To test the performance of APMA in this book and demonstrate its generality and adaptability in integrating the performance advantages of different algorithms and solving various satellite task scheduling problems, this section sets up several comparison algorithms. In the following conventional scheduling experiments, the algorithms compared with APMA include four different types of algorithms: heuristic algorithm, local search algorithm, EA, and memetic algorithm:

(1) Heuristic algorithm

Heuristic algorithm is a commonly used algorithm in solving satellite task scheduling problems and various industrial optimization problems as well as constructing initial solution in various optimization algorithms. In this chapter, two common heuristic algorithms are used in the experiments, including the first-in-first-service algorithm (FIFS) and the random allocation algorithm (referred to as random algorithm), which are used for the "construction" strategy in APMA.

(2) Local search algorithm

Local search algorithm is an important part of APMA, which is commonly used to address various combinational optimization problems, especially complex industrial optimization problems. In this chapter, five common heuristic algorithms are used in the experiments, that is, HC, TS, SA, LA, and ISL, which are involved in the parallel and competition strategies of APMA. The parameters of each algorithm are set as described in the previous section. The total number of iterations is set to $|T| \times 500$ generations (T is the task set).

(3) Evolutionary algorithm (EA)

EA is also a common algorithm in solving various combinational optimization problems, and its evolutionary strategy is also an important part of APMA algorithm. In this chapter, two common EAs are used in the experiments: GA and differential evolution (DE) algorithm. The population sizes are both set to 100, and all initial individuals are generated by 1,000 iterations of HC. To improve the optimization performance as much as possible, the numbers of population iterations are both set to 100 generations.

(4) Memetic algorithm (MA, hybrid algorithm)

MA is a hybrid algorithm of a local search algorithm and EA, offsetting the weaknesses by learning from the strong points of both algorithms. In this chapter, three kinds of MAs are used in the experiments, including traditional MA with GA as the main loop, APMA proposed in this book, and its simplified version, that is, the parallel memetic algorithm (PMA) without "competition" and "evolution" strategies. The population sizes of the algorithms are all set to 100. The initial solution of traditional MA is generated by 1,000 iterations of HC, and the number of population iterations is set to 100 (consistent with the EAs such as GA and DE); the initial solutions of APMA and PMA in this book are both generated by their "construction" strategy, and the two algorithms are terminated by the total number of iterations of the above local search algorithms, during which the number of evolutionary operations is set to 10 times.

The main parameters of the above algorithms are summarized in Table 6.3. It should be noted that in the table, the number of algorithm iterations of GA, DE, and

MA with EAs as the main loops is 100, while that of the other five local search algorithms, as well as PMA and APMA with local searches as the main loops, is $|T| \times 500$. During the experiments of each scenario, since the running time of each algorithm with local search as the main loop is basically the same (can be regarded as terminated at the same time) and is much smaller than that of the algorithms with EAs as the main loops, the running times of the algorithms in the conventional scheduling experiments in this chapter are not compared.

Table 6.3: Summary of main algorithm parameters of satellite conventional task scheduling experiments.

No.	Algorithm		Initial solution	Main parameter	Number of iterations
1	Local search algorithms	HC	–	–	$|T| \times 500$
2		TS		Tabu length coefficient $a_T = 0.1$	
3		SA		Annealing method is geometric annealing	
4		LA		Late length coefficient $a_L = 2.0$	
5		ISL		Frequency of disturbance and repair is 10	
6	EAs	GA	HC 1,000 iterations	Population size is 100, and crossover and mutation probability is 1.0	100
7		DE			100
8	Hybrid algorithm	MA	HC 1,000 iterations	Population size is 100, and crossover and mutation probability is 1.0	100
9		PMA	"Construction" Strategy	Using HC, TS, SA, LA, and ISL, population size is 100, and number of evolutionary operations is 10 times	$|T| \times 500$
10		APMA			

In summary, this section carries out the experiments on algorithm parameter setting for the task scheduling problems of "SuperView-1" commercial remote-sensing satellite as an example. It should be noted that the same algorithm parameters may produce very different optimization results in different problems and scenarios, and this section aims at reasonable parameter analysis and setting to provide a reasonable parameter setting basis for the experiments in this book. Next, this chapter will, respectively, carry out various types of satellite conventional and emergency task scheduling experiments, conduct algorithm comparisons, and test the feasibility and effectiveness of the algorithms in this book.

6.2 Remote-sensing satellite task scheduling experiments

In December 2017, "SuperView-1" 03 and 04 satellites were successfully launched and networked with the 01 and 02 satellites already in orbit, marking the completion of the first commercial agile remote-sensing satellite constellation domestically. This section takes the four remote-sensing satellites in the "SuperView-1" constellation as the experimental objects, carries out conventional and emergency task scheduling experiments for remote-sensing satellites, examines the satellite task scheduling engine in this book, and realizes the control objectives of fully utilizing satellite resources and enhancing the economic benefits of commercial constellations.

6.2.1 Experimental scenarios

6.2.1.1 Experimental scenarios for conventional scheduling

The "SuperView-1" commercial remote-sensing satellite constellation is the first commercial constellation domestically composed of agile remote-sensing satellites. It currently contains four optical remote-sensing satellites in sun-synchronous orbit with a resolution of 0.5 m and an orbital period of about 97 min. The orbital parameters are listed in Table 6.4, and the schematic diagram of the orbits is shown in Figure 6.3 (There is a certain deviation from the ideal orbits under the influence of perturbation.) Users of the "SuperView-1" cover many traditional sectors such as land surveying and mapping, urban construction, agriculture, forestry and water conservancy, geology and minerals, environmental monitoring, national defense security, and emergency disaster reduction. The constellation also has great application potential in emerging sectors such as the Internet, location services, and smart cities. The "SuperView-1" 01 and 02 satellites were launched in December 2016, and 03 and 04 satellites were launched in December 2017. In the next few years, more than 20 satellites will be launched to form a new "SuperView-1" constellation on the basis of the existing 4 satellites.

The current market price of commercial remote-sensing satellite images is about ¥120/km^2. The imaging width of the "SuperView-1" remote-sensing satellites is 12 km, and the moving speed of the sub-satellite point is about 7 km/s. In other words, "SuperView-1" remote-sensing satellites can take about 84 km^2 images per second, which generates about ¥10,000 economic income. It can be seen that reasonable and efficient task scheduling can bring huge economic benefits to "SuperView-1."

In this section, the conventional scheduling experiment uses the real data set provided by the control agency of "SuperView-1," and the data set is summarized in Table 6.5. Different from the simulation data set used in Benchmark and other remote-sensing satellite task scheduling studies, the data set used by this experiment has the following characteristics: 1) authenticity: since the date set is provided by the control agency, it truly reflects the optimization requirements of the "SuperView-1"

Table 6.4: Orbit parameters of "SuperView-1" commercial remote-sensing satellites.

Satellite	Orbit semi-major axis (km)	Orbit eccentricity	Orbit inclination (°)	Argument of perigee (°)	Right ascension of ascending node (°)	Average anomaly (°)
1	6,923.140	0.00157	97.510	163.089	118.707	222.281
2	6,923.188	0.00145	97.428	159.290	106.742	18.052
3	6,923.463	0.00073	97.487	64.352	112.454	177.262
4	6,923.279	0.00136	97.488	53.241	112.644	34.562

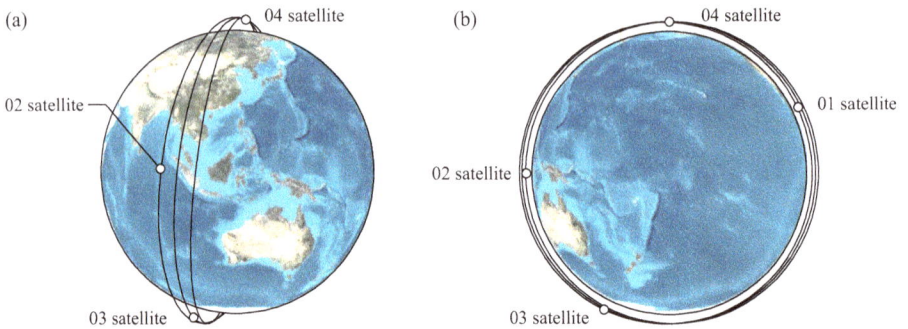

Figure 6.3: Schematic diagram of orbits of "SuperView-1" commercial remote-sensing satellites.

task scheduling problems; 2) integrity: in the "SuperView-1" task scheduling problems, it is necessary to decide the imaging, downlink, and memory-erasing events of the satellites at the same time; 3) constraint complexity: the data set covers more than 100 constraints of the "SuperView-1" task scheduling problems; and 4) scenario diversity: it includes different task scheduling scenarios such as single-satellite scheduling, multi-satellite scheduling, single-day scheduling, and multi-day scheduling.

6.2.1.2 Experimental scenarios for emergency scheduling

Based on the experimental scenarios for conventional task scheduling of "SuperView-1," this section designs the following experimental scenarios for emergency scheduling: Given the conventional scheduling results in each scenario, the control agency of "SuperView-1" receives a certain number of emergency task requirements at regular intervals, including those that have to be executed (scheduled) and those that do not have to be executed, with the probability that the emergency tasks have to be executed is 50%. Specifically 1) in the single-satellite single-day task scheduling scenario, that is, scenarios 1, 9, 10, and 11, the control agency randomly receives 0–4 emergency task requirements every 1 h; 2) in the single-satellite multi-day scenario, that is, scenario 2, the control agency randomly receives 0–4 emergency task requirements every 2 h; 3) in the multi-satellite single-day scenarios, that is, scenarios 3, 5, 6, and 8, the control agency randomly receives 0–16 emergency task requirements every 1 h; and 4) in the multi-satellite multi-day scenarios, that is, scenarios 4 and 7, the control agency randomly receives 0–16 emergency task requirements every 2 h.

Table 6.5: Summary of task scheduling data set of "SuperView-1" commercial remote-sensing satellites.

Scenario	Satellite number	Scheduling cycle	Total number of tasks	Average number of imaging event EOs	Average number of downlink event EOs
1	01	24 h	113	131.9	2,022.6
2	01	48 h	190	130.0	1,539.2
3	01, 02, 03, 04	24 h	345	121.5	1,860.3
4	01, 02, 03, 04	48 h	617	114.3	2,085.0
5	01, 02, 03, 04	24 h	272	102.1	1,968.7
6	01, 02, 03, 04	24 h	580	116.1	1,909.4
7	01, 02, 03, 04	48 h	1266	121.0	1,950.7
8	01, 02, 03, 04	24 h	686	123.8	1,931.6
9	02	24 h	148	126.4	1,877.2
10	03	24 h	55	86.8	1,888.6
11	04	24 h	137	122.6	1,970.3

After receiving the emergency tasks, the control agency of "SuperView-1" immediately calls DDRO to address the scheduling problems, triggering task allocation and single-satellite window rolling and rescheduling. In the single-day scenarios, the real-time window length is set to 5 h, that is, only the tasks within 5 h from the current moment are involved in rescheduling. In the multi-day scenarios, the real-time window length is set to 10 h. In this process, to avoid drastic changes to the original scheme, PMA is used to address the rescheduling problems of nonmandatory tasks, and the number of iterations is $|T| \times 250$, that is, half of that of the conventional scheduling experiments. The rest of the parameters are the same as those of the conventional scheduling experiments.

6.2.2 Conventional scheduling experiments

Based on the general-purpose modeling method for satellite task scheduling and the satellite conventional task scheduling algorithm APMA in this book, this section carries out conventional task scheduling experiments for the "SuperView-1" commercial remote-sensing satellites.

After calculation, the comparison results of the algorithms for conventional task scheduling experiments of "SuperView-1" commercial remote-sensing satellites are shown in Table 6.6 and Figure 6.4. Table 6.6 lists the average values and standard deviations of the experimental results of 12 algorithms in 4 categories, including heuristic algorithms, local search algorithms, EAs and hybrid algorithms. Figure 6.4 presents the box plots of the experimental results of each algorithm and the iteration curves of the median values during the experimental process. Among the 11 examples, APMA proposed in this book obtains the best average value 10 times and the lowest standard deviation 3 times, indicating that APMA has the best optimization performance and good robustness.

The statistical results in Table 6.6 and Figure 6.4 show that FIFS and random have poor optimization performance. Among them, Random has extremely poor optimization performance. The reason is that the constraints of the remote-sensing satellite task scheduling problems in this section are complex, and it is difficult for this heuristic algorithm to obtain high-quality feasible solutions. It can be seen that the heuristic algorithms can quickly generate feasible solutions in a short time, but due to the lack of an iterative optimization mechanism, it is difficult to guarantee the solution quality. Therefore, the algorithms are usually used to construct the initial solution of combinational optimization problems. It is worth noting that FIFS is a commonly used heuristic strategy for solving complex task scheduling problems in industrial sectors, and it is widely used in many in-service satellite task scheduling systems. However, its actual optimization performance is far inferior to that of meta-heuristic algorithm, so it is of great significance to popularize and apply meta-heuristic algorithm in industrial problems such as satellite task scheduling. In this chapter, the standard deviation of FIFS results is not listed in the table because the results of multiple runs of FIFS are consistent.

The convergence rate and optimization performance of EAs such as GA and DE are generally lower than those of local search algorithms, which is due to the following reasons: the task scheduling constraints of remote-sensing satellites are complex, the crossover and mutation operators in EAs (especially the differential mutation operators in DE) are very likely to produce solutions that violate the constraints, the probability of high-quality solutions in the population iteration process is small, and the optimization efficiency is low. In the scheduling problems with scheduling cycle of 48 h such as scenarios 3–8, the optimization performance of GA is even inferior to that of heuristic algorithms. In all scenarios, the iteration curves of DE always stay at the level of the initial solution, with no obvious optimization sign. It can be seen that traditional EAs such as GA and DE are not suitable for satellite task scheduling problems with complex constraints such as remote-sensing satellite task scheduling, and their optimization performance and robustness need to be further improved.

For the local search algorithms HC, TS, SA, LA and ISL, the optimization performance and robustness of each algorithm are generally high. Among them, the optimization performance and convergence rate of SA and LA are outstanding, and SA ob-

Table 6.6: Comparison results of algorithms for conventional task scheduling experiments of "SuperView-1" commercial remote-sensing satellites.

Scenario	Heuristic algorithms		Local search algorithms					EAs			Hybrid algorithms	
	FIFS	Random	HC	TS	SA	LA	ISL	GA	DE	MA	PMA	APMA
1	35	8.1 (1.4)	74.3 (1.3)	70.4 (4.2)	78.3 (1.6)	78.6 (1.3)	68.7 (3.5)	52.8 (3.5)	44.0 (1.5)	63.7 (2.3)	78.3 (1.6)	**78.7 (1.3)**
2	67	10.9 (1.7)	96.4 (6.3)	101.4 (5.3)	98.0 (0.0)	98.8 (2.9)	95.9 (5.1)	61.2 (2.3)	47.6 (2.3)	79.7 (1.8)	102.9 (5.1)	**109.4 (5.8)**
3	113	29.1 (2.7)	240.5 (4.1)	236.0 (4.9)	260.9 (2.5)	261.2 (3.0)	234.2 (6.6)	71.5 (3.4)	48.5 (1.8)	124.5 (6.4)	261.8 (3.1)	**262.8 (3.4)**
4	182	54.6 (6.5)	474.0 (5.9)	464.5 (5.8)	451.5 (13.0)	468.8 (14.5)	461.6 (11.3)	65.4 (2.9)	43.7 (2.2)	146.5 (6.5)	489.8 (5.6)	**493.9 (6.2)**
5	96	24.7 (2.8)	221.7 (4.5)	222.3 (3.5)	235.1 (3.1)	237.0 (1.5)	219.1 (4.3)	76.8 (4.5)	55.9 (2.6)	121.0 (4.7)	238.1 (1.6)	**238.9 (1.5)**
6	234	13.3 (2.1)	281.8 (6.3)	278.2 (5.8)	300.9 (3.8)	297.4 (3.5)	277.8 (8.0)	56.0 (3.4)	38.1 (2.0)	120.2 (5.0)	298.1 (1.7)	**301.3 (3.1)**
7	357	18.2 (1.4)	521.2 (3.6)	525.0 (10.0)	470.1 (27.6)	519.5 (20.4)	512.7 (10.6)	44.6 (0.7)	29.0 (0.7)	111.7 (7.4)	539.8 (6.4)	**546.6 (6.2)**
8	239	14.7 (1.7)	307.7 (7.5)	310.0 (3.5)	**335.4 (2.3)**	331.8 (3.7)	298.2 (8.8)	53.7 (3.2)	35.1 (1.1)	117.0 (6.7)	332.3 (3.2)	334.5 (2.5)
9	29	7.3 (1.9)	78.3 (3.0)	77.1 (3.0)	**87.5 (0.7)**	86.8 (1.4)	70.0 (10.5)	54.7 (3.0)	42.9 (3.0)	66.4 (2.0)	87.0 (1.1)	**87.5 (1.5)**
10	30	6.7 (1.6)	51.9 (0.7)	51.5 (0.8)	52.4 (0.8)	52.7 (1.1)	49.8 (3.0)	46.8 (1.3)	42.4 (0.8)	50.7 (0.8)	52.8 (0.9)	**53.5 (0.7)**
11	29	8.9 (2.0)	76.6 (5.1)	78.4 (2.8)	83.6 (0.9)	83.7 (1.7)	72.2 (6.5)	56.9 (1.6)	45.8 (1.8)	66.8 (1.8)	84.5 (1.1)	**85.7 (1.3)**
Statistics	0/11	0 (0)/11	0 (2)/11	0 (0)/11	2 (3)/11	0 (2)/11	0 (0)/11	0 (1)/11	0 (4)/11	0 (0)/11	0 (1)/11	10 (3)/11
ranking	9	12	5	6	2	4	7	10	11	8	3	1

Note: 1) Data format in the table: bold average values (standard deviations) are the best results; 2) Algorithm ranking principle: the more times of the best average values and the higher the average values, the more prioritized the ranking.

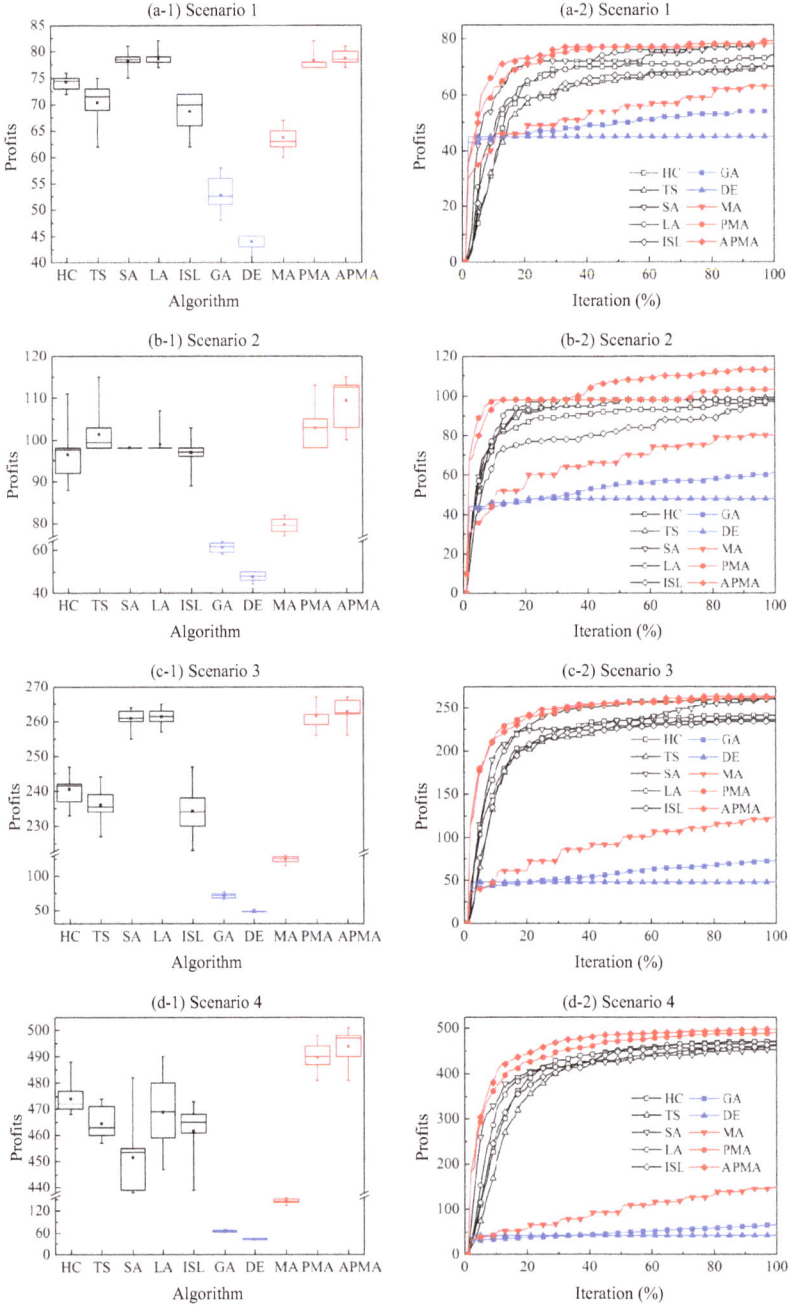

Figure 6.4: Comparison results of algorithms for conventional task scheduling experiments of "SuperView-1" commercial remote-sensing satellites: (a) scenario 1; (b) scenario 2; (c) scenario 3; (d) scenario; 4 (e) scenario 5; (f) scenario 6; (g) scenario 7; (h) scenario 8; (i) scenario 9; (j) scenario 10; and (k) scenario 11.

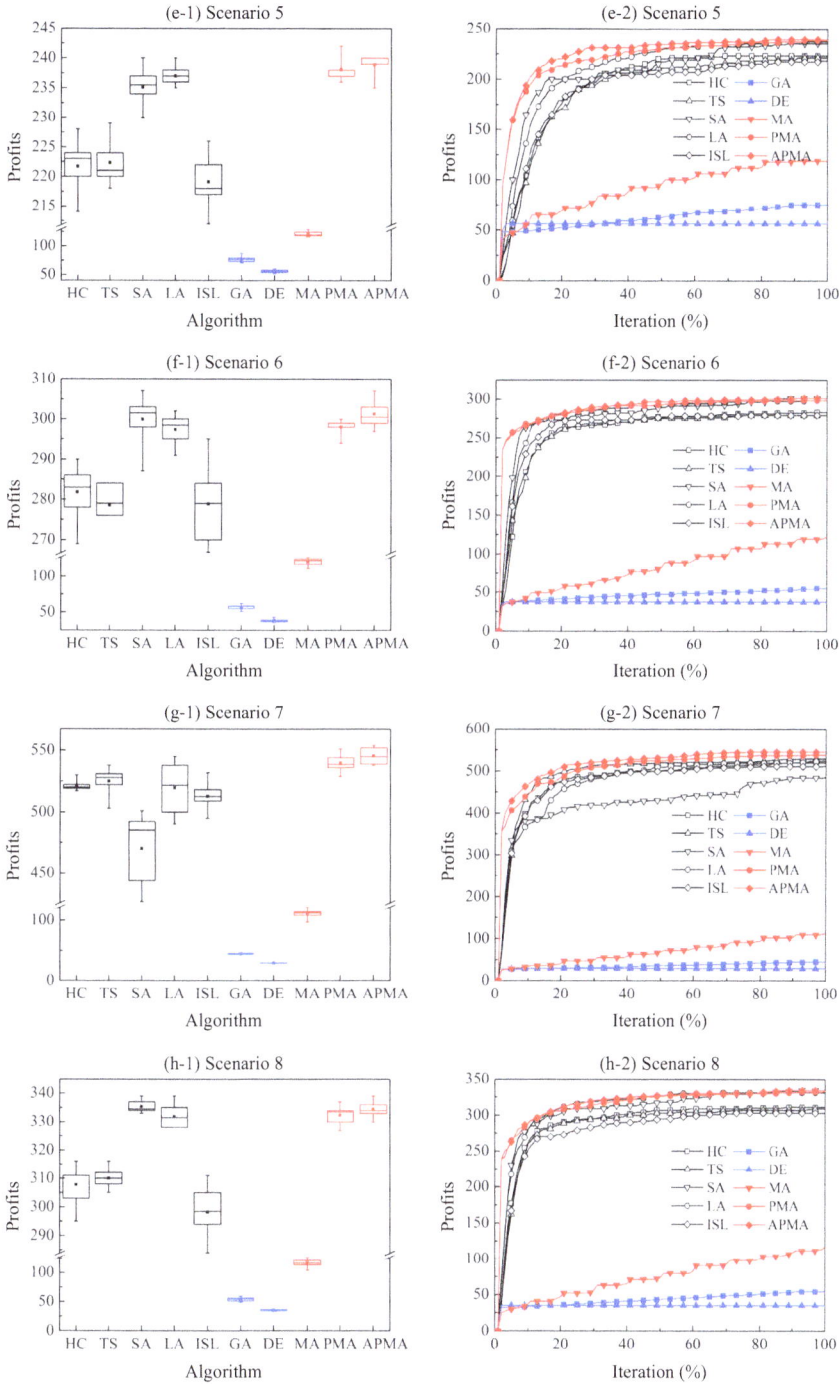

Figure 6.4 (continued)

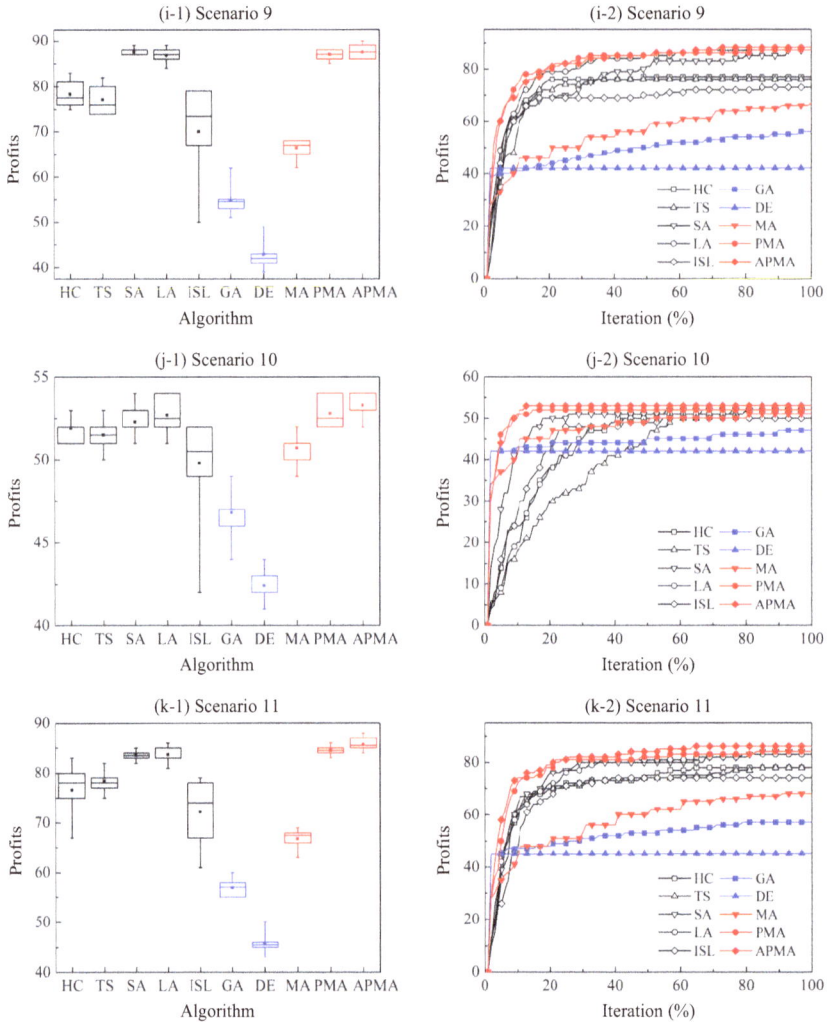

Figure 6.4 (continued)

tains the best average value twice in the 11 examples. Notably, as the simplest and most basic local search algorithm, HC also has good optimization performance, robustness, and convergence rate in the 11 examples, which shows the application potential of local search algorithms in complex industrial problems such as remote-sensing satellite task scheduling. In addition, the optimization performance of ISL is poor, and the convergence rate is also low, which is due to the following reasons: disturbance and repair mechanisms are introduced into ISL to skip from local optima. However, under the actual situation that HC can achieve good optimization results for the problems, the disturbance and repair mechanism of ISL generate many solutions that violate the constraints, which reduces the convergence rate and optimization efficiency of the algorithm. It can be seen that the local search algorithms based on HC are very suitable for solving remote-sensing satellite task scheduling problems, and certain meta-heuristic mechanisms in TS, SA, and LA are helpful to improve their optimization performance, but the disturbance and repair mechanism in ISL destroys the advantages and continuity of local search, which is not conducive to solving remote-sensing satellite task scheduling problems.

Regarding hybrid algorithms MA, PMA and APMA, the optimization performance and convergence rate of MA are better than those of EAs such as GA and DE, but they are not as good as those of local search algorithms, which is due to the fact that traditional MA takes GA as the main loop, and there is the same problem of insufficient constraint optimization efficiency as that of EAs such as GA. On the contrary, APMA and its simplified version PMA proposed in this book use the local search algorithms as the main loop, which can fullly exploit the rapid optimization and convergence ability of the local search algorithms in the constraint solution space, and at the same time quickly obtain high-quality initial solutions by the "construction" strategy, and help the algorithms escape from local optima through the "evolution" strategy, showing the best convergence rate and optimization performance. Moreover, the experimental results show that the optimization performances of local search algorithms are different, but APMA can fully utilize the computing power of the computers and the performance advantages of the local search algorithms through the "parallel" and "competition" strategies to realize the "survival of the fittest" algorithms and operators and achieve the best overall optimization performance, whereas the optimization performance of the simplified version PMA is slightly inferior to that of APMA due to the lack of "competition" mechanism.

In summary, the experiments of "SuperView-1" commercial remote-sensing satellite conventional task scheduling in this section practice the general-purpose modeling method for satellite task scheduling in this book, and fully examine the algorithm performance of APMA. The results show that APMA can make full use of computing power of the computers and the performance advantages of various algorithms to realize outperformance and robustness in the process of solving the conventional task scheduling problems of "SuperView-1" commercial remote-sensing satellites.

6.2.3 Emergency scheduling experiments

Based on the median results of conventional scheduling experiments in Section 6.2.2 and the satellite emergency task scheduling algorithm DDRO, this section further conducts emergency task scheduling experiments of "SuperView-1" commercial remote-sensing satellites.

After calculation, the experimental results of emergency task scheduling of "Super-View-1" commercial remote-sensing satellites are shown in Figure 6.5. During the emergency scheduling process, a different number of emergency tasks, including mandatory and nonmandatory ones, are input every 1 h (or 2 h). It is worth noting that under the complex constraints of remote-sensing satellite task scheduling, to schedule mandatory emergency tasks, some tasks currently executed may be forced to be canceled, resulting in a reduction in the profits of the tasks.

As shown in Figure 6.5, under the effect of DDRO emergency scheduling in this book, the profits of satellite task scheduling fluctuate to varying degrees in each scenario. Among them, the profits of satellite task scheduling in scenarios 3, 5, 6, 7, and 8 show an overall trend of decreasing first and increasing later, which is due to the following reasons: in the early stage of rolling scheduling, the tasks to be executed by the satellites are close to saturation again, and to schedule the mandatory emergency tasks, some tasks are forced to be canceled, and the profits decrease correspondingly; in particular, there are many tasks involved in memory-related constraints in remote-sensing satellite task scheduling, and the tasks that are forced to be canceled due to this constraint lead to a significant decrease in the profits. At the same time, within the DDRO dynamic optimization framework and short computation time, the satellite task scheduling problems are optimized to a limited extent, and it is difficult to recover the profits. However, with the passage of time, under the effect of DDRO rolling optimization for many consecutive times, the optimization degree of the current satellite task scheduling problems gradually improves, and the profits then increase and approach (slightly lower than) the initial profits, and the tasks to be executed are close to the saturation state again. Considering that the remote-sensing satellites have executed a large number of emergency tasks during this period, which meets the special requirements of remote-sensing satellites in terms of emergency response, and at the same time, the satellite task scheduling profit basically maintains the original level, reflecting the effectiveness of this book's DDRO in solving emergency task scheduling problems of "SuperView-1" remote-sensing satellites.

At the same time, it is easy to find that during the dynamic optimization process of DDRO, the task profits of some satellites always remain at a low level and the fluctuation is not obvious, such as satellite 03 in scenario 3 and satellites 02 and 03 in scenarios 4 and 5. The reason for this phenomenon is that the above satellites are close to task saturation at the beginning of rolling scheduling, and their ability to perform new tasks is low. DDRO has reasonably predicted this information and allocated the emergency tasks (especially the mandatory ones) to other capable satellites, which reflects the reasonableness of DDRO in solving the emergency task scheduling problems of SuperView-1 remote-sensing satellites. However, it is undeniable that, under the influence of visibility, timeliness and other constraints and the actual situation, sometimes the satellites with low profits still need to sacrifice part of the current task profits to execute the emergency tasks, for example, the profits of satellites 01, 02, and 04 in scenario 6 are obviously reduced during the 7th–9th hours, and the profit of satellite 03 in scenario 7 is significantly reduced during the 2nd-4th hours. This phe-

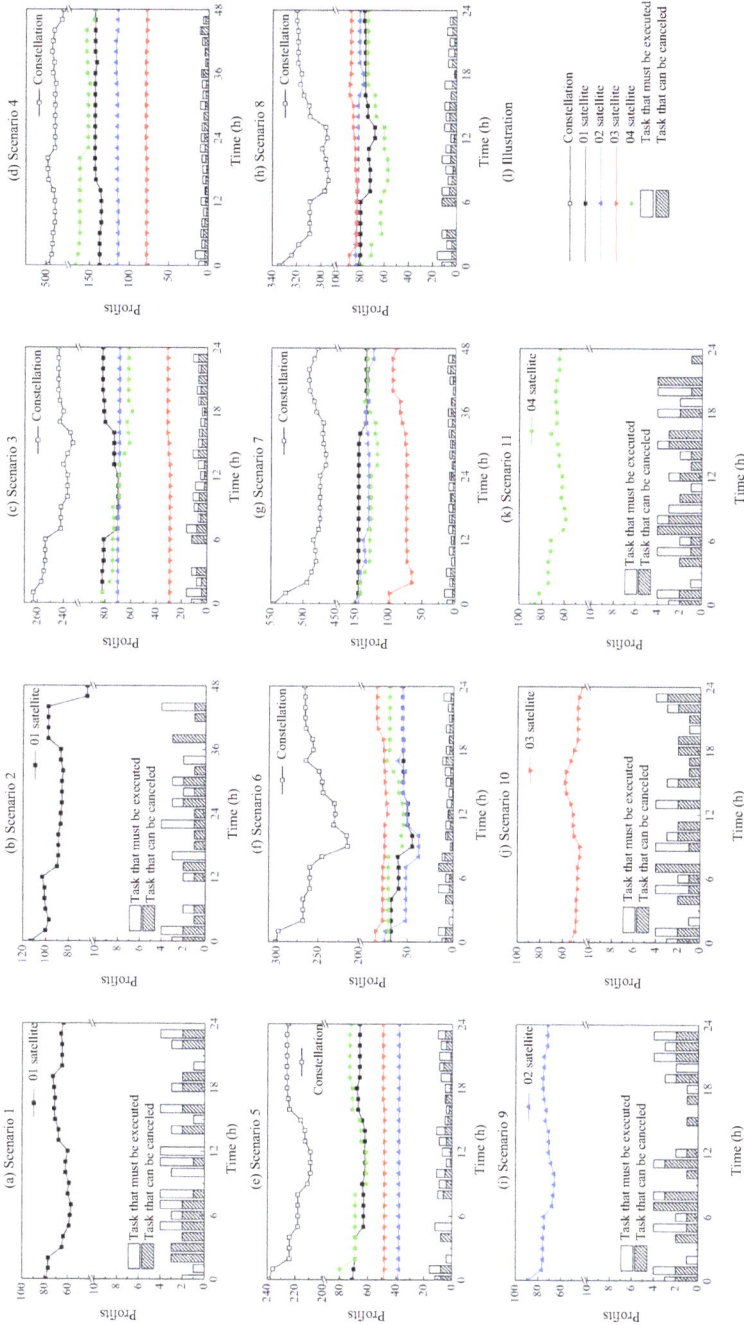

Figure 6.5: Experimental results of emergency task scheduling of "SuperView-1" commercial remote-sensing satellites: (a) scenario 1; (b) scenario 2; (c) scenario 3; (d) scenario 4; (e) scenario 5; (f) scenario 6; (g) scenario 7; (h) scenario 8; (i) scenario 9; (j) scenario 10; (k) scenario 11; and (l) illustrations.

Table 6.7: Comparison of profits and running time between conventional and emergency task scheduling of "SuperView-1" commercial remote-sensing satellites.

Scenario	Conventional task scheduling		Emergency task scheduling experiments		Result comparison	
	Profit (%)	Running time (s)	Profit (%)	Running time (s)	Profit (%)	Running time (s)
1	78.7 (1.3)	32.1	66.4 (6.3)	11.1	−15.6%	−65.3%
2	109.4 (5.8)	126.5	91.3 (10.9)	47.3	−16.5%	−62.6%
3	262.8 (3.4)	288.3	245.1 (8.5)	50.6	−6.7%	−82.5%
4	493.9 (6.2)	742.9	490.5 (4.5)	168.4	−0.7%	−77.3%
5	238.9 (1.5)	154.0	221.2 (7.9)	21.9	−7.4%	−85.8%
6	301.3 (3.1)	387.4	255.7 (21.2)	83.7	−15.1%	−78.4%
7	546.6 (6.2)	783.6	483.1 (18.0)	209.3	−11.6%	−73.3%
8	334.5 (2.5)	506.3	310.6 (10.9)	80.4	−7.1%	−84.1%
9	87.5 (1.5)	43.6	73.3 (4.5)	16.3	−16.2%	−62.6%
10	53.5 (0.7)	23.8	49.4 (3.9)	10.1	−7.7%	−57.4%
11	85.7 (1.3)	34.2	68.0 (6.5)	13.1	−20.7%	−61.7%
Average	–	–	–	–	−11.4%	−71.9%

Note: 1) The statistics of conventional task scheduling scenarios are the results of 10 independent operations; and 2) the statistics of emergency task scheduling scenarios are the results of 24 consecutive emergency scheduling.

nomenon is also in line with the actual situation of the long-term process of remote-sensing satellite control.

Based on the above experimental results of conventional and emergency task scheduling of "SuperView-1" commercial remote-sensing satellites, the comparison of profits and running time between the conventional and emergency task scheduling is listed in Table 6.7. It can be seen that, compared to the conventional task scheduling results (based on APMA), DDRO can reduce the response time by 71.9% on average, which meets the requirement for fast response. The average reduction of the profits is 11.4%, which basically maintains the profit level, and the optimization of emergency task scheduling is obvious. As a result, the emergency task scheduling experiments in this section test the rationality, flexibility, and timeliness of DDRO, showing that it can quickly schedule and execute emergency tasks in a long-term dynamic environment, maintain the profit of remote-sensing satellite task scheduling, achieve the design purpose, and meet the dynamic and real-time emergency task scheduling needs in the long-term process of remote-sensing satellite control.

In summary, this section practices the general-purpose modeling method for satellite task scheduling in this book through the conventional and emergency task scheduling experiments of "SuperView-1" commercial remote-sensing satellites, examines the general-purpose solving algorithms APMA and DDRO, and provides a practical basis for the application of this book's satellite task scheduling engine in remote-sensing satellite task scheduling problems.

6.3 Relay satellite task scheduling experiments

In November 2016, the "Tianlian-1" 04 satellite was successfully launched and successfully networked with the 01–03 satellites already in orbit, marking the full completion of the first domestically developed geosynchronous orbit relay satellite system. This section takes the four relay satellites in the "Tianlian-1" constellation as the experimental objects to carry out the conventional and emergency task scheduling experiments of the relay satellites for seven consecutive days. The experiments have examined the satellite task scheduling engine of this book and realized the control goal of fully utilizing the advantages of space-based relay communication to meet the relay communication needs of medium- and low-orbit satellites.

6.3.1 Experimental scenarios

6.3.1.1 Experimental scenarios for conventional scheduling
"Tianlian-1" is the first relay satellite system of China, consisting of 4 relay satellites in geosynchronous orbit. The orbital parameters are listed in Table 6.8, and the schematic diagram of the orbits is shown in Figure 6.6 (There is a certain deviation from the ideal orbits under the influence of perturbation.) "Tianlian-1" has long been providing relay communication services for medium- and low-orbit satellites, space stations, manned spacecraft, and other aerospace equipment, and is known as the "space-based ground station," which is important for the construction of the multidimensional TT&C network, alleviating domestic TT&C pressure, and improving the robustness and emergency response capabilities of satellite systems. In recent years, with the increasing number of spacecraft in orbit, the scale of relay communication tasks of "Tianlian-1" has increased dramatically, and the request for task scheduling has become increasingly urgent. In this section, the conventional scheduling experiments select a 7-day task scheduling simulation data of "Tianlian-1" as the experimental data set, as shown in Table 6.9.

Table 6.8: Orbital parameters of "Tianlian-1" relay satellites.

Satellite	Semimajor axis (km)	Eccentricity	Inclination (°)	Argument of perigee (°)	Right ascension of ascending node (°)	Average anomaly (°)
1	42,314.97	0.00341	4.616	261.445	71.571	298.521
2	42,314.35	0.00124	2.832	119.497	75.272	167.768
3	42,315.01	0.00060	0.769	263.985	98.871	209.060
4	42,313.98	0.00175	0.146	187.083	264.076	277.204

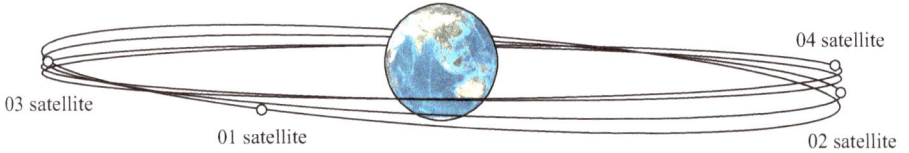

04 satellite

03 satellite

01 satellite

02 satellite

Figure 6.6: Schematic diagram of orbits of "Tianlian-1" relay satellites.

Table 6.9: Overview of task scheduling data set of "Tianlian-1" relay satellites.

Scenario	Satellite number	Scheduling cycle	Total number of tasks	Average number of downlink event executable opportunities
1	01, 02, 03, 04	24 h	1,000	347.7
2	01, 02, 03, 04	24 h	1,000	345.1
3	01, 02, 03, 04	24 h	1,000	343.0
4	01, 02, 03, 04	24 h	1,000	343.0
5	01, 02, 03, 04	24 h	1,000	344.5
6	01, 02, 03, 04	24 h	1,000	355.1
7	01, 02, 03, 04	24 h	1,000	343.4

6.3.1.2 Experimental scenarios for emergency scheduling

Based on the experimental scenarios for conventional task scheduling of "Tianlian-1," this section designs the following experimental scenarios for emergency scheduling. Given the conventional scheduling results in each scenario, the control agency of "Tianlian-1" randomly receives 0–16 emergency tasks every 1 h, and the probability that the emergency tasks have to be executed is 50%.

After receiving the emergency tasks, the control agency of "Tianlian-1" immediately calls DDRO to address the scheduling problems, triggering task allocation and single-satellite window rolling and rescheduling. In this process, the real-time window length is set to 5 h, that is, only the tasks within 5 h from the current moment are involved in rescheduling. To avoid drastic changes to the original scheme, PMA is used to address the rescheduling problems of nonmandatory tasks, and the number of iterations is $|T| \times 250$, that is, half of that of the conventional scheduling experiments. The rest of the parameters are the same as those of the conventional scheduling experiments.

6.3.2 Conventional scheduling experiments

Based on the general-purpose modeling method for satellite task scheduling and the satellite conventional task scheduling algorithm APMA in this book, this section carries out the conventional task scheduling experiments of the "Tianlian-1" relay satellites.

After calculation, the comparison results of the algorithms for conventional task scheduling experiments of "Tianlian-1" relay satellites are shown in Table 6.10 and Figure 6.7. The table lists the average values and standard deviations of the experimental results of 12 algorithms in 4 categories, including heuristic algorithms, local search algorithms, EAs and hybrid algorithms. The figure presents the box plots of the experimental results of each algorithm and the iteration curves of the median values during the experimental process. Among the 7 examples, APMA proposed in this book obtains all seven best average values and four lowest standard deviations, indicating that APMA has the best optimization performance and robustness.

From the statistical results in Table 6.10 and Figure 6.7, it can be seen that, similar to the remote-sensing satellite task scheduling experiments: the heuristic algorithm FIFS has a general optimization performance; local search algorithms generally have good optimization performance; however, the optimization performance of EAs such as GA and DE is generally inferior to that of local search algorithms, even inferior to heuristic algorithms such as FIFS and Random. The iteration curves of DE always stay at the level of initial solution, with no obvious optimization sign.

For the local search algorithms HC, TS, SA, LA, and ISL, the optimization performance and robustness of each algorithm are relatively close to each other, and the convergence rate is basically the same. Among them, the optimization performance of TS and HC is more outstanding, and their performance is very close to that of APMA. However, SA, which performs well in remote-sensing satellite task scheduling experiments, performs obviously poorly in the experiments of this section, and its optimization performance and convergence rate are not as good as other local search algorithms, which is attributed to the following reasons: compared with remote-sensing satellite task scheduling, the constraint complexity of relay satellite task scheduling is relatively low, and the difficulty of optimization is reduced so that the local search algorithms can converge faster and obtain high-quality solutions, but the meta-heuristic mechanism of probabilistic acceptance of inferior solutions in SA affects its convergence and optimization performance. This phenomenon also objectively shows that the algorithms may exhibit different optimization performances when solving different satellite task scheduling problems.

Table 6.10: Comparison results of algorithms for conventional task scheduling experiments of "Tianlian-1" relay satellite.

Scenario	Heuristic algorithms			Local search algorithms				EAs			Hybrid algorithms	
	FIFS	Random	HC	TS	SA	LA	ISL	GA	DE	MA	PMA	APMA
1	822	817.4 (6.7)	963.8 (1.8)	965.6 (2.8)	946.5 (3.7)	965.2 (1.6)	964.9 (2.4)	605.4 (5.9)	504.5 (6.0)	820.5 (4.6)	965.4 (1.4)	**968.7 (1.3)**
2	831	805.7 (6.9)	964.6 (3.5)	965.4 (1.8)	943.7 (2.3)	965.7 (1.6)	963.4 (2.5)	597.7 (2.4)	497.7 (3.6)	810.7 (5.2)	964.7 (1.5)	**968.6 (2.0)**
3	841	811.3 (6.1)	960.7 (1.6)	960.6 (2.0)	935.9 (6.0)	960.1 (3.9)	958.1 (2.2)	597.2 (5.4)	497.5 (4.6)	798.4 (6.0)	959.8 (2.5)	**964.0 (1.8)**
4	827	803.2 (7.7)	964.1 (3.1)	963.4 (2.2)	942.0 (4.5)	963.2 (2.3)	962.5 (2.2)	594.4 (5.7)	497.1 (4.9)	803.5 (4.7)	962.1 (2.3)	**965.9 (2.1)**
5	832	807.9 (7.0)	960.2 (2.8)	961.0 (2.6)	936.2 (4.4)	959.5 (2.7)	959.2 (4.1)	599.3 (7.3)	498.0 (5.5)	801.4 (5.1)	959.7 (1.3)	**963.6 (2.1)**
6	827	819.5 (5.3)	963.7 (2.3)	963.9 (2.0)	945.7 (4.8)	962.0 (1.8)	961.9 (2.4)	601.4 (6.1)	500.2 (3.9)	797.0 (3.9)	963.9 (3.1)	**966.9 (1.7)**
7	799	798.0 (8.2)	956.0 (2.0)	957.0 (1.8)	933.6 (3.7)	956.4 (3.0)	955.2 (2.6)	592.0 (4.7)	493.8 (2.9)	789.3 (4.8)	957.1 (1.9)	**959.0 (1.8)**
Statistics	0/7	0 (0)/7	0 (1)/7	0 (1)/7	0 (0)/7	0 (0)/7	0 (0)/7	0 (0)/7	0 (0)/7	0 (0)/7	0 (2)/7	7 (4)/7
Ranking	8	9	3	2	7	5	6	11	12	10	4	1

Note: 1) Data format in the table: bold average values (standard deviations) are the best results; and 2) algorithm ranking principle: the more times of the best average values and the higher the average values, the more prioritized the ranking.

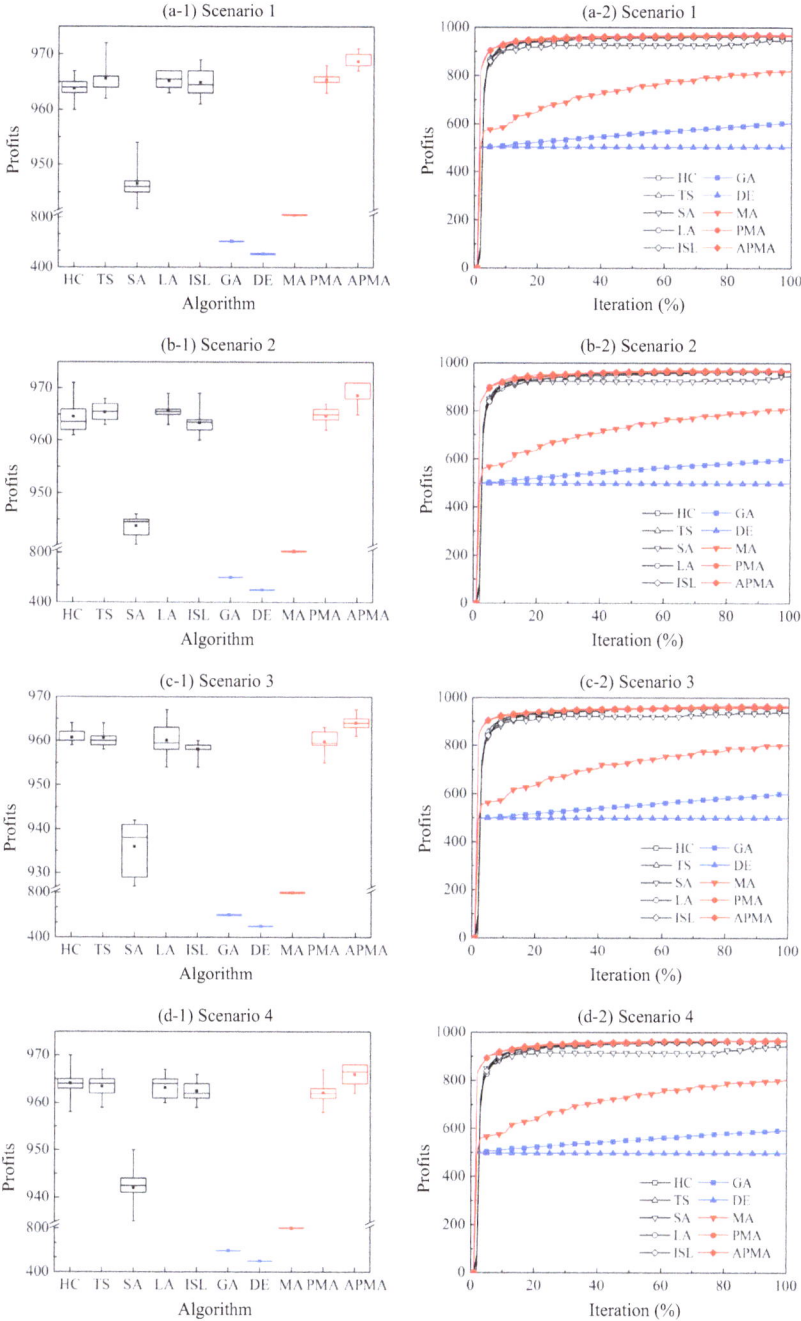

Figure 6.7: Comparison results of algorithms for conventional task scheduling experiments of "Tianlian-1" relay satellite: (a) scenario 1; (b) scenario 2; (c) scenario 3; (d) scenario 4; (e) scenario 5; (f) scenario 6; and (g) scenario 7.

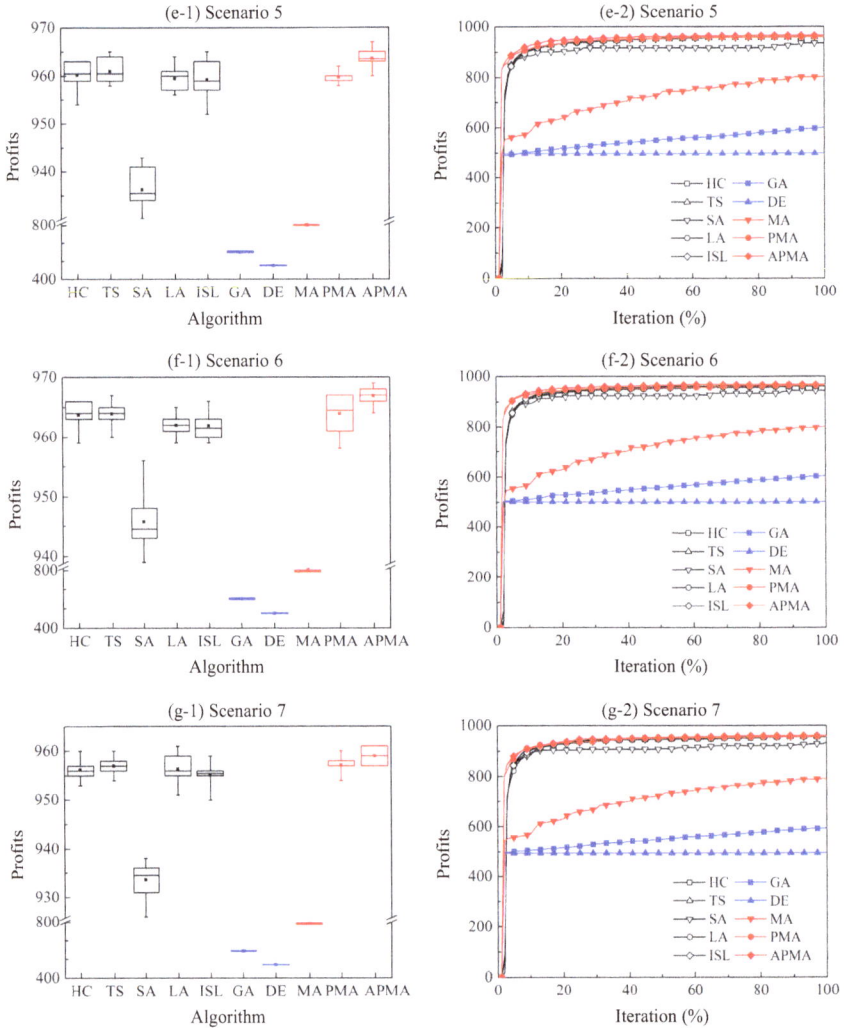

Figure 6.7 (continued)

For the hybrid algorithms MA, PMA, and APMA, similar to the experimental results of remote-sensing satellite task scheduling, the optimization performance and convergence rate of MA are better than those of EAs such as GA and DE, but not as good as those of local search algorithms, which is still attributed to the inefficient constraint optimization efficiency of traditional MA with GA as the main loop. On the contrary, even though the local search algorithms may show very different optimization performances in solving different satellite task scheduling problems, APMA proposed in this book can make full use of the computing power of the computers, integrate the performance advantages of different algorithms, and achieve the best comprehensive convergence optimization performance and good robustness.

To sum up, the experiments of "Tianlian-1" relay satellite conventional task scheduling in this section practice the general-purpose modeling method for satellite task scheduling in this book, showing its feasibility and generality. The algorithm performance of APMA is fully examined, which shows that APMA can make full use of computing power of computers and the performance advantages of various algorithms to realize outperformance and robustness in the process of solving the conventional task scheduling problems of "Tianlian-1" relay satellites.

6.3.3 Emergency scheduling experiments

Based on the medians of the conventional scheduling experiments in Section 6.3.2 and the satellite emergency task scheduling algorithm DDRO, this section further carries out the "Tianlian-1" relay satellite emergency task scheduling experiments.

After calculation, the results of the emergency task scheduling experiments of "Tianlian-1" relay satellites are shown in Figure 6.8. Similar to the conventional task scheduling experiments of remote-sensing satellites, different numbers of emergency tasks, including mandatory and nonmandatory tasks, are randomly input every 1 h. At the same time, under the influence of constraints, to schedule the mandatory emergency tasks, some executed tasks are forced to be canceled, and the task profits decrease accordingly.

As seen in Figure 6.8, under the DDRO optimization framework, the profits of satellite task scheduling fluctuate to varying degrees in each scenario; different from the experimental results of conventional task scheduling of remote-sensing satellites, the fluctuation range of task scheduling profits of relay satellites is small, and there is an obvious increase in the profits in scenarios 1, 2, and 6. The reasons for this phenomenon are: the complexity of relay satellite task scheduling constraints is relatively low, and relatively few tasks are forced to be canceled due to the execution of emergency tasks. At the same time, with the passage of time, under the effect of DDRO rolling optimization for many consecutive times, the optimization degree of the current satellite task scheduling problems gradually improves. Considering that the relay satellites have executed a large number of emergency tasks during this period, which meets the special requirements of relay satellites in terms of emergency response, and at the same time, the satellite task scheduling profit maintains the original level and increases to a certain extent, reflecting the effectiveness of DDRO in solving emergency task scheduling problems of "Tianlian-1" relay satellites.

Based on the above experimental results of conventional and emergency task scheduling of "Tianlian-1" relay satellites, the comparison of profits and running time between the conventional and emergency task scheduling is listed in Table 6.11. It can be seen that, compared with conventional task scheduling results (based on APMA), DDRO can reduce the response time by 86.8% on average, which meets the requirement

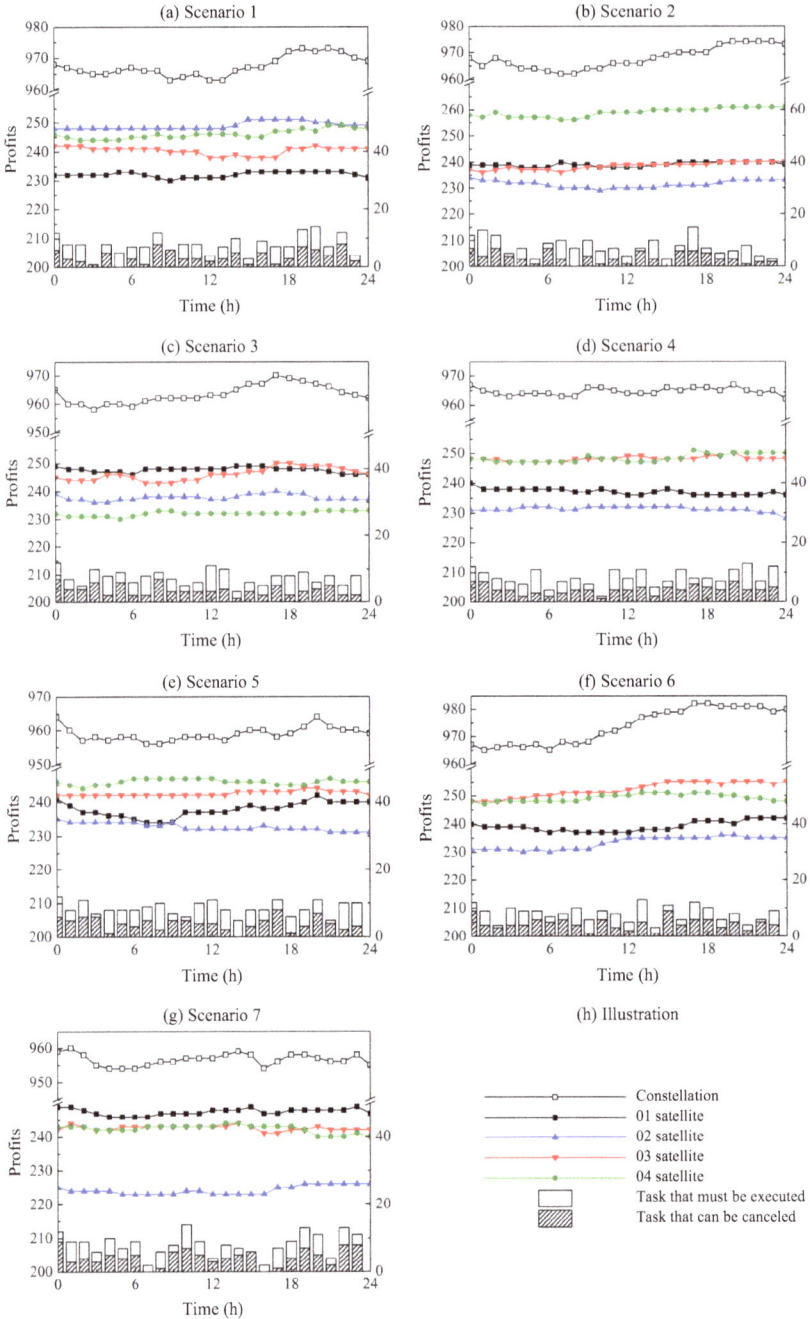

Figure 6.8: Experimental results of emergency task scheduling of "Tianlian-1" relay satellites: (a) scenario 1; (b) scenario 2; (c) scenario 3; (d) scenario 4; (e) scenario 5; (f) scenario 6; (h) scenario 7; and (i) illustrations.

for fast response. The average reduction of the profits is 0.1%, which maintains the profit level, and the optimization of emergency task scheduling is obvious. As a result, the emergency task scheduling experiments in this section test the generality, flexibility and timeliness of DDRO, showing that it can quickly schedule and execute emergency tasks in a long-term dynamic environment, maintain the profit of relay satellite task scheduling, achieve the design purpose, and meet the dynamic and real-time emergency task scheduling needs in the long-term process of relay satellite control.

Table 6.11: Comparison of profits and running time between conventional and emergency task scheduling of "Tianlian-1" relay satellites.

Scenario	Conventional task scheduling		Emergency task scheduling experiments		Result comparison	
	Profit	Running time (s)	Profit	Running time (s)	Profit (%)	Running time (%)
1	968.7 (1.3)	154.2	967.4 (3.1)	22.6	−0.1	−85.4
2	968.6 (2.0)	178.6	967.9 (4.0)	24.2	−0.1	−86.4
3	964.0 (1.8)	185.1	963.4 (3.3)	24.3	−0.1	−86.9
4	965.9 (2.1)	180.4	964.7 (1.3)	22.9	−0.1	−87.3
5	963.6 (2.1)	192.7	958.9 (2.1)	24.3	−0.5	−87.4
6	966.9 (1.7)	173.5	973.7 (6.5)	23.1	0.7	−86.7
7	959.0 (1.8)	188.2	956.6 (1.7)	23.3	−0.3	−87.6
Average	–	–	–	–	−0.1	−86.80

In summary, this section practices the general-purpose modeling method for satellite task scheduling in this book through the conventional and emergency task scheduling experiments of "Tianlian-1" relay satellites, examines the general-purpose solving algorithms APMA and DDRO, and provides practical basis for the application of this book's satellite task scheduling engine in relay satellite task scheduling problems.

6.4 Navigation satellite task scheduling experiments

On June 23, 2020, the last satellite of the third-generation "BDS" satellite navigation system ("BDS-3") was successfully launched, marking the full completion of the deployment of the global satellite navigation system constellation domestically and a new era of domestic navigation system. This section takes 30 navigation satellites in "BDS-3" system as experimental objects, carries out conventional and emergency task scheduling experiments of the navigation satellites for 7 consecutive days to test the satellite task scheduling engine of this book, and to realize the control goals of reducing navigation system delay and guaranteeing the accuracy of the system.

6.4.1 Experimental scenarios

6.4.1.1 Experimental scenarios for conventional scheduling

"BDS-3" is the third-generation global satellite navigation system independently built domestically. It consists of 24 medium Earth orbit (MEO) satellites, 3 geosynchronous orbit (GEO) satellites, and 3 inclined geosynchronous orbit (IGEO) satellites, totaling 30 satellites. Their orbital parameters are listed in Table 6.12 [99, 100], and the schematic diagram of the orbits is shown in Figure 6.9 (There is a certain deviation from the ideal orbits under the influence of perturbation.) Since the main control station of "BDS-3" is Beijing Station, the navigation satellites in the system that are visible to Beijing are "anchor satellites," and the rest of the satellites are "non-anchor satellites."

In this section, a 7-day task scheduling simulation data of "BDS-3" is selected as the experimental data set for the conventional scheduling experiments, and the total scheduling period is 7 days. Among them, the duration of each scheduling scenario (superframe) is 1 min, with a total of 10,080 scenarios (superframes); each superframe is further divided into 20 timeslots with a duration of 3 s. Meanwhile, since this test set contains many scenarios, to quantitatively test the performance of the algorithms in this book, this section selects seven scenarios (i.e., the first scenarios every day) to conduct algorithm performance comparison experiments, and the data are summarized in Table 6.13.

Table 6.12: Summary of orbital parameters of "BDS-3" navigation satellites.

Satellite type	No.	Orbital parameter
MEO	01, 02, . . ., 24	24/3/1 Walker constellation, altitude approximately 21,528 km, inclination 55°
GEO	25, 26, 27	Altitude approximately 35,786 km, longitude 80°, 110.5°, and 140°
IGSO	28, 29, 30	Altitude approximately 35,786 km, phase angle 120°, and inclination 55°

6.4.1.2 Experimental scenarios for emergency scheduling

Due to the short scheduling cycle of BDS-3 navigation satellite task scheduling itself, the duration of each scheduling scenario (superframe) is only 1 min, and each scenario is naturally connected in the time domain, which has the natural conditions for the implementation of rolling scheduling. Therefore, in this section, the real-time window length in the emergency scheduling experiments is set to be the same as 1 min, that is, the lengths in the emergency and conventional scheduling experiments are consistent. Different from the emergency task scheduling experiments of remote-sensing satellites and relay satellites, the number of tasks in the navigation satellite task scheduling problems is a fixed value, that is, the number of timeslots, as shown in Table 6.13. The objective function is the average delay of the navigation system, which is an index for overall

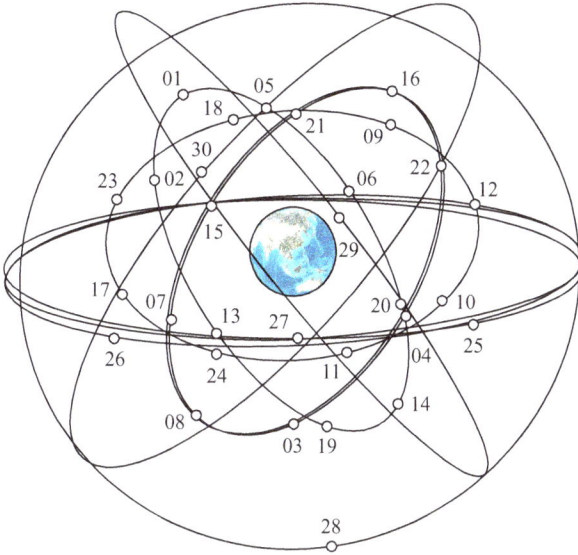

Figure 6.9: Schematic diagram of orbits of "BDS-3" navigation satellites.

Table 6.13: Summary of task scheduling data set of "BDS-3" navigation satellites.

Scenario	Satellite number	Scheduling cycle (min)	Total number of tasks (i.e., number of timeslots)	Number of downlink events (i.e., number of satellites)	Average number of executable opportunities (i.e., number of visible satellites)
1	01, 02, . . ., 30	1	20	30	17.8
2	01, 02, . . ., 30	1	20	30	17.7
3	01, 02, . . ., 30	1	20	30	17.4
4	01, 02, . . ., 30	1	20	30	17.3
5	01, 02, . . ., 30	1	20	30	17.4
6	01, 02, . . ., 30	1	20	30	17.4
7	01, 02, . . ., 30	1	20	30	17.5

evaluation. Therefore, the problems of task allocation and insertion are not involved in the navigation satellite emergency task scheduling experiments.

Based on the conventional task scheduling experimental scenarios of "BDS-3," to simulate the failure of some inter-satellite links of "BDS-3" under unexpected influences such as signal interference and satellite failure, this section designs the following emergency scheduling experimental scenarios: In the scenarios (superframes) of "BDS-3" for 7 consecutive days, given the visible inter-satellite links and conventional scheduling results, 15–80 random activated inter-satellite links of non-anchor satellites temporarily fail for some reasons and cannot downlink data in the current scenario. In this case, the control agency of "BDS-3" immediately calls DDRO to address the problem, triggering task rescheduling. Meanwhile, to avoid drastic changes to the original scheme, PMA is used to address the task rescheduling problems, and the number of iterations is $|T| \times 250$, that is, half of that of the conventional scheduling experiments. The rest of the parameters are the same as those of the conventional scheduling experiments.

6.4.2 Conventional scheduling experiments

6.4.2.1 Experimental results for 7 consecutive days

Based on the general-purpose modeling method for satellite task scheduling and the satellite conventional task scheduling algorithm APMA in this book, this section carries out the conventional task scheduling experiments of "BDS-3" navigation satellites.

After calculation, the results of the conventional task scheduling experiments of "BDS-3" navigation satellites for 7 consecutive days are shown in Table 6.14 and Figure 6.10. Here, this book has stipulated that satellite task scheduling is a maximization problem, so the objective function of navigation satellite task scheduling is the negative value of the average system delay (the unit is timeslot, denoted as ts). The tables and figures in this section show the positive values of the average system delay (ts). The smaller the value is, the better the optimization performance.

From Table 6.14 and Figure 6.10(a) and (b), it can be seen that in the conventional scheduling experiments for 7 consecutive days, the minimum delay of (non-anchor) navigation satellites in each scenario is 1 ts (i.e., 3 s), accounting for 84.6%, while the maximum value is 4 ts (i.e., 12 s), accounting for only 0.1%. The average delay of each scenario for the 7 days fluctuates in the range of 1.045–1.605 ts, with an average value of 1.164 ts (i.e., 3.492 s). It is worth noting that the average delay in each scenario is the objective function of the experiments in this section. Since there are always a certain number of non-anchor satellites in BDS-3, the minimum delay is 1 ts. While the average value of the optimization results of the experiments in this section is close to this minimum value, and the delay of most non-anchor satellites has been reduced to 1–2 ts, which indicates that APMA in this book is effective in solving navigation satellite task scheduling problems, and achieves the purpose of reducing the average delay of the navigation system and improving the system accuracy.

On the other hand, the minimum value of the number of links built by the navigation satellites in each scenario is 7, and the maximum value is 20, both accounting for 0.1%. The most common numbers of links built by the satellites is 12–13, accounting for 20.5% and 21.1%, respectively. The average number of links built in each scenario within the 7 days fluctuates within the range of 11.3–13.8, with an average value of 12.5, meaning that each navigation satellite establishes inter-satellite links with 41.7% of other satellites in the system in average, achieving the goal of enhancing the diversity of the links and guaranteeing the accuracy of inter-satellite ranging, and meeting the basic requirements

Table 6.14: Experimental results of conventional task scheduling of "BDS-3" navigation satellites.

Index	Delay (ts)			Number of links built		
	Maximum delay	Average delay	Minimum delay	Maximum number of links built	Average number of links built	Minimum number of links built
Maximum	4	1.605	1	20	13.8	11
Average	3.018	**1.164**	1	16.1	12.5	8.8
Minimum	2	1.045	1	14	11.3	7

Note: 1) The average delay in the table is the objective function of each scenario, and the average value of this index (bold) is the 7-day average delay of the system; and 2) the average number of links built in the table is the secondary objective function of each scenario.

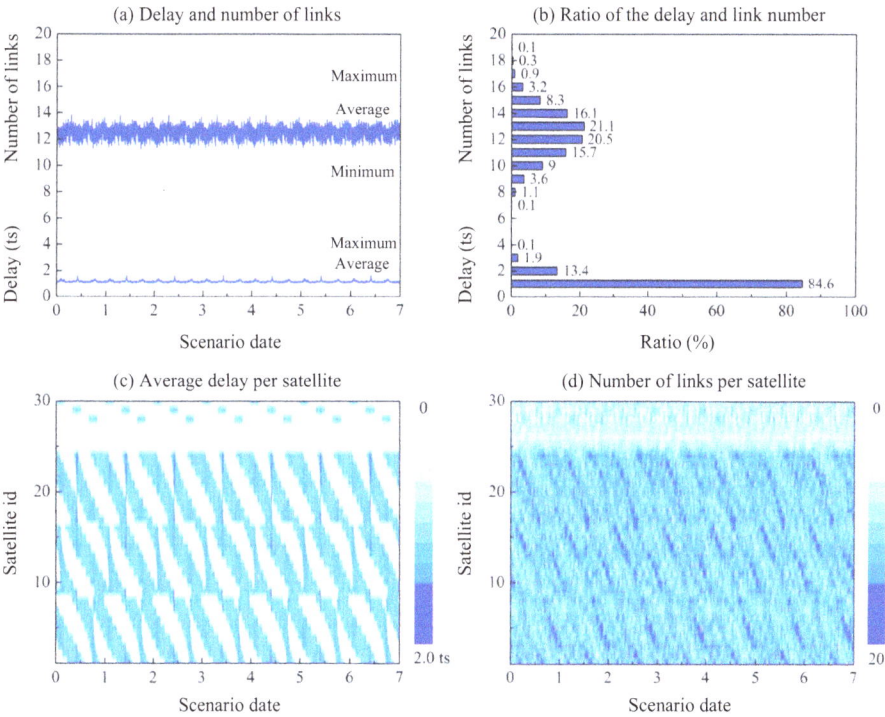

Figure 6.10: Experimental results of conventional task scheduling problems of "BDS-3" navigation satellites: (a) delay and number of links built; (b) delay versus number of links built; (c) average delay of single satellite; and (d) number of links built by single satellite.

of navigation satellite control. The average delay and the number of links built by each navigation satellite in the experimental results of each scenario is shown in Figure 6.10(c) and (d). The results in the figures show a certain periodicity pattern, which is in line with the actual situation of orbit periodicity of "BDS-3" navigation satellites.

6.4.2.2 Experimental results of algorithm comparison

Furthermore, the comparison results of the algorithms for the conventional task scheduling experiments in seven scenarios of the "BDS-3" navigation satellites are shown in Table 6.15 and Figure 6.11. The table lists the average values and standard deviations of the experimental results of 12 algorithms in 4 categories, including heuristic algorithms, local search algorithms, EAs and hybrid algorithms. The figure presents the box plots of the experimental results of each algorithm and the iteration curves of the median values during the experimental process. Similarly, Table 6.15 and Figure 6.11 show the positive values of the average system delay (ts), with smaller values indicating better optimization.

From the statistical results in Table 6.15 and Figure 6.11, it can be seen that, similar to the task scheduling experiments of remote-sensing satellites and relay satellites: the heuristic algorithms have a general optimization performance. The local search algorithms have good optimization performance. However, the EAs have poor optimization performance, in which the optimization performance of GA is close to that of ISL, and the iteration curves of DE always stays at the level of the initial solution, with no obvious optimization sign. Among the seven examples, APMA proposed in this book obtains all 7 best average values and 1 lowest standard deviation, indicating that APMA has the best optimization performance. The other algorithms don't obtain any best average value, showing that the optimization performance of APMA in this book has obvious advantages.

For the heuristic algorithms FIFS and Random, although FIFS can address the task scheduling problems of remote-sensing satellites and relay satellites to a certain extent, it is extremely ineffective in solving task scheduling problems of navigation satellites in this section, and the gap with the rest of the algorithms is extremely large, which is attributed to the fact that FIFS blindly activates the inter-satellite links according to the sequences of tasks and satellite numbers, resulting in a single topology of the inter-satellite links and a sharp increase in system delay. On the contrary, Random randomly activates the inter-satellite links, which leads to more diverse inter-satellite link topologies, and the system delay can be reduced to a certain extent. It can be seen that compared with the regular algorithms commonly used in many complex industrial optimization problems, the algorithm with randomness mechanism is more suitable for solving navigation satellite task scheduling problems.

For the local search algorithms HC, TS, SA, LA, and ISL, LA has the most outstanding optimization performance and good robustness, and its performance is very close to that of APMA in this book; HC and TS have the second best optimization performance, but the early convergence rate is on the low side, and the their performances are more or less the same; SA, which performs well in the remote-sensing satellite task scheduling experiments, has poor optimization performance in this section, but its early convergence rate is the highest; ISL has the worst optimization performance and convergence rate, and its performance is close to that of GA. This phenomenon also shows that the algorithms may exhibit different optimization performances when solving different satellite task scheduling problems. Here, the reason for the poor optimization performance and convergence rate of ISL may be: due to the particularity of the objective function of navigation satellite task scheduling, the disturbance and repair mechanism of ISL produces a large number of solutions that violate the constraints, and the benefit value decreases significantly, further reducing its optimization performance.

For the hybrid algorithms MA, PMA and APMA, the optimization performance of MA is better than that of EAs such as GA and DE, but it is not as good as that of local search algorithms such as LA, HC, and TS, which is still due to the insufficient constrained optimization efficiency of the traditional MA with GA as the main loop. As for APMA proposed in this book, even though the local search algorithms may show very different optimization performances and convergence rates when solving different satellite task scheduling problems, it can make full use of computing power of the computers, integrate the performance advantages of local search algorithms, as well as heuristic and evolutionary algorithms and other types of algorithms, and achieve the best comprehensive convergence optimization performance and good robustness.

To sum up, the experiments of "BDS-3" navigation satellite conventional task scheduling in this section practice the general-purpose modeling method for satellite task scheduling in this book, showing its feasibility and generality. The algorithm performance of APMA is fully examined, which shows that APMA can make full use of the computing power of the computers and the performance advantages of various algorithms to realize outperformance in the process of solving the conventional task scheduling problems of "BDS-3" navigation satellites.

6.4.3 Emergency scheduling experiment

Based on the results of conventional scheduling experiments for 7 consecutive days in Section 6.4.2 and the satellite emergency task scheduling algorithm DDRO, this section further carries out the "BDS-3" navigation satellite emergency task scheduling experiment.

After calculation, the experimental results of emergency task scheduling of "BDS-3" navigation satellites for 7 consecutive days are shown in Figure 6.12, and the comparison results with conventional task scheduling are shown in Table 6.16. From Figure 6.12(a), it can be seen that in each scenario (superframe), some activated inter-satellite links in the navigation system fail, and data cannot be downlinked in the current scenario, resulting in a sharp increase in the average delay of the navigation sys-

Table 6.15: Comparison results of algorithms for conventional task scheduling experiments of "BDS-3" navigation satellites.

Scenario	Heuristic algorithm			Local search algorithms				EAs		Hybrid algorithm		
	FIFS	Random	HC	TS	SA	LA	ISL	GA	DE	MA	PMA	APMA
1	6.070	1.738 (0.073)	1.361 (0.051)	1.336 (0.037)	1.404 (0.040)	1.242 (0.040)	1.499 (0.031)	1.494 (0.028)	1.702 (0.041)	1.441 (0.020)	1.238 (0.046)	**1.223** (0.030)
2	6.432	1.574 (0.055)	**1.253** (**0.015**)	1.254 (0.032)	1.305 (0.028)	1.165 (0.036)	1.383 (0.056)	1.393 (0.024)	1.564 (0.038)	1.308 (0.025)	1.163 (0.046)	**1.143** (0.021)
3	6.432	1.563 (0.118)	1.242 (0.023)	1.244 (0.023)	1.317 (0.034)	1.173 (0.030)	1.365 (0.032)	1.385 (0.029)	1.575 (0.027)	1.323 (**0.014**)	1.199 (0.040)	**1.147** (0.027)
4	6.432	1.584 (0.080)	**1.245** (**0.018**)	1.246 (0.024)	1.293 (0.029)	1.176 (0.032)	1.359 (0.030)	1.377 (0.036)	1.581 (0.027)	1.317 (0.019)	1.165 (0.031)	**1.142** (0.022)
5	6.432	1.591 (0.077)	1.255 (0.033)	1.249 (**0.023**)	1.299 (0.027)	1.168 (0.035)	1.384 (0.032)	1.394 (0.033)	1.586 (0.028)	1.316 (0.028)	1.190 (0.042)	**1.146** (**0.023**)
6	6.432	1.596 (0.063)	1.240 (0.023)	1.249 (0.030)	1.323 (0.023)	1.170 (0.025)	1.342 (0.022)	1.384 (0.022)	1.569 (0.022)	1.313 (**0.015**)	1.195 (0.049)	**1.195** (0.030)
7	6.432	1.565 (0.056)	1.253 (0.024)	1.243 (0.029)	1.317 (0.029)	1.153 (**0.010**)	1.385 (0.030)	1.383 (0.018)	1.557 (0.034)	1.316 (0.031)	1.174 (0.032)	**1.140** (0.015)
Statistics	0/7	0 (0)/7	0 (2)/7	0 (1)/7	0 (0)/7	0 (1)/7	0 (0)/7	0 (0)/7	0 (0)/7	0 (3)/7	0 (0)/7	7 (1)/7
Ranking	12	11	5	4	6	2	8	9	10	7	3	1

Note: 1) The table shows the average system delay, that is, the negative profit. The lower the value, the better the performance; 2) data format in the table: Bold average values (standard deviations) are the best results; and 3) algorithm ranking principle: the more instances of the best average values and the higher the average values, the more prioritized the ranking.

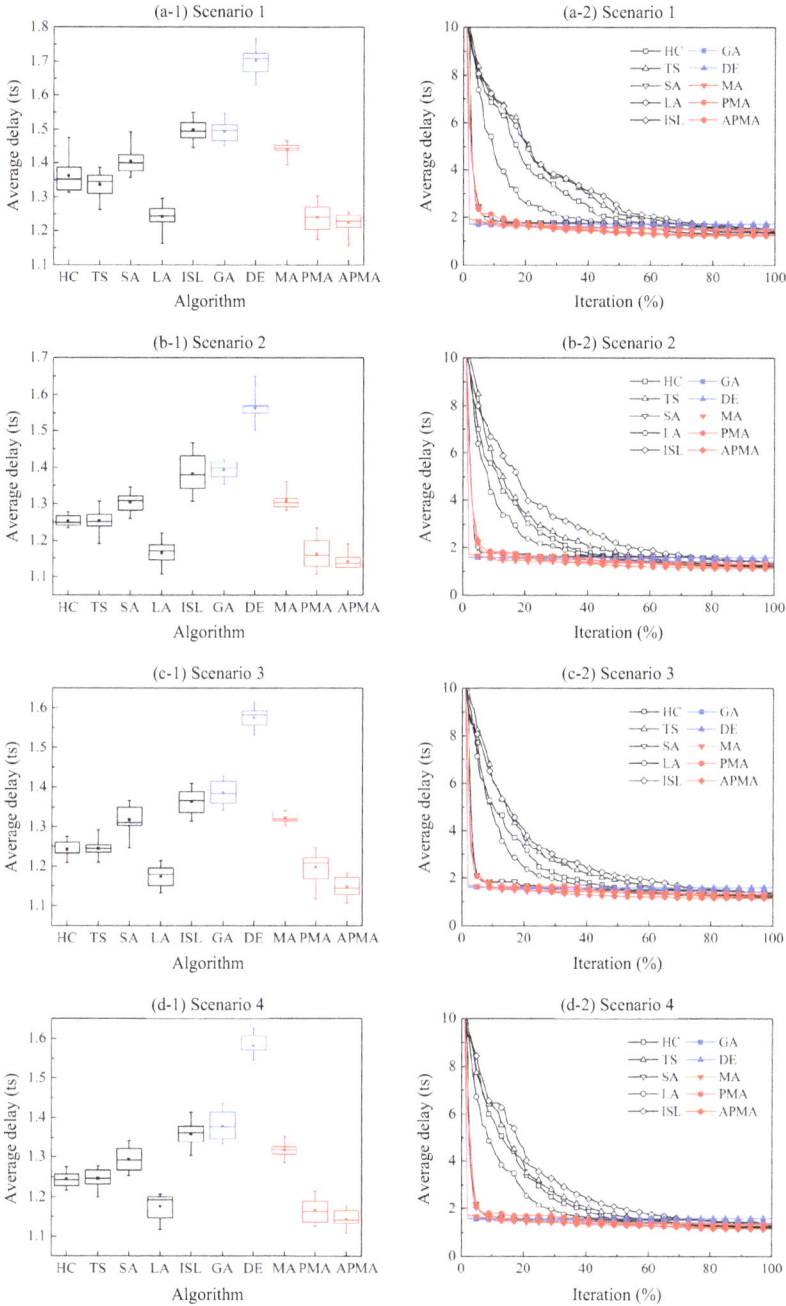

Figure 6.11: Comparison results of algorithms for conventional task scheduling experiments of "BDS-3" navigation satellites: (a) scenario 1; (b) scenario 2; (c) scenario 3; (d) scenario 4; (e) scenario 5; (f) scenario 6; and (g) scenario 7.

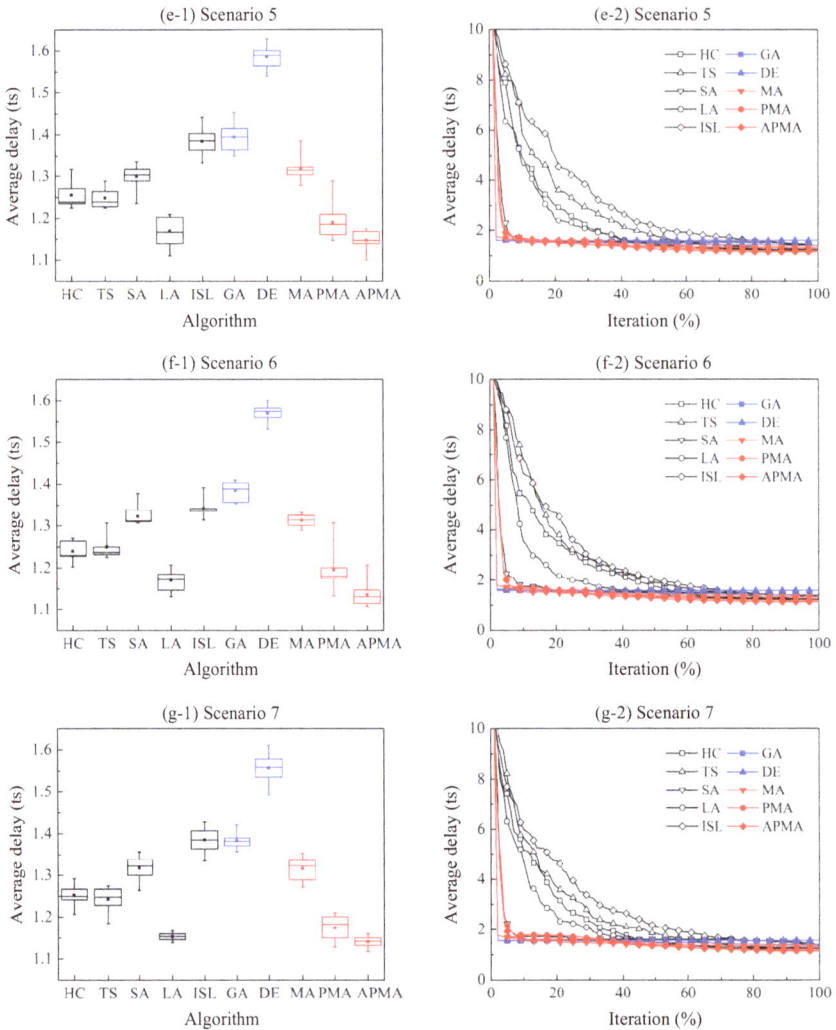

Figure 6.11 (continued)

tem. In this case, DDRO in this book carries out dynamic optimization to minimize the delay increment caused by the link failure within the limited time to ensure the accuracy and robustness of the navigation system. After calculation, the average delay of "BDS-3" for 7 consecutive days in the emergency scheduling experiments is 1.347 ts, which is 15.7% higher than the conventional scheduling result of 1.164 ts, and the response time of each super frame (scenario) is reduced by 45.7%. A certain periodicity is shown in Figure 6.12(b), which is consistent with the results of the conventional task scheduling in Figure 6.10(c). In this experiment, despite the adverse effects of frequent visible inter-satellite link failures in the navigation system, the average delay of the system is still maintained at a normal and reasonable level, which meets the special needs of emergency response of the navigation satellite system and reflects the effectiveness of DDRO in solving "BDS-3" navigation satellite emergency task scheduling problems.

Table 6.16: Comparison of profits and running time between conventional and emergency task scheduling of "BDS-3" navigation satellites.

Conventional task scheduling		Emergency task scheduling experiments		Result comparison	
Average delay	Running time	Average delay	Running time	Average delay	Running time
1.164 (0.069)	31.7 s	1.347 (0.102)	17.2 s	15.7%	−45.7%

Above all, the "BDS-3" navigation satellite emergency task scheduling experiment in this section tests the rationality and effectiveness of DDRO, showing that it can cope with the unexpected events of link failure in a long-term dynamic environment, basically maintain the average delay of the navigation system, guarantee the accuracy and robustness of the system, and achieve the design purpose, and meet the dynamic and real-time emergency task scheduling needs in the long-term process of navigation satellite control.

In summary, this section practices the general-purpose modeling method for satellite task scheduling in this book through the conventional and emergency task scheduling experiments of "BDS-3" navigation satellites, examines the solving algorithms APMA and DDRO, and provides a practical basis for the application of this book's satellite task scheduling engine in navigation satellite task scheduling problems.

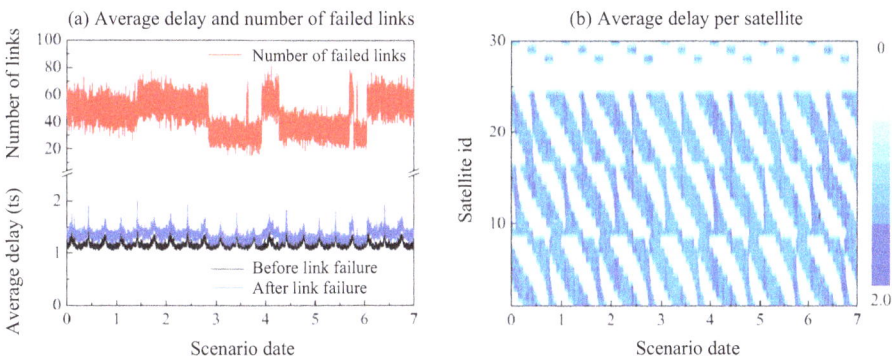

Figure 6.12: Experimental results of emergency task scheduling of "BDS-3" navigation satellites: (a) average delay and number of failed links per scenario and (b) average delay per single satellite.

6.5 Satellite range scheduling experiments

Based on the public data set of satellite range scheduling of the U.S. Air Force, this section carries out conventional and emergency satellite range scheduling experiments for 7 consecutive days to test the satellite task scheduling engine in this book and achieve the control goals of timely tracking, telemetry and command of the target satellites.

6.5.1 Experimental scenarios

6.5.1.1 Experimental scenarios for conventional scheduling

This section selects the Benchmark dataset [150] for satellite measurement and control network (AFSCN) task scheduling released by the U.S. Air Force in 2003, which is the most commonly used dataset in satellite range scheduling studies. The data set contains 7 (7-day) satellite range scheduling scenarios, each scenario contains 19 ground stations, and the task scales of each scenario are listed in Table 6.17. For this data set, Luo et al. [133] proposed a conflict-resolution heuristic algorithm to implement efficient and accurate task scheduling by pre-calculating the possible conflicts between the TT&C tasks, achieving outperformance and refreshing several known optimal solutions in the data set.

Table 6.17: Summary of dataset for AFSCN range scheduling.

Scenario	Ground station number	Scheduling cycle (h)	Total number of tasks	Average number of downlink event executable opportunities	Maximum number of scheduled tasks of the literature
1	01, 02, . . ., 19	24	322	121.6	316
2	01, 02, . . ., 19	24	302	137.3	299
3	01, 02, . . ., 19	24	310	131.2	308
4	01, 02, . . ., 19	24	318	131.5	316
5	01, 02, . . ., 19	24	305	145.5	303
6	01, 02, . . ., 19	24	299	137.5	294
7	01, 02, . . ., 19	24	297	147.1	293

6.5.1.2 Experimental scenarios for emergency task scheduling

Based on the experimental scenarios for AFSCN conventional scheduling, this section designs the following emergency scheduling scenarios: the satellite TT&C agency randomly receives 0–60 emergency tasks every 1 h, and the probability that the emergency tasks have to be executed is 50%.

After receiving the emergency tasks, the satellite TT&C agency immediately calls DDRO to address the scheduling problems, triggering task allocation and single-platform (ground station) window rolling and rescheduling. In this process, the real-time window length is set to 5 h, that is, only the tasks within 5 h from the current moment are involved in rescheduling. To avoid drastic changes to the original scheme, PMA is used to address the rescheduling problems of nonmandatory tasks, and the number of iterations is $|T| \times 250$, that is, half of that of the conventional scheduling experiments. The rest of the parameters are the same as those of the conventional scheduling experiments.

6.5.2 Conventional scheduling experiments

Based on the general-purpose modeling method for satellite task scheduling and the satellite conventional task scheduling algorithm APMA in this book, this section carries out the conventional task scheduling experiments of AFSCN satellites.

After calculation, the comparison results of the algorithms for conventional task scheduling experiments of AFSCN satellites are shown in Table 6.18 and Figure 6.13. Table 6.18 lists the average values and standard deviations of the experimental results of 12 algorithms in 4 categories, including heuristic algorithms, local search algorithms, EAs and hybrid algorithms. Figure 6.13 presents the box plots of the experimental results of each algorithm and the iteration curves of the median values during the experimental process. Among the seven examples, APMA proposed in this book obtains all seven best average values and five lowest standard deviations, and the best optimization results in each scenario can reach the best known values given in literature [133], showing the best optimization performance and good robustness.

From the statistical results in Table 6.18 and Figure 6.13, it can be seen that, similar to the task scheduling experiments of other satellites, the heuristic algorithm FIFS has a general optimization performance. Local search algorithms generally have good optimization performance. However, the optimization performance of EAs such as GA and DE is generally inferior to that of local search algorithms, even inferior to heuristic algorithms such as Random. The iteration curves of DE always stay at the level of initial solution in all scenarios, with no obvious optimization sign.

For the local search algorithms HC, TS, SA, LA, and ISL, TS has outstanding optimization performance. In seven examples, it obtains five best average values and seven lowest standard deviations, showing good optimization results and best robustness. Its performance is very close to that of APMA in this book. Meanwhile, other local search algorithms such as HC, LA and ISL show good optimization performance as well. However, it is worth noting that the SA has low early convergence rate and poor optimization performance, and the average value of its optimization results is even inferior to that of Random. The early convergence rate of LA is also low, which are due to the following reasons: the complexity of constraints of AFSCN satellite range scheduling is relatively low, the scale of the scheduling problems is relatively small, and the difficulty of optimization is reduced, so the local search algorithms can converge faster, but the meta-heuristic mechanism of probabilistic acceptance of inferior solutions in SA and backtracking in LA affects their convergence rates. This phenomenon again shows that the algorithms may exhibit different optimization performances when solving different satellite task scheduling problems.

For the hybrid algorithms MA, PMA, and APMA, the optimization performance of MA is better than that of EAs such as GA and DE, but it is not as good as that of local search algorithms such as TS and HC; its convergence rate is better than that of SA and is close to that of LA, which is still due to the insufficient constrained optimization efficiency of the traditional MA with GA as the main loop. APMA proposed in this book can make full use of computing power of the computers, integrate the performance advantages of various algorithms such as heuristic algorithms, local search algorithms, and EAs, and its best optimization results in each scenario can reach the best known values given in the literature [133], showing the best optimization performance and good robustness.

To sum up, the experiments of AFSCN satellite conventional range scheduling in this section practice the general-purpose modeling method for satellite task scheduling in this book, showing its feasibility and generality. The algorithm performance of APMA is fully examined, which shows that APMA can make full use of computing power of the computers and the performance advantages of various algorithms to realize outperformance and robustness in the process of solving the conventional range scheduling problems of AFSCN satellites.

6.5.3 Emergency scheduling experiment

Based on the medians of conventional scheduling experiments in Section 6.5.2 and the algorithm DDRO for satellite emergency task scheduling, this section further carries out AFSCN satellite emergency range scheduling experiment.

After calculation, the experimental results of AFSCN satellite emergency range scheduling are shown in Figure 6.14. Here, for the sake of clarity and conciseness of the pictures in Figure 6.14, the 19 ground stations are divided into 4 groups, including group 01–05, group 06–10, group 11–15 and group 16–19. Similar to the emergency task scheduling experiments of remote-sensing satellites and relay satellites, different numbers of emergency tasks, including mandatory and nonmandatory tasks, are randomly input every 1 h. Under the influence of constraints, some executed tasks are forced to be canceled, and the task profits decrease accordingly.

As shown in Figure 6.14, under the dynamic optimization framework of DDRO, the profits of satellite task scheduling fluctuate to varying degrees in each scenario. Different from the experimental results of emergency task scheduling of relay satellites, though the fluctuation range of profits of AFSCN satellite range scheduling is also small, there is a significant decrease in the profits in scenarios 3, 4, 5, and 6, which are due to the following reasons: despite the complexity of the constraints of

Table 6.18: Comparison results of algorithms for conventional range scheduling experiments of AFSCN satellites.

Scenario	Heuristic algorithm			Local search algorithms				EAs			Hybrid algorithm	
	FIFS	Random	HC	TS	SA	LA	ISL	GA	DE	MA	PMA	APMA
1	269	308.8 (1.4)	313.4 (0.7)	315.0 (0.0)	305.2 (2.4)	311.3 (0.7)	314.6 (0.7)	292.7 (2.6)	264.8 (2.0)	307.7 (1.2)	315.0 (0.0)	**315.3 (0.5)**
2	262	292.1 (2.3)	295.9 (0.7)	298.4 (0.5)	290.5 (3.0)	295.6 (0.5)	297.2 (0.8)	280.4 (1.4)	255.8 (2.4)	293.4 (1.3)	298.4 (0.5)	**298.6 (0.5)**
3	264	300.9 (2.0)	306.4 (0.5)	**308.0 (0.0)**	299.1 (2.3)	306.2 (0.4)	307.3 (0.7)	288.1 (2.0)	260.5 (2.1)	300.6 (1.0)	**308.0 (0.0)**	**308.0 (0.0)**
4	265	309.0 (2.3)	313.0 (1.2)	315.1 (0.1)	304.4 (2.0)	312.8 (0.9)	314.6 (0.8)	291.9 (1.5)	265.1 (2.6)	309.1 (0.9)	315.3 (0.5)	**315.4 (0.5)**
5	255	297.0 (0.7)	300.3 (0.5)	**303.0 (0.0)**	294.7 (1.5)	302.2 (0.4)	302.0 (0.7)	286.1 (1.6)	260.9 (2.3)	298.3 (1.2)	**303.0 (0.0)**	**303.0 (0.0)**
6	239	288.4 (1.8)	291.3 (1.2)	**294.0 (0.0)**	285.7 (1.8)	292.5 (0.7)	293.2 (0.6)	277.3 (1.3)	252.4 (1.8)	288.9 (1.1)	**294.0 (0.0)**	**294.0 (0.0)**
7	257	291.2 (0.9)	292.7 (0.5)	**293.0 (0.0)**	288.3 (2.2)	292.6 (0.5)	292.7 (0.5)	280.4 (2.3)	254.1 (2.0)	291.3 (0.7)	**293.0 (0.0)**	**293.0 (0.0)**
Statistics	0/7	0 (0)/7	0 (0)/7	4 (7)/7	0 (0)/7	0 (1)/7	0 (0)/7	0 (0)/7	0 (0)/7	0 (0)/7	4 (6)/7	7 (5)/7
Ranking	11	8	5	3	9	6	4	10	12	7	2	1

Note: 1) Data format in the table: bold average values (standard deviations) are the best results; and 2) algorithm ranking principle: the more times of the best average values and the higher the average values, the more prioritized the ranking.

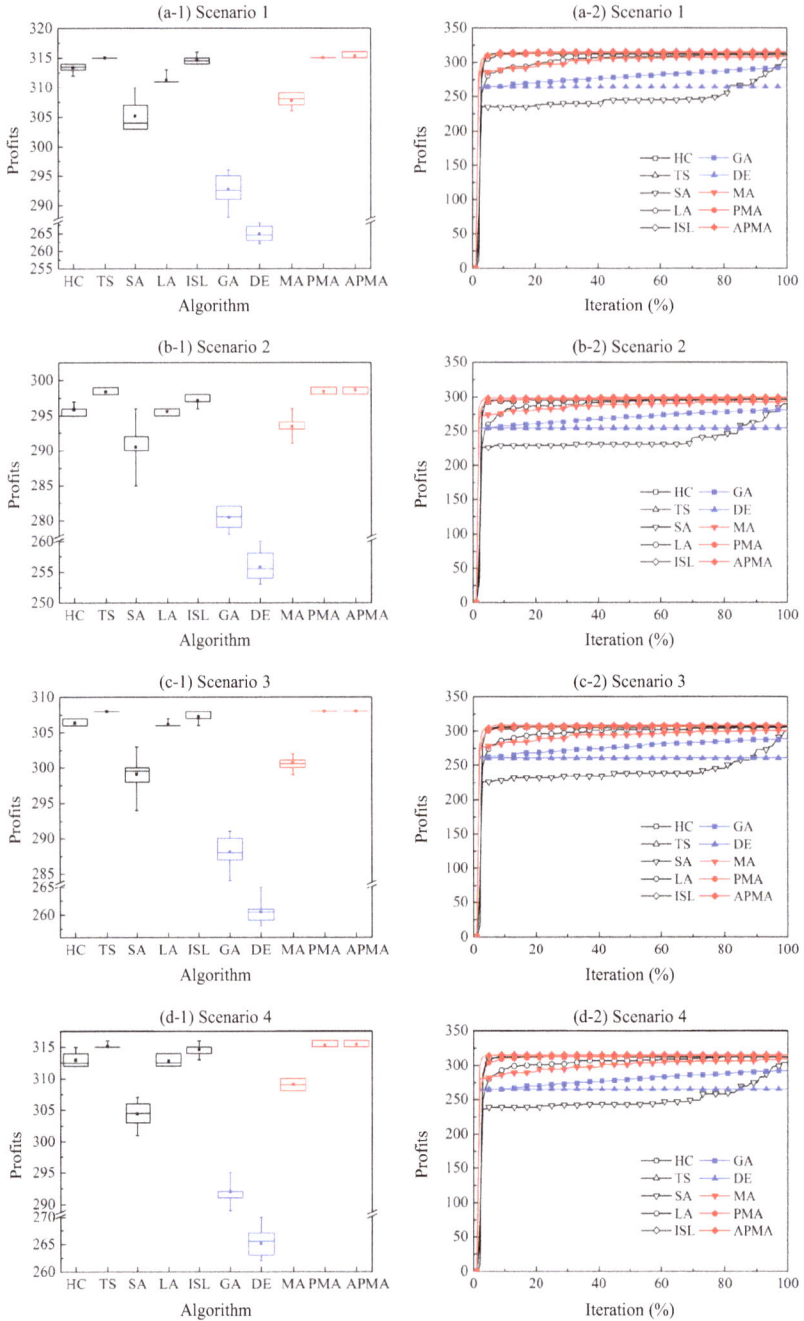

Figure 6.13: Comparison results of algorithms for conventional range scheduling experiments of AFSCN satellites: (a) scenario 1; (b) scenario 2; (c) scenario 3; (d) scenario 4; (e) scenario 5; (f) scenario 6; and (g) scenario 7.

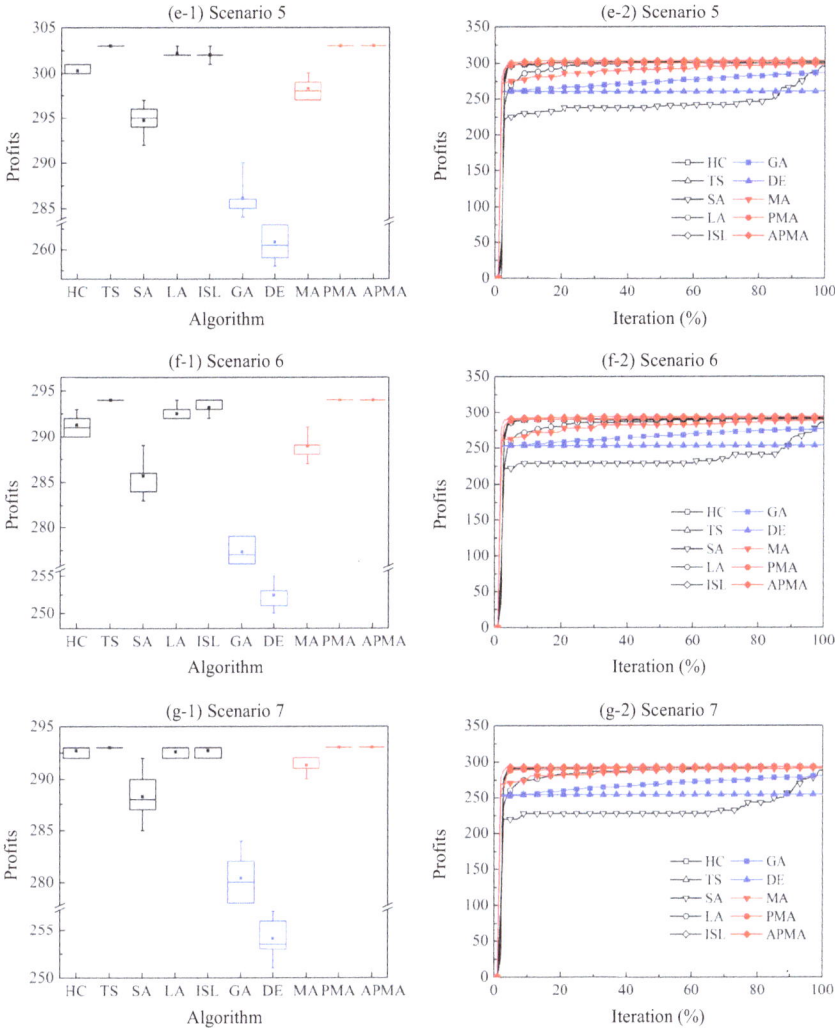

Figure 6.13 (continued)

the problems is relatively low, the task scheduling has reached the saturation state at the beginning of rolling scheduling (which has reached the best level in the literature [133]), and the number of tasks forced to be canceled for executing emergency tasks is relatively high. In this experiment, considering the fact that each ground station still performs a large number of emergency tasks in the long-term process when the task scheduling is close to saturation, it meets the special requirements of satellite range scheduling in terms of emergency response, which reflects the effectiveness of DDRO in solving the AFSCN satellite emergency range scheduling problems.

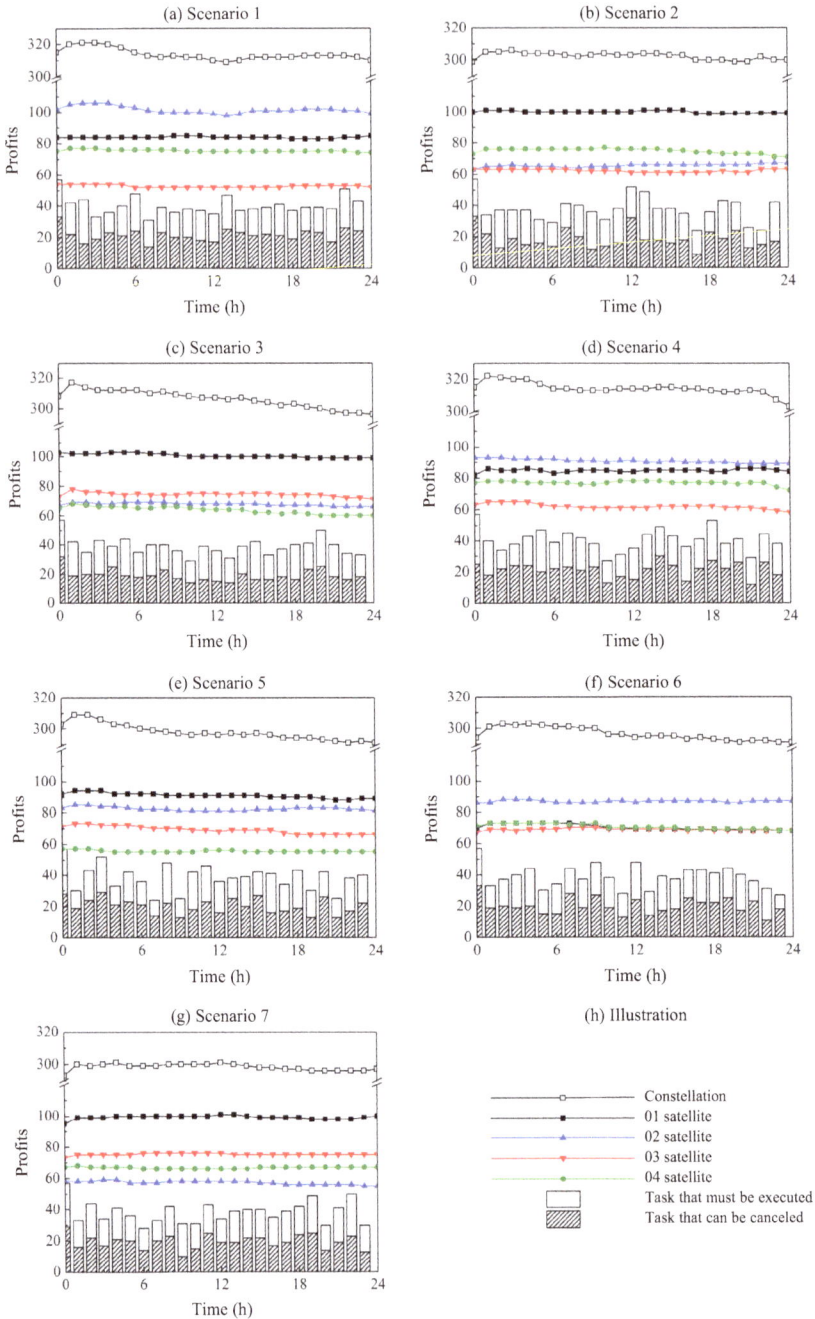

Figure 6.14: Experimental results of local optimization of AFSCN satellite range scheduling problems: (a) scenario 1; (b) scenario 2; (c) scenario 3; (d) scenario 4; (e) scenario 5; (f) scenario 6; (g) scenario 7; and (h) illustrations.

Based on the above experimental results of conventional and emergency range scheduling of AFSCN satellites, the comparison of profits and running time between the conventional and emergency task scheduling is listed in Table 6.19. It can be seen that, compared with conventional task scheduling results (based on APMA), DDRO can reduce the response time by 89.7% on average, which meets the requirement for fast response; the average increase in profits reaches 0.1%, which maintains the profit level, and the optimization of emergency task scheduling is obvious. As a result, the emergency task scheduling experiments in this section test the generality, flexibility and timeliness of DDRO, showing that it can quickly schedule and execute emergency tasks in a long-term dynamic environment, maintain the profit of satellite range scheduling, achieve the design purpose, and meet the dynamic and real-time emergency task scheduling needs in the long-term process of satellite TT&C.

In summary, this section practices the general-purpose modeling method for satellite task scheduling in this book through the conventional and emergency range scheduling experiments of AFSCN satellites, examines the general-purpose solving algorithms APMA and DDRO, and provides a practical basis for the application of this book's satellite task scheduling engine in satellite range scheduling problems.

Table 6.19: Comparison of profits and running time between conventional and emergency range scheduling of AFSCN satellites.

Scenario	Conventional task scheduling		Emergency task scheduling experiments		Result comparison	
	Profit	Running time	Profit	Running time	Profit	Running time
1	315.3 (0.5)	76.2 s	313.7 (3.5)	8.3 s	−0.5%	−89.1%
2	298.6 (0.5)	72.5 s	302.4 (2.1)	7.5 s	1.3%	−89.6%
3	308.0 (0.0)	80.4 s	306.2 (5.8)	7.9 s	−0.6%	−90.2%
4	315.4 (0.5)	78.6 s	314.2 (4.0)	8.2 s	−0.4%	−89.5%
5	303.0 (0.0)	73.5 s	297.7 (5.2)	7.2 s	−1.7%	−90.2%
6	294.0 (0.0)	75.2 s	296.3 (4.2)	7.9 s	0.8%	−89.5%
7	293.0 (0.0)	76.8 s	298.3 (2.0)	7.8 s	1.7%	−89.9%
Average	–	–	–	–	0.1%	−89.7%

Note: 1) The statistics of conventional task scheduling scenarios are the results of 10 independent operations; and 2) the statistics of emergency task scheduling scenarios are the results of 24 consecutive emergency scheduling.

6.6 Applications of satellite task scheduling engine

Besides the practical applications of the satellite task scheduling engine in the above task scheduling experiments such as remote-sensing satellite scheduling, relay satellite scheduling, navigation satellite scheduling, and satellite range scheduling, the engine in this book further supports the development of a number of satellite task scheduling services or simulation systems, as shown in Figure 6.15, which have shown good performance. Next, this section gives a brief introduction to the four main application systems supported by the satellite task scheduling engine:

(1) **The second-generation task scheduling system of "SuperView-1" commercial remote-sensing satellites.** The satellite task scheduling engine in this book has contributed to the development of the second-generation task scheduling system of "SuperView-1" commercial remote-sensing satellites. The system can significantly enhance the joint control profits of the four current "SuperView-1" remote-sensing satellites, and meet the daily conventional scheduling requirements of the control agency for imaging, downlink and memory-erasing events, as well as the emergency scheduling requirements under special dynamic environments such as adding or deleting tasks, satellite failures, manual interventions, and constraint changes. Moreover, the system has reserved task scheduling model interfaces for more than 20 satellites planned to be launched in the future for the "SuperView-1" constellation, which can

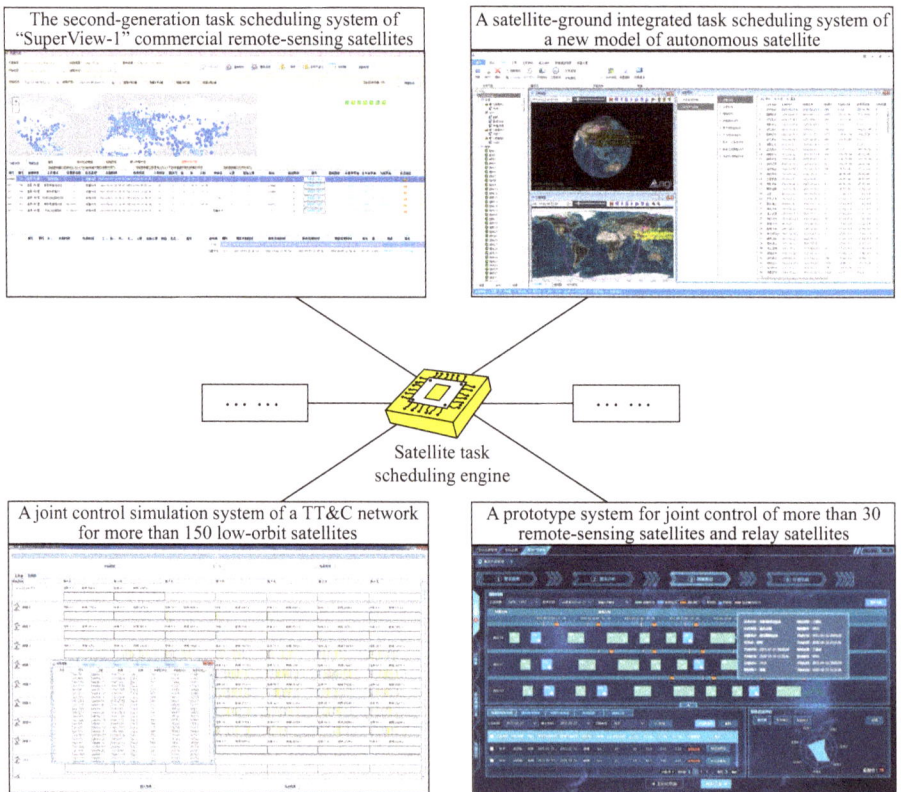

Figure 6.15: Examples of application systems supported by the satellite task scheduling engine in this book.

support system updates, iterations and upgrades in the long-term operation process, and provide a long-term solution for the joint control of "SuperView-1."

(2) A satellite-ground integrated task scheduling system of a new model of autonomous satellite. The satellite task scheduling engine in this book not only has supported the development of the satellite's ground-based control subsystem to meet the scheduling requirements of conventional and emergency tasks (including imaging, downlink, and memory-erasing events) implemented by the control agency but also has supported the development of the onboard autonomous task scheduling subsystem of the satellite to meet the emergency task scheduling requirements spontaneously implemented by the satellite. The system practices technologies of the satellite task scheduling engine in this book, demonstrates the feasibility of autonomous satellite control, and provides technical support and practical basis for the autonomous and intelligent development of the satellite system.

(3) A joint control simulation system of a TT&C network for more than 150 low-orbit satellites. The satellite task scheduling engine in this book has supported the development of a joint control simulation system for more than 150 low-orbit satellites (and dozens of sets of TT&C equipment). The system can significantly improve the TT&C efficiency of low-orbit satellites, the completion rate of conventional tasks is close to 100% and the response time is less than 5 min. The emergency response time is less than 1 min in case of task additions and deletions, equipment failure, etc. Moreover, the system reserves task scheduling model interfaces for medium- and high-orbit satellites, which can support the joint control of low-, medium-, and high-orbit all-space satellite TT&C networks. Because the complexity of satellite range scheduling is relatively low, the system also opens up the interface for constraint and algorithm customization, which can provide users with autonomous and flexible ways of constraint management and algorithm design.

(4) A prototype system for joint control of more than 30 remote-sensing satellites and relay satellites. Against the background of the demand for joint control of cross agencies and cross satellite models, the satellite task scheduling engine in this book has supported the development of a prototype system for joint control of more than 10 types of remote-sensing satellites and relay satellites, with a total of more than 30 satellites. The system can well meet the basic daily scheduling requirements of remote-sensing satellites and relay satellites, and basically meet their emergency scheduling requirements in dynamic environments such as adding or deleting tasks, satellite failures, manual interventions, and constraint changes. For the first time, the system incorporates satellites of different agencies, models, and types into a joint control system, indicating the feasibility of joint control of cross agencies and cross satellite models, providing important technical support and practical basis for breaking the control barrier of "one satellite, one system," enhancing the flexibility and robustness of the satellite system, and transforming the way of control and application of satellites in the future.

To sum up, with strong application value and promotion prospects, the satellite task scheduling engine in this book can support the development of a number of satellite task scheduling services or simulation systems and provide practical and efficient task scheduling tools for satellite control agency.

6.7 Chapter summary

Based on the general-purpose modeling method for satellite task scheduling in this book and the general-purpose solving methods APMA and DDRO for conventional and emergency task scheduling, this chapter takes four types of actual satellite task scheduling problems such as "SuperView-1" commercial remote-sensing satellite scheduling, "TianLian-1" relay satellite scheduling, "BDS-3" navigation satellite scheduling, and AFSCN satellite range scheduling as examples to carry out conventional and emergency task scheduling experiments, respectively. The main contents and conclusions are as follows:

(1) The general-purpose modeling method for satellite task scheduling is practiced. Based on the method, this chapter successfully completes the modeling and optimization experiments of "SuperView-1" remote-sensing satellite task scheduling, "Tianlian-1" relay satellite task scheduling, "BDS-3" navigation satellite task scheduling and AFSCN satellite range scheduling, and verifies the feasibility and generality of the method.

(2) The APMA for satellite conventional task scheduling is comprehensively tested. Compared with traditional MA, APMA in this book uses heuristic algorithms to construct initial solutions, local search algorithms as the main loop, and EAs as the improvement strategy, which significantly improves the convergence optimization performance and robustness. Compared with other algorithms, although the performance of the algorithms in different satellite task scheduling problems and scenarios varies, APMA in this book can always can make full use of computing power of the computers and the performance advantages of local search algorithms and achieve the best convergence optimization performance and good robustness, providing a general-purpose and efficient means of solving various kinds of satellite conventional task scheduling problems.

(3) The distributed dynamic rolling optimization algorithm DDRO for satellite emergency task scheduling is experimented. The algorithm can respond quickly in the long-term dynamic environment, maintain the profit level of satellite task scheduling, achieve the design purpose, and provide a general-purpose and flexible means of solving various kinds of satellite emergency task scheduling problems.

(4) The feasibility and application prospect of the satellite task scheduling engine are illustrated. Based on the satellite task scheduling engine, this chapter successfully completes the experiments to illustrate that the engine can fully meet the actual needs of satellite task scheduling and introduces the development of related systems to demonstrate the application value and promotion prospects of the satellite task scheduling engine in this book.

Chapter 7
Summary and outlook

7.1 Summary

Since the space industry is in a period of rapid development with an explosive increase in satellites, satellite task scheduling is particularly important. In response to the satellite control status quo of "one satellite, one system," the contradiction between the needs of conventional control and emergency response, some countries' technological blockade of satellites, and facing the urgent needs for the development of a satellite task scheduling engine and its application, this book designs the top-level framework of a satellite task scheduling engine, proposes the general-purpose modeling method for satellite task scheduling, puts forward the adaptive parallel memetic algorithm (APMA) for satellite conventional task scheduling, explores the distributed dynamic rolling optimization (DDRO) algorithm for satellite emergency task scheduling, and experiments with their applications. The following conclusions are drawn based on the experiments:

(1) **The satellite task scheduling engine framework can meet the actual needs of modeling and optimizing satellite task scheduling problems, providing a general-purpose and modular modeling and solving process for satellite task scheduling problems.** For remote-sensing satellite scheduling, relay satellite scheduling, navigation satellite scheduling, and satellite range scheduling, this book decouples the satellite task scheduling model and algorithms and establishes the satellite task scheduling engine framework decoupling "model-conventional algorithms-emergency algorithms." Practical results show that the framework can provide a general-purpose and modular modeling and solving process for various kinds of satellite task scheduling problems, meet the complex and practical needs of satellite control at this stage, and set up a rational and feasible top-level framework for the satellite task scheduling engine.

(2) **The general-purpose modeling method of satellite task scheduling can be applied to remote-sensing satellite scheduling, relay satellite scheduling, navigation satellite scheduling, and satellite range scheduling, providing a general-purpose model for the satellite task scheduling engine.** To address the problems of insufficient generality of satellite task scheduling models and the satellite control status quo of "one satellite, one system," this book proposes a general-purpose modeling method for satellite task scheduling that decouples "decision-constraints-profits" decoupling and constructs a general-purpose 0–1 mixed integer decision model, a constraint model, and a profit model in sequence. The practical results show that the method proposed successfully integrates remote-sensing satellite scheduling, relay satellite scheduling, navigation satellite scheduling, and satellite range scheduling into a unified modeling system, opening up a new approach for mathematical and

https://doi.org/10.1515/9783111537191-007

general-purpose modeling of satellite task scheduling problems and providing an important general-purpose model for the satellite task scheduling engine in this book.

(3) The APMA can address various kinds of conventional satellite task scheduling problems generally and efficiently, providing a core algorithm for the satellite task scheduling engine. For the current daily and weekly conventional task scheduling requirements of the satellite control agency, this book proposes the APMA and successively designs a heuristic-based fast initial solution construction strategy, a parallel search-based general-purpose local optimization strategy, a competition-based algorithm and operator adaptive selection strategy, and a population evolution-based global optimization strategy, which overcome the deficiencies of the traditional memetic algorithm. The experimental results show that APMA can make full use of the computing power of computers and the collaboration of the multiple strategies, achieving the best convergence optimization performance and good robustness in all experiments, providing a general-purpose and efficient means of solving various kinds of satellite conventional task scheduling problems, and a core algorithm for the satellite task scheduling engine.

(4) The DDRO algorithm can address various kinds of satellite emergency task scheduling problems generally and flexibly, providing another algorithm for the satellite task scheduling engine. To meet the emergency task scheduling needs of satellite control agency under dynamic influences such as task additions and deletions, and satellite failures, this book designs a DDRO algorithm and also successively designs a task negotiation and allocation strategy based on a dynamic contract net, a single-platform task rescheduling strategy based on a rolling time domain, a task rapid insertion strategy based on schedulable prediction, and a real-time conflict-avoidance strategy based on a constraint network. The experimental results show that DDRO can respond quickly in the long-term dynamic environment and maintain the profit level of satellite task scheduling, providing a flexible and effective means of solving various kinds of satellite emergency task scheduling problems, as well as another algorithm for the satellite task scheduling engine.

To sum up, the framework of the satellite task scheduling engine proposed in this book is rational and feasible, the general-purpose modeling method is applicable and effective, and the conventional and emergency task algorithms are efficient and flexible. The study in this book satisfies the practical and urgent needs of the satellite control agency and has important engineering application prospects and transformation value.

7.2 Outlook for future study

Due to the limited time, there are still many inadequacies in this book, and many contents are open to further studies. Based on the authors' knowledge, this section briefly states the study contents that are not covered in the study concept of the satellite task scheduling engine in this book and puts forward the following ideas and prospects for future study:

(1) **The task scheduling problems of experimental satellites, early warning satellites, astronomical satellites, in-orbit service satellites, space stations, deep space probes, and other spacecraft can be incorporated into the study system of this book so that the satellite task scheduling engine can be developed into a more comprehensive, systematic, and general space-level task scheduling solver.** In this book, the satellite task scheduling engine is mainly for solving four typical satellite task scheduling problems, including remote-sensing satellite scheduling, relay satellite scheduling, navigation satellite scheduling, and satellite range scheduling. With the rapid development of the space industry, more and more various kinds of spacecraft are launched, such as experimental satellites, astronomical satellites, in-orbit service satellites, space stations, and deep space probes, and the task scheduling problems of such spacecrafts have gradually become prominent. Among the task scheduling problems of such spacecrafts, since the task scheduling problems of experimental satellites and astronomical satellites have relatively fixed orbits, they are similar to those of remote-sensing satellites; since the task scheduling problems of in-orbit service satellites and deep space probes involve decisions on the timing of orbit change as well as specialized knowledge of orbital transfer, the complexity is relatively high; and since the task scheduling problems of space stations involve the scheduling of the daily plan of spacecraft as well as the scheduling of astronaut life resources such as electricity and oxygen, the problems are highly coupled with the problems related to in-orbit service satellites and rocket launches, and the complexity is relatively high.

In this regard, future studies need to further improve the general-purpose modeling method for satellite task scheduling in this book, address the modeling problems of the task scheduling problems of the abovementioned spacecrafts in a targeted manner, form a set of more comprehensive, systematic, and general-purpose satellite and spacecraft task scheduling modeling system, and develop the satellite task scheduling engine in this book into a space-grade general-purpose solver for task scheduling. This will break through the software technology blockade of some countries in all aspects, provide core software technical support for the research and development of domestic general-purpose space task scheduling software, and comprehensively promote the integrated control of the space system.

(2) **The APMA can be extended into a new multi-objective algorithm to meet the optimization requirements of multi-spacecraft task scheduling and to provide a multi-objective decision basis for the control agency.** The satellite task scheduling problems in this book are all single-objective optimization problems, aiming to meet the specialized optimization needs of the control agency and give the required task scheduling scheme. However, in the real process of spacecraft control, multi-objective optimization is often required. For example, in the remote-sensing satellite task scheduling problems, there exist the dual optimization requirements of the quantity and quality of the imaging tasks; in the navigation satellite task scheduling

problems, there exist the dual optimization requirements of the average delay of the system and the average number of links built; in the relay satellite task scheduling and satellite range scheduling problems, there exist the dual optimization requirements of the total task profit and the payload balance of the satellites and the ground stations. The general-purpose modeling method for satellite task scheduling in this book can model the above optimization requirements, but it is still necessary to have a special multi-objective optimization algorithm for solving the problems.

In view of this, the "parallel" and "evolutionary" strategies in this book's APMA both produce a large number of diverse solution sets, which satisfy the basic needs of the multi-objective evolutionary algorithm (MOEA) to perform the relevant operations. Compared with the traditional MOEA with EA as the main loop, APMA also has the advantages of fast search and constrained optimization with a local search algorithm as the main loop and shows better optimization performance in solving complex spacecraft task scheduling problems. As a result, APMA has the feasibility and performance advantages to be extended into an MOEA. In future studies, the authors will try to follow this direction to address the multi-objective spacecraft task scheduling problems with a new set of MOEA and combine the elitism strategy, competition strategy, and co-evolutionary strategy, and so on, to output the set of non-dominated multi-objective Pareto scheme sets, so as to provide the control agency with a diversified set of candidate schemes.

(3) **The DDRO algorithm in this book can be extended into a new robust optimization algorithm to guarantee the fast response of the space system and enhance the robustness and invulnerability of the system at the same time.** In this book, in the process of solving satellite emergency task scheduling problems, the objective function of the model is consistent with conventional scheduling problems. However, considering the complexity and uncertainty of the space environment and the subjective demand of the users, the randomness and uncertainty of the task scheduling problems in the process of real spacecraft control are much stronger than that of the simulation experiments designed in this book. In this case, robustness becomes another important index to measure the task scheduling scheme. Although the DDRO algorithm in this book can effectively address the task scheduling problems in emergency scenarios, it does not consider the robustness of the emergency scheduling scheme and cannot yet guarantee to maintain the profit level of task scheduling under the influence of other uncertainties, which needs to be further improved. In this regard, in future studies, the authors will try to design a new robust optimization algorithm on the basis of DDRO to improve the robustness of the solutions while guaranteeing the fast solution of the spacecraft emergency task scheduling problems, thus providing an important algorithmic basis for improving the robustness of the space system. In addition, military struggles may occur in space in the future, so when designing relevant algorithms and strategies, it is also necessary to take into account the effects of spacecraft de-orbiting, destruction, and other extreme dynamic situations, to minimize the adverse effects of military strikes on the space system, and

to safeguard the basic response and counter-strike capability of the system after the strike.

(4) More machine learning techniques can be introduced into the model construction and algorithm design process in this book to enhance the intelligence and autonomy level of the satellite task scheduling engine and to form a new feature of "aerospace + AI + combinational optimization." In recent years, "machine learning + combinational optimization" has become a new study hotspot, and more and more scholars have introduced machine learning technology into the study of combinational optimization problems, which has provided many good solutions for classical optimization problems such as traveling salesman problems and vehicle routing problems. In this regard, in future studies, the authors will try to introduce more up-to-date machine learning techniques in the process of model construction and algorithm design for spacecraft task scheduling problems. For example, computer-assisted constraint checking can be introduced in the process of model construction to further reduce the complexity of constraint computation and to improve the overall iterative efficiency of the algorithms. The search direction of local search algorithms can be regulated in the design of conventional scheduling algorithms to reduce the randomness of blind searching and to help the algorithms to obtain high-quality solutions more efficiently and accurately. In the design of emergency scheduling algorithms, the rapid task insertion algorithm based on schedulability can be further improved to enhance the precision and foresight of task insertion and to reduce the possible adverse effects of the task insertion on the subsequent long-term scheduling process. As a result, the intelligence and autonomy of the satellite task scheduling engine in this book are greatly improved, and the deep integration of machine learning and combinational optimization technology in task scheduling problems in the space field is comprehensively promoted, forming a new feature of "aerospace + AI + combinational optimization."

(5) The satellite task scheduling engine can be deployed in the cloud, and a new framework for the engine can be designed based on the cloud platform to explore a new mode for the design, application, and maintenance of remote and online satellite task scheduling. Orbit Logic Company of the United States has opened up the remote access of its universal satellite task scheduling software STK/Scheduler, providing a new service mode of online task scheduling, which makes users' access more convenient, system deployment more efficient, and function iteration and fault repair timelier. Moreover, cloud computing can also further leverage the performance advantages of the APMA in this book and obtain better solution results through more diverse algorithm competition. Therefore, in future studies, the authors intend to move the study results of this book to the cloud, design a new framework of the satellite/spacecraft task scheduling engine based on the cloud platform, and explore a new mode for the design, application, and maintenance of remote and online satellite task scheduling systems.

References

[1]   Union of Concerned Scientists. Satellite Database [DB/OL]. 2020 [2020-3-1]. https://www.ucsusa.
        org/resources/satellite-database, 2020-12-31.
[2]   Wolfe W J, Sorensen S E. Three scheduling algorithms applied to the earth observing systems
        domain [J]. Management Science, 2000, 46(1): 148–168.
[3]   Cordeau J F, Laporte G. Maximizing the value of an Earth observation satellite orbit [J]. Journal of
        the Operational Research Society, 2005, 56(8): 962–968.
[4]   Cordeau J F, Laporte G, Mercier A. A unified Tabu search heuristic for vehicle routing problems with
        time windows [J]. Journal of the Operational Research Society, 2001, 52(8): 928–936.
[5]   Bianchessi N, Cordeau J F, Descrosiers J. et al., A heuristic for the multi-satellite, multi-orbit and
        multi-user management of Earth observation satellites [J]. European Journal of Operational
        Research, 2007, 177(2): 750–762.
[6]   Li J F, Tan Y J. VRP and JSP Models of coordinate scheduling problem for observing satellites [J].
        Systems Engineering, 2006, 24(6): 111–115. In Chinese.
[7]   Guo Y H, Li J, Zhao K. et al., A heuristic method for earth observing satellites united imaging
        scheduling [J]. Journal of Astronautics, 2009, 30(2): 652–658. In Chinese.
[8]   Cai D R. Research-based the "Ant Colony" algorithm multi-satellite mission planning joint imaging
        problems [D]. Chengdu: University of Electronic Science and Technology of China, 2012, In Chinese.
[9]   Gabrel V, Vanderpooten D. Enumeration and interactive selection of efficient paths in a multiple
        criteria graph for scheduling an earth observing satellite [J]. European Journal of Operational
        Research, 2002, 139(3): 533–542.
[10]   Bianchessi N, Righini G. Planning and scheduling algorithms for COSMO-SkyMed constellation [J].
        Aerospace Science and Technology, 2008, 12(7): 535–544.
[11]   Chen H, Li J, Jing N. et al., Scheduling model and algorithms for autonomous electromagnetic
        detection satellites [J]. Acta Aeronautica et Astronautica Sinica, 2010, 31(5): 1045–1053. In Chinese.
[12]   Chen X G, Wu X Y. ACO algorithm of satellite data transmission scheduling based on solution
        construction graph [J]. Systems Engineering and Electronics, 2010, 32(3): 592–597. In Chinese.
[13]   Xu B, Wang D H, Liu W X, et al. A hybrid navigation constellation inter-satellite link assignment
        algorithm for the integrated optimization of the inter-satellite observing and communication
        performance [C]. Proceedings of the China Satellite Navigation Conference. Berlin: Springer-Verlag,
        2015: 283–296.
[14]   Zhang F, Wang J, Li J. et al., Multicriteria optimal imaging scheduling based on time ordered acyclic
        directed graph [J]. Journal of National University of Defense Technology, 2005, 27(6): 61–66. In
        Chinese.
[15]   Wang J. Research on modeling and optimization techniques in united mission scheduling of
        imaging satellites [D]. Changsha: National University of Defense Technology, 2007, In Chinese.
[16]   Hall N G, Magazine M J. Maximizing the value of a space mission [J]. European Journal of
        Operational Research, 1994, 78(2): 224–241.
[17]   Gu Z S, Chen Y W. MIP Model and algorithm for resolving scheduling of earth observation satellites
        [J]. Journal of Spacecraft TT&C Technology, 2007, 26(1): 19–24. In Chinese.
[18]   He R J, Gu Z S. Multi-satellite scheduling model and column generation algorithm for imaging
        satellites [J]. Journal of Spacecraft TT&C Technology, 2008, 27(4): 5–9. In Chinese.
[19]   Xiao Y Y, Zhang S Y, Yang P. et al., A two-stage flow-shop scheme for the multi-satellite observation
        and data-downlink scheduling problem considering weather uncertainties [J]. Reliability
        Engineering and System Safety, 2019, 188: 263–275.
[20]   Lemaître M, Verfaillie G, Jouhaud F. et al., Selecting and scheduling observations of agile satellites
        [J]. Aerospace Science and Technology, 2002, 6(5): 367–381.

https://doi.org/10.1515/9783111537191-008

[21]   Xu R, Chen H P, Liang X L. et al., Priority-based constructive algorithms for scheduling agile earth observation satellites with total priority maximization [J]. Expert Systems with Applications, 2016, 51(1): 195–206.

[22]   Lin W C, Liao D Y, Liu C Y. et al., Daily imaging scheduling of an earth observation satellite [J]. IEEE Transactions on Systems Man and Cybernetics, 2005, 35(2): 213–223.

[23]   Cheng S W, Zhang H, Shen L C. Research on satellite operation planning using mip method [J]. Computer Engineering and Applications, 2011, 47(3): 229–232. In Chinese.

[24]   Ren X H, Yang L P, Zhu Y W. Mission planning for multiple geo satellite proximity inspection [J]. Flight Dynamics, 2011, 29(3): 76–79. In Chinese.

[25]   Sun K, Yang Z Y, Wang P. et al., Mission planning and action planning for agile earth-observing satellite with genetic algorithm [J]. Journal of Harbin Institute of Technology, 2013, 20(5): 51–56.

[26]   Cui K K, Xiang J H, Zhang Y L. Mission planning optimization of video satellite for ground multi-object staring imaging [J]. Advances in Space Research, 2018, 61(6): 1476–1489.

[27]   Bensana E, Lemaitre M, Verfaillie G. Earth observation satellite management [J]. Constraints, 1999, 4(3): 293–299.

[28]   Gabrel V. Strengthened 0–1 linear formulation for the daily satellite mission planning [J]. Journal of Combinatorial Optimization, 2006, 11(3): 341–346.

[29]   Jin X S, Li J, Liu X H. et al., An algorithm for satellite imaging scheduling based on Lagrangian relaxation and max weighted component algorithm [J]. Journal of Astronautics, 2008, 29(2): 310–315. In Chinese.

[30]   Wang P, Reinelt G, Gao P. et al., A model, a heuristic and a decision support system to solve the scheduling problem of an earth observing satellite constellation [J]. Computers & Industrial Engineering, 2011, 61(2): 322–335.

[31]   Liu X L, Jiang W, Li Y J Mutation particle swarm optimization for earth observation satellite mission planning [C]. Proceedings of International Conference on Management Science & Engineering, Dallas, USA: IEEE, 2012: 236–243.

[32]   Jiang J, Choi J, Bae H J. et al., Image collection planning for Korea Multi-Purpose SATellite-2 [J]. European Journal of Operational Research, 2013, 230(1): 190–199.

[33]   Wu G, Liu J, Ma M. et al., A two-phase scheduling method with the consideration of task clustering for earth observing satellites [J]. Computers & Operations Research, 2013, 40(7): 1884–1894.

[34]   Lemaître M, Verfaillie G, Jouhaud F, et al. How to manage the new generation of agile earth observation satellites [C]. Proceedings of the 6th International SpaceOps Conference, Paris, France: ESA: AIAA, 2000: 1–8.

[35]   He Y M, He L, Wang Y. et al., Autonomous mission replanning method for imaging satellites considering real-time weather conditions [J]. Journal of Computational and Theoretical Nanoscience, 2016, 13(10): 6967–6973.

[36]   He Y M, Chen Y W, Lu J M. et al., Scheduling multiple agile earth observation satellites with an edge computing framework and a constructive heuristic algorithm [J]. Journal of Systems Architecture, 2019, 95: 55–66.

[37]   Mok S H, Jo S, Bang H. et al., Heuristic-based mission planning for an agile earth observation satellite [J]. International Journal of Aeronautical and Space Sciences, 2019, doi: 10.1007/s42405-018-0105-4.

[38]   Chu X G, Chen Y N, Tan Y J. An anytime branch and bound algorithm for agile earth observation satellite onboard scheduling [J]. Advances in Space Research, 2017, 60(9): 2077–2090.

[39]   Song B Y, Yao F, Chen Y N. et al., A hybrid genetic algorithm for satellite image downlink scheduling problem [J]. Discrete Dynamics in Nature and Society, 2018, 1531452.

[40]   Zhang C, Li Y B. Planning and scheduling method for multi agile satellite coordinated mission [J]. Science Technology and Engineering, 2017, 17(22): 276–282. In Chinese.

[41] Liu X L, Laporte G, Chen Y W. et al., An adaptive large neighborhood search metaheuristic for agile satellite scheduling with time-dependent transition time [J]. Computers & Operations Research, 2017, 86: 41–53.

[42] Kim H, Chang Y K. Mission scheduling optimization of SAR satellite constellation for minimizing system response time [J]. Aerospace Science and Technology, 2015, 40(1): 17–32.

[43] Wu G H, Wang H L, Pedrycz W. et al., Satellite observation scheduling with a novel adaptive simulated annealing algorithm and a dynamic task clustering strategy [J]. Computers & Industrial Engineering, 2017, 113: 576–588.

[44] She Y C, Li S, Zhao Y B. Onboard mission planning for agile satellite using modified mixed-integer linear programming [J]. Aerospace Science and Technology, 2018, 72: 204–216.

[45] Chen X Y, Reinelt G, Dai G M. et al., A mixed integer linear programming model for multi-satellite scheduling [J]. European Journal of Operational Research, 2019, 275(2): 694–707.

[46] Frank J, Do M, Tran T T Scheduling ocean color observations for a GEO-stationary satellite [C]. Proceedings of the 26th International Conference on International Conference on Automated Planning and Scheduling, California, USA: AAAI, 2016: 376–384.

[47] Sarkheyli A, Bagheri A, Ghorbani-Vaghei B, Askari-Moghadamd R. Using an effective Tabu search in interactive resources scheduling problem for LEO satellites missions [J]. Aerospace Science and Technology, 2013, 29(1): 287–295.

[48] Niu X N, Zhai X J, Tang H, et al. Multi-satellite scheduling approach for dynamic areal tasks triggered by emergent disasters [C]. Proceedings of International Archives of the Photogrammetry Remote-sensing and Spatial Information Sciences, Prague, Czech: ISPRS, 2016: 475–481.

[49] Chen X Y, Reinelt G, Dai G M. et al., Priority-based and conflict-avoidance heuristics for multi-satellite scheduling [J]. Applied Soft Computing, 2018, 69: 177–191.

[50] Valicka C G, Garcia D, Staid A. et al., Mixed-integer programming models for optimal constellation scheduling given cloud cover uncertainty [J]. European Journal of Operational Research, 2019, 275(2): 431–445.

[51] Nag S, Li A S, Merrick J H. Scheduling algorithms for rapid imaging using agile Cubesat constellations [J]. Advances in Space Research, 2018, 61: 891–913.

[52] Zhu W M, Hu X X, Xia W. et al., A two-phase genetic annealing method for integrated Earth observation satellite scheduling problems [J]. Soft Computing, 2019, 23(1): 181–196.

[53] He L, Liu X L, Laporte G. et al., An improved adaptive large neighborhood search algorithm for multiple agile satellites scheduling [J]. Computers & Operations Research, 2018, 100: 12–25.

[54] Li J, Li J, Chen H. et al., A data transmission scheduling algorithm for rapid-response earth-observing operations [J]. Chinese Journal of Aeronautics, 2014, 27(2): 349–364.

[55] Du Y H, Xing L N, Chen Y G. et al., Unified modeling and multi-strategy collaborative optimization for satellite task scheduling [J]. Control and Decision, 2019, 34(9): 1847–1856. In Chinese.

[56] Bensana E, Verfaillie G, Agnese J C, et al. Exact and inexact methods for the daily management of an Earth observation satellite [C]. Proceeding of the 4th SpaceOps Conference, Paris, France: ESA, 1996: 507–514.

[57] Vasquez M, Hao J K. A "Logic-constrained" knapsack formulation and a Tabu search algorithm for the daily photograph scheduling of an Earth observation satellite [J]. Computational Optimization and Applications, 2001, 20(2): 137–157.

[58] Spangelo S, Cutler J, Gilson K. et al., Optimization-based scheduling for the single-satellite, multi-ground station communication problem [J]. Computers & Operations Research, 2015, 57: 1–16.

[59] Gao L, Zhou L A, Sha J C. Task allocation model and algorithm for dss cooperation mechanism [J]. Journal of Systems Engineering, 2009, 24(4): 445–450. In Chinese.

[60] Hao H C, Jiang W, Li Y J. et al., Research on agile satellite dynamic mission planning based on multi-agent [J]. Journal of National University of Defense Technology, 2013, 35(1): 53–59. In Chinese.

[61] Skobelev P O, Simonova E V, Zhilyaev A. et al., Application of multi-agent technology in the scheduling system of swarm of earth remote-sensing satellites [J]. Procedia Computer Science, 2017, 103: 396–402.

[62] Du B, Li S. A new multi-satellite autonomous mission allocation and planning method [J]. Acta Astronautica, 2018, doi: 10.1016/j.actaastro.2018.11.001.

[63] Zheng Z X, Guo J, Gill E. Distributed onboard mission planning for multi-satellite systems [J]. Aerospace Science and Technology, 2019, 89: 111–122.

[64] Xing L N, Wang Y, He Y M. et al., An earth observation satellite task schedulability prediction method based on BP artificial network [J]. Chinese Journal of Management Science, 2015, S1: 117–124. In Chinese.

[65] Wang C, Jing N, Li J. et al., An algorithm of cooperative multiple satellites mission planning based on multi-agent reinforcement learning [J]. Journal of National University of Defense Technology, 2011, 33(1): 53–58. In Chinese.

[66] Wang C, Li J, Jing N. et al., A distributed cooperative dynamic task planning algorithm for multiple satellites based on multi-agent hybrid learning [J]. Chinese Journal of Aeronautics, 2011, 24(4): 493–505.

[67] Wang H J, Yang Z, Zhou W G. et al., Online scheduling of image satellites based on neural networks and deep reinforcement learning [J]. Chinese Journal of Aeronautics, 2019, 32(4): 1011–1019.

[68] Reddy S C, Brown W L. Single processor scheduling with job priorities and arbitrary ready and due times [R]. Beltsville, USA: Computer Sciences Corporation, 1986.

[69] Reddy S C Scheduling of tracking and data relay satellite system (TDRSS) antennas: scheduling with sequence dependent setup times [C]. Proceedings of the ORSA/TIMS Joint National Meeting, Denver, USA: Military Operations Research Society, 1988.

[70] Rojanasoonthon S, Bard J F, Reddy S D. Algorithms for parallel machine scheduling: a case study of the tracking and data relay satellite system [J]. Journal of the Operational Research Society, 2003, 54(8): 806–821.

[71] Rojanasoonthon S, Bard J. A GRASP for parallel machine scheduling with time windows [J]. Informs Journal on Computing, 2005, 17(1): 32–51.

[72] Zhuang S F, Yin Z D, Wu Z L, et al. The relay satellite scheduling based on artificial bee colony algorithm [C]. Proceedings of the 17th International Symposium on Wireless Personal Multimedia Communications, Sydney, Australia: IEEE, 2014: 635–640.

[73] Zhuang S F. Research on resource scheduling technology of tracking and data relay satellite system [D]. Harbin: Harbin Institute of Technology, 2017, In Chinese.

[74] He C, Meng X G, Zhu Z M. et al., Design of mission programming algorithm for tdrs based on execution time slide adjustment strategy [J]. Journal of Spacecraft TT&C Technology, 2015(3): 246–253.

[75] Liu R Z, Sheng M, Tang C Y. et al., Tasking planning based on task splitting and merging in relay satellite network [J]. Journal on Communications, 2017, 38(Z1): 110–117. In Chinese.

[76] Guo C, Xiong W, Hao L Y. Relay satellite system task scheduling algorithm based on double-layer priority [J]. Application Research of Computers, 2018(5): 1506–1510. In Chinese.

[77] Gu Z S. Research on the relay satellite dynamic scheduling problem modeling and optimizational technology [D]. Changsha: National University of Defense Technology, 2007, In Chinese.

[78] Zhao J, Zhao W H, Li Y J. et al., Scheduling algorithm for data relay satellite with microwave and laser hybrid links [J]. Chinese Journal of Lasers, 2013, 40(10): 1005005. In Chinese.

[79] Zhao J, Zhao W H, Li Y J. et al., Multi-objective resources scheduling algorithm for microwave and laser hybrid links data relay satellite based on improved NSGA-II algorithm [J]. Chinese Journal of Lasers, 2013, 40(12): 1205003. In Chinese.

[80] Zhao W H, Zhao J, Zhao S H. et al., Resources scheduling for data relay satellite with microwave and optical hybrid links based on improved niche genetic algorithm [J]. Optik, 2014, 125(3): 3370–3375.

[81] Fang Y S, Chen Y W, Zou K. et al., Research on relay satellite scheduling problem with CSP [J]. Operations Research and Management Science, 2005, 14(4): 74–79. In Chinese.

[82] Fang Y S, Chen Y W, Gu Z S. CSP model of the relay satellite scheduling [J]. Journal of National University of Defense Technology, 2005(2): 6–10. In Chinese.

[83] Fang Y S, Chen Y W, Wang J M. Constraint programming model and algorithms for multiple access links scheduling of tracking and data relay satellite system (TDRSS) [J]. Spacecraft Recovery & Remote-sensing, 2006, 27(4): 62–67. In Chinese.

[84] He C, Li Y J, Qiu Z. Task programming models and algorithms of tracking and data relay satellite in application-on-demand [J]. Chinese Space Science and Technology, 2017, 37(6): 46–55. In Chinese.

[85] He M F, Zhu Y Q, Jia X Q. Scheduling model and heuristic algorithm for tracking and data relay satellite considering multiple slide windows [J]. Journal of Zhengzhou University (Engineering Science), 2018, 39(5): 15–25.

[86] Tang M. The Research on strategies of satellite system data transmission serving. Changsha: National University of Defense Technology, 2013, In Chinese.

[87] Luba O, Boyd L, Gower A. et al., GPS III system operations concepts [J]. IEEE Aerospace and Electronic Systems Magazine, 2005, 20(1): 10–18.

[88] Zhao S. The game between major powers in satellite navigation systems: the obstruction of Russia's construction of GLONASS ground station in the United States [J]. Space International, 2014, 36(10): 36–40. In Chinese.

[89] Gershman R, Buxbaum K L, Ludwinski J M, et al. Galileo mission planning for Low Gain Antenna based operations [C]. Proceedings of the 3rd International Symposium on Space Mission Operations and Ground Data Systems, Washington, USA: NASA, 1994: 279–286.

[90] Iv J C M Performing the Galileo mission using the S-band low-gain antenna [C]. Proceedings of Aerospace Applications Conference, Vail, USA: IEEE, 1994: 145–183.

[91] Toribio S G: Mission planning [C]. Proceedings of the SpaceOps 2004 Conference, Montreal, Canada: AIAA, 2004.

[92] Hall S, Moreira F, Franco T Operations planning for the Galileo constellation [C]. Proceedings of the SpaceOps 2008 Conference, Heidelberg, Germany: AIAA, 2008.

[93] Marinelli F, Nocella S, Rossi F. et al., A Lagrangian heuristic for satellite range scheduling with resource constraints [J]. Computers & Operations Research, 2011, 38(11): 1572–1583.

[94] Long Y J, Chen Y W, Xing L N. et al., Uplink task scheduling model and two-phase heuristic algorithm of navigation satellites [J]. Journal of National University of Defense Technology, 2013, 35(2): 34–39. In Chinese.

[95] Tang Y Y, Wang Y K, Chen J Y, et al. Uplink scheduling of navigation constellation based on genetic algorithm [C]. Proceedings of the 13th International Conference on Signal Processing, Chengdu, China: IEEE, 2017: 1124–1129.

[96] Yan J G, Xing L N, Zhang Z S. et al., Dual time window constrained job-shop scheduling algorithm [J]. Science Technology and Engineering, 2016, 16(26): 85–92. In Chinese.

[97] Zhang Z S, Wang P, He R J. et al., A heuristic algorithm for the scheduling of clock synchronization and uplink tasks between satellites and ground stations based on inter-satellite links [J]. Gnss World of China, 2012(5): 38–45. In Chinese.

[98] Yang D N, Yang J, Xu P J. Timeslot scheduling of inter-satellite links based on a system of a narrow beam with time division [J]. GPS Solutions, 2017, 21(3): 999–1011.

[99] Sun L Y, Wang Y K, Huang W D. et al., Inter-satellite communication and ranging link assignment for navigation satellite systems [J]. GPS Solutions, 2018, 22(2): 38.

[100] Sun L Y, Huang W D, Zhou Y F. et al., Monitor link assignment for reentry users based on BeiDou inter-satellite links [J]. Advances in Space Research, 2019, 64(3): 747–758.

[101] Liu S Y, Yang J, Guo X Y. et al., Inter-satellite link assignment for the laser/radio hybrid network in navigation satellite systems [J]. GPS Solutions, 2020, 24(2): 49.

[102] Parish D A. A genetic algorithm approach to automating satellite range scheduling [D]. Ohio, USA: Air Force Institute of Technology, 1994.

[103] Barbulescu L, Howe A E, Watson J P, et al. Satellite range scheduling: A comparison of genetic, heuristic and local search [C]. Proceedings of the International Conference on Parallel Problem Solving from Nature, Berlin, Germany: Springer-Verlag, 2002: 611–620.

[104] Barbulescu L, Howe A, Whitley D. AFSCN scheduling: How the problem and solution have evolved [J]. Mathematical and Computer Modelling, 2006, 43(9–10): 1023–1037.

[105] Barbulescu L, Howe A E, Whitley L D. et al., Understanding algorithm performance on an oversubscribed scheduling application [J]. Journal of Artificial Intelligence Research, 2006, 27(1): 577–615.

[106] Zheng J J, Zhang N T, Zhang L Y. Dynamical schedule model for rational utilize of TT&C system [J]. Chinese High Technology Letters, 2002, 12(7): 22–27. In Chinese.

[107] Zhang P, Feng X X, Ge X Q. A hardware resource allocation method for multi-antenna ground station based on improved genetic algorithm [J]. Computer Engineering and Science, 2017, 39(6): 1155–1163. In Chinese.

[108] Barbulescu L, Watson J P, Whitley L D. et al., Scheduling space-ground communications for the air force satellite control network [J]. Journal of Scheduling, 2004, 7(1): 7–34.

[109] Li Y Q, Wang R X, Liu Y. et al., Satellite range scheduling with the priority constraint: An improved genetic algorithm using a station ID encoding method [J]. Chinese Journal of Aeronautics, 2015, 28(3): 789–803.

[110] Arbabi M, Garate J A, Kocher D F Interactive real time scheduling and control [C]. Proceedings of the Summer Simulation Conference, San Diego, USA: Society for Computer Simulation, 1985: 271–277.

[111] Zufferey N, Amstutz P, Giaccari P. Graph colouring approaches for a satellite range scheduling problem [J]. Journal of Scheduling, 2008, 11(4): 263–277.

[112] Blöchliger I, Zufferey N. A graph coloring heuristic using partial solutions and a reactive Tabu scheme [J]. Computers & Operations Research, 2008, 35(3): 960–975.

[113] Zhang Y, Dang Q, Huang Y X. A memetic algorithm with predictive selection for multi-satellite TT&C resources scheduling problems [J]. Journal of Xi'an Jiaotong University, 2009, 43(10): 37–41. In Chinese.

[114] Xu X H. Study on Land-based Satellite TT&C scheduling models and algorithms [D]. Beijing: Tsinghua University, 2012, In Chinese.

[115] Zhang N, Feng Z R Cooperative ant colony optimization for multisatellite resource scheduling problem [C]. Proceedings of the 2007 IEEE Congress on Evolutionary Computation, Singapore, Singapore: IEEE, 2007: 2822–2828.

[116] Zhang N, Feng Z R, Ke L J. Guidance-solution based ant colony optimization for satellite control resource scheduling problem [J]. Applied Intelligence, 2011, 35(3): 436–444.

[117] Zhang Z J, Zhang N, Feng Z R. Multi-satellite control resource scheduling based on ant colony optimization [J]. Expert Systems with Applications, 2014, 41(6): 2816–2823.

[118] Zhang Z J, Hu F N, Zhang N. Ant colony algorithm for satellite control resource scheduling problem [J]. Applied Intelligence, 2018, 48(10): 3295–3305.

[119] Chen X G, Wu X Y. Model of satellite data transmission resource workload balance scheduling and ant colony optimization algorithm [J]. Systems Engineering, 2008, 26(12): 91–97. In Chinese.

[120] Chen X G. Research on model and algorithm of ant colony optimization for satellite data transmission scheduling [D]. Changsha: National University of Defense Technology, 2010, In Chinese.

[121] Wang H B, Xu M Q, Wang R X. et al., Solving space and ground TT&C resources integrated scheduling problem with ant colony optimization-simulated annealing algorithm [J]. Journal of Astronautics, 2012, 33(11): 1636–1645. In Chinese.

[122] Vazquez A J, Erwin R S. On the tractability of satellite range scheduling [J]. Optimization Letters, 2015, 9(2): 311–327.

[123] Vazquez A J, Erwin R S Robust fixed interval satellite range scheduling [C]. Proceedings of the 2015 IEEE Aerospace Conference, Big Sky, USA: IEEE, 2014.

[124] Vázquez A J, Erwin R S. An introduction to optimal satellite range scheduling [M]. Berlin, Germany: Springer-Verlag, 2015.

[125] Wang Y Z, Zhao J, Nie C. Study on Petri net model for multi-satellites-ground station system [J]. Journal of Air Force Engineering University, 2003, 4(2): 7–11. In Chinese.

[126] Wang Y Z, Zhao J, Nie C. Study on optimal scheduling for multi-satellites-ground stations system [J]. Computer Simulation, 2003, 20(7): 7–19,54. In Chinese.

[127] Jin G, Wu X Y, Gao W B. Simulation-based study on resource deployment of satellite ground station [J]. Journal of System Simulation, 2004, 16(11): 2401–2403. In Chinese.

[128] Wang Y Z, Gao W B, Nie C. Summary of the resource configuration optimization for a multi-satellite ground station system [J]. Systems Engineering and Electronics, 2004, 26(4): 437–439. In Chinese.

[129] Gooley T D. Automating the Satellite Range Scheduling Process [D]. Ohio, USA: Air Force Institute of Technology, 1993.

[130] Gooley T D, Borsi J J, Moore J T. Automating air force satellite control network (AFSCN) scheduling [J]. Mathematical and Computer Modelling, 1996, 24(2): 91–101.

[131] He R J, Tan Y J. Apply constraint satisfaction to optimal allocation of satellite ground station resource [J]. Computer Engineering and Applications, 2004, 40(18): 229–232. In Chinese.

[132] Liu Y, He R J, Tan Y J. Modeling the scheduling problem of multi-satellites based on the constraint satisfaction [J]. Systems Engineering and Electronics, 2004, 26(8): 1076–1079. In Chinese.

[133] Luo K P, Wang H H, Li Y J. et al., High-performance technique for satellite range scheduling [J]. Computers & Operations Research, 2017, 85: 12–21.

[134] Xhafa F, Sun J, Barolli A. et al., Genetic algorithms for satellite scheduling problems [J]. Mobile Information Systems, 2012, 8(4): 351–377.

[135] Xhafa F, Herrero X, Barolli A. et al., Evaluation of struggle strategy in Genetic Algorithms for ground stations scheduling problem [J]. Journal of Computer and System Sciences, 2013, 79(7): 1086–1100.

[136] Xhafa F, Herrero X, Barolli A, et al. A simulated annealing algorithm for ground station scheduling problem [C]. Proceedings of the 16th International Conference on Network-based Information Systems, Gwangju, South Korea: IEEE, 2013: 24–30.

[137] Xhafa F, Herrero X, Barolli A, et al. A Tabu search algorithm for ground station scheduling problem [C]. Proceedings of the 28th International Conference on Advanced Information Networking & Applications, Victoria, Canada: IEEE, 2014: 1033–1040.

[138] Valicka C G, Garcia D, Staid A, et al. Space surveillance network scheduling under uncertainty: Models and benefits [C]. Proceedings of the Advanced Maui Optical and Space Surveillance Technologies Conference, Hawaii, USA: The Maui Economic Development Board, 2016: 124.

[139] Liu Z B, Feng Z R, Ren Z G. Route-reduction-based dynamic programming for large-scale satellite range scheduling problem [J]. Engineering Optimization, 2019, doi: 10.1080/0305215X.2018.1558445.

[140] Greve G H, Hopkinson K M, Lamont G B. Evolutionary sensor allocation for the space surveillance network [J]. Journal of Defense Modeling and Simulation, 2018, 15(3): 303–322.

[141] Liu C F, Yang J. Research of decision support system for grand-based tracking telemeter and command resource distribution [J]. Aeronautical Computing Technique, 2003, 33(4): 80–83. In Chinese.

[142] Liu C F. The construct and realization for tracking telemeter and command resource scheduling management system [J]. Aeronautical Computing Technique, 2005, 35(3): 68–71. In Chinese.

[143] Ling X D, Liu B, Wu X Y. et al., An ontology-based model of multi-satellite TT&C scheduling problem [J]. Computer and Digital Engineering, 2010, 38(8): 62–66. In Chinese.

[144] Jayaweera S K, Erwin R S, Carty J. Distributed space situational awareness (D-SSA) with a satellite-assisted collaborative space surveillance network [J]. IFAC Proceedings Volumes, 2011, 44(1): 8792–8798.

[145] Shen D, Jia B, Chen G S, et al. Game optimal sensor management strategies for tracking elusive space objects [C]. Proceedings of the 2017 IEEE Aerospace Conference, Big Sky, USA: IEEE, 2017: 1–8.

[146] Ling X D, Wu X Y, Liu Q. Analysis of modeling of TT&C resource scheduling problem based on agent technology [J]. Systems Engineering and Electronics, 2008, 30(11): 2220–2223. In Chinese.

[147] Du H M, Liu M S. The method of satellite TT&C resources dynamic scheduling problem based on the technique of multi-agent collaboration [J]. Journal of The Academy of Equipment Command & Technology, 2010, 21(3): 76–80. In Chinese.

[148] Feng H S, Chen Y, Wu X Y. SVM regression model for satellite ground station resources allocation [J]. Journal of Spacecraft TT&C Technology, 2011, 30(2): 15–19. In Chinese.

[149] Ahn H S, Jung O, Choi S. et al., An optimal satellite antenna profile using reinforcement learning [J]. IEEE Transactions on Systems, Man, and Cybernetics, 2011, 41(3): 393–406.

[150] Air Force Office of Scientific Research. Exploiting elementary landscapes for search (AFSCN scheduling problems) [DB/OL]. 2003 [2024-9-30]. http://www.cs.colostate.edu/sched/data.html.

[151] Beaumet G, Verfaillie G, Charmeau M C. Feasibility of autonomous decision making on board an agile earth-observing satellite [J]. Computational Intelligence, 2015, 27(1): 123–139.

[152] Liu X D, Li J, Chen H. et al., How to solve the collision in multi-satellite imaging mission planning [J]. Electronic Optics & Control, 2008, 15(10): 10–15. In Chinese.

[153] Ran C X, Xiong G Y, Wang H L. et al., Study of electronic reconnaissance satellites mission scheduling model and algorithm [J]. Communication Countermeasures, 2009, 30(1): 3–8. In Chinese.

[154] Liu B B, Li H, Zhao M. et al., Imaging satellite mission planning based on task compression [J]. Radio Engineering, 2017, 47(11): 77–82. In Chinese.

[155] Jin G, Wu X Y, Gao W B. Ground station resource scheduling optimization model and its heuristic algorithm [J]. Systems Engineering and Electronics, 2004, 26(12): 1839–1841. In Chinese.

[156] Yang P, Yang F, Wu B. et al., Heuristic algorithm and conflict-based backjumping algorithm for satellite TT&C resource scheduling [J]. Journal of Astronautics, 2007, 28(6): 1609–1613. In Chinese.

[157] Tsatsoulis C, Van Dyne M Integrating artificial intelligence techniques to generate ground station schedules [C]. Proceedings of the 2014 IEEE Aerospace Conference, Big Sky, USA: IEEE, 2014: 1–9.

[158] Lemaître M, Verfaillie G, Fargier H, et al. Equitable allocation of earth observing satellites resources [C]. Proceedings of the 5th ONERA-DLR Aerospace Symposium, Toulouse, French: ONERA, 2003.

[159] Du Y H, Wang T, Xin B. et al., A data-driven parallel scheduling approach for multiple agile earth observation satellites [J]. IEEE Transactions on Evolutionary Computation, 2020, 24(4): 679–693.

[160] Zhou J S. Research on task allocation for multiple satellites based on multiple agents [D]. Changsha: National University of Defense Technology, 2009, In Chinese.

[161] Qiu D S, Huang W, Huang X J. et al., Task merging and detecting with hybrid scheduling for electronic reconnaissance satellites [J]. Systems Engineering and Electronics, 2011, 33(9): 2012–2018. In Chinese.

[162] Sun K, Xing L N, Chen Y W. Agile earth observing satellites mission scheduling based on decomposition optimization algorithm [J]. Computer Integrated Manufacturing Systems, 2013, 19(1): 127–136. In Chinese.

[163] Land A H, Doig A G. An automatic method of solving discrete programming problems [J]. Econometrica, 1960, 28(3): 497–520.

[164] Little J D C, Murty K G, Sweeney D W. et al., An algorithm for the traveling salesman problem [J]. Operations Research, 1963, 11(6): 972–989.

[165] Wang P, Tan Y J. Column generation for the earth observation satellites scheduling problem [J]. System Engineering Theory and Practice, 2011, 31(10): 1932–1939. In Chinese.

[166] Ribeiro G M, Constantino M F, Lorena L A N. Strong formulation for the Spot 5 daily photograph scheduling problem [J]. Journal of Combinatorial Optimization, 2009, 20(4): 385–398.

[167] Wang J J. Research on the scheduling of optical earth observation satellites under uncertainties of clouds [D]. Changsha: National University of Defense Technology, 2015, In Chinese.

[168] Jin X S, Li J, Wang J. et al., Research on optimizing scheduling of multi-satellite joint imaging based on stochastic search and relaxation methods [J]. Acta Armamentarii, 2009, 30(1): 49–55. In Chinese.

[169] Bellman R E, Kalaba R E. Dynamic programming and modern control theory [M]. New York, USA: Academic Press, 1965.

[170] Bai B C, He R J, Li J F. et al., Satellite orbit task merging problem and its dynamic programming algorithm [J]. Systems Engineering and Electronics, 2009, 31(7): 1738–1742. In Chinese.

[171] Damiani S, Verfaillie G, Charmeau M C A continuous anytime planning module for an autonomous earth watching satellite [C]. Proceedings of the 15th International Conference on Automated Planning and Scheduling, California, USA: AAAI, 2005: 19–28.

[172] Liu Y, Chen Y W, Tan Y J. The method of mission planning of the ground station of satellite based on dynamic programming [J]. Chinese Space Science and Technology, 2005, 25(1): 44–47. In Chinese.

[173] Qin L, Zhang Q. Distribution strategy of satellite remote sensing data based on dynamic programming [J]. Remote-sensing Information, 2016, 31(5): 30–35. In Chinese.

[174] Peng G S, Dewil R, Verbeeck C. et al., Agile earth observation satellite scheduling: An orienteering problem with time-dependent profits and travel times [J]. Computers & Operations Research, 2019, 111: 84–98.

[175] Holland J H. Adaptation in natural and artificial systems to biology, control, and artificial intelligence [M]. Ann Arbor, USA: University of Michigan Press, 1975.

[176] Zhou Y R, Chen H, Li L M. et al., Immune genetic algorithm for satellite data transmission scheduling [J]. Journal of Chinese Mini-Micro Computer Systems, 2015, 36(12): 2725–2729. In Chinese.

[177] Chen H, Zhou Y, Du C, et al. A satellite cluster data transmission scheduling method based on genetic algorithm with rote learning operator [C]. Proceedings of the Congress on Evolutionary Computation, Vancouver, Canada: IEEE, 2016: 5076–5083.

[178] Li Y F, Wu X Y. Application of genetic algorithm in satellite data transmission scheduling problem [J]. System Engineering Theory and Practice, 2008, 28(1): 124–131. In Chinese.

[179] Han C Q, Liu Y R, Li H. Mission planning for small satellite constellations based on improved genetic algorithm [J]. Chinese Journal of Space Science, 2019, 39(1): 129–134. In Chinese.

[180] Niu X N, Tang H, Wu L X. Satellite scheduling of large areal tasks for rapid response to natural disaster using a multi-objective genetic algorithm [J]. International Journal of Disaster Risk Reduction, 2018, 28: 813–825.

[181] Hosseinabadi S, Ranjbar M, Ramyar S. et al., Scheduling a constellation of agile earth observation satellites with preemption [J]. Journal of Quality Engineering and Production Optimization, 2017, 2(1): 47–64.

[182] Du Y H, Xing L N, Zhang J W. et al., MOEA based memetic algorithm for multi-objective satellite range scheduling problems [J]. Swarm and Evolutionary Computation, 2019, 50: 100576.

[183] Colorni A, Dorigo M, Maniezzo V Distributed optimization by ant colonies [C]. Proceedings of the 1st European Conference on Artificial Life, London England: The MIT Press, 1991: 134–142.

[184] Qiu D S, Guo H E, He C. et al., Intensive task scheduling method for multi-agile imaging satellites [J]. Acta Aeronauticaet AstronauticaSinica, 2013, 34(4): 882–889. In Chinese.

[185] Geng Y Z, Guo Y N, Li C J. et al., Optimal mission planning with task clustering for intensive point targets observation of staring mode agile satellite [J]. Control and decision, 2019, 10.13195/j. kzyjc.2018.0800. In Chinese.

[186] Yan Z Z, Chen Y W, Xing L N. Agile satellite scheduling based on improved ant colony algorithm [J]. System Engineering Theory and Practice, 2014, 34(3): 793–801. In Chinese.

[187] Chen Y N, Xing L N, Chen Y W. Scheduling of agile satellites based on ant colony algorithm [J]. Science Technology and Engineering, 2011, 11(3): 484–489. In Chinese.

[188] Zhu X X, Tan Y J, Deng H Z. et al., The improved ant colony algorithm solving the scheduling problem of imaging satellites. Science Technology and Engineering, 2012, 12(31): 8322–8326. In Chinese.

[189] Gao K B, Wu G H, Zhu J H. Multi-satellite observation scheduling based on a hybrid ant colony optimization [J]. Advanced Materials Research, 2013, 765–767: 532–536.

[190] Wu G H, Ma M H, Zhu J H. et al., Multi-satellite observation integrated scheduling method oriented to emergency tasks and common tasks [J]. Journal of Systems Engineering and Electronics, 2012, 23(5): 723–733.

[191] Xing L N, Chen Y W. Mission planning of satellite ground station system based on the hybrid ant colony optimization [J]. Acta Utomatica Sinica, 2008, 34(4): 414–418. In Chinese.

[192] Yao F, Xing L N. Learnable ant colony optimization algorithm for solving satellite ground station scheduling problems [J]. Systems Engineering and Electronics, 2012, 34(11): 2270–2274. In Chinese.

[193] Huang S L, Ma D Q, Fang D M. et al., Satellite data transmission scheduling based on improved ant colony system [J]. Radio Engineering, 2015, 45(7): 27–30. In Chinese.

[194] Li Z X, Li J, Mu W T Space-ground TT&C resources integrated scheduling based on the hybrid ant colony optimization [C]. Proceedings of the 28th Conference of Spacecraft TT&C Technology, Singapore, Singapore: Springer-Verlag, 2016: 179–196.

[195] Kennedy J, Eberhart R Particle swarm optimization [C]. Proceedings of the IEEE International Conference on Neural Networks, Perth, Australia: IEEE, 1995: 1942–1948.

[196] Kennedy J, Eberhart R C A discrete binary version of the particle swarm algorithm [C]. Proceedings of the IEEE International Conference on Systems, Man, and Cybernetics. Computational Cybernetics and Simulation, Orlando, USA: IEEE, 1997: 4104–4108.

[197] Tang S X, Yi X Q, Luo X S. An improved particle swarm optimization algorithm for early warning satellites scheduling problems [J]. Systems Engineering, 2012, 30(1): 116–121. In Chinese.

[198] Chang F, Wu X Y. Satellite data transmission task scheduling based on advanced particle swarm optimization [J]. Systems Engineering and Electronics, 2009, 31(10): 2404–2408. In Chinese.

[199] Chen H, Li L M, Zhong Z N. et al., Approach for earth observation satellite real-time and playback data transmission scheduling [J]. Journal of Systems Engineering and Electronics, 2015, 26(5): 982–992.

[200] Chen Y, Zhang D Y, Zhou M Q, et al. Multi-satellite observation scheduling algorithm based on hybrid genetic particle swarm optimization [C]. Proceedings of Advances in Information Technology and Industry Applications, Berlin, Germany: Springer-Verlag, 2012: 441–448.

[201] Guo X B, Liu J C, Zhou H B. Research on transmission task scheduling for distributed satellite systems [J]. Radio Communications Technology, 2016, 42(4): 29–32. In Chinese.

[202] Liu J Y, Wang Z W. Research on the tasks scheduling algorithm for imaging satellite observing forest area [J]. Journal of Central South University of Forestry & Technology, 2018, 38(10): 41–46. In Chinese.

[203] Glover F. Tabu search – part I [J]. ORSA Journal on Computing, 1989, 1(3): 190–205.

[204] Glover F. Tabu search – part II [J]. ORSA Journal on Computing, 1990, 2(1): 4–32.

[205] He R J, Tan Y J. On parallel machine scheduling problem with time windows [J]. System Engineering, 2004, 22(5): 18–22. In Chinese.

[206] Zuo C R, Wang H Y. Research on scheduling of earth observing satellites based on taboo search algorithm [J]. Computer Engineering and Applications, 2010, 46(1): 215–217. In Chinese.

[207] Chen Y W, Fang Y S, Li J F. et al., Constraint programming model of satellite mission scheduling [J]. Journal of National University of Defense Technology, 2006, 28(5): 126–132. In Chinese.

[208] Li J F, He R J, Yao F. et al., Variable neighborhood tabu search algorithm for integrated imaging satellites scheduling problem [J]. System Engineering Theory and Practice, 2013, 33(12): 3040–3044. In Chinese.

[209] Habet D, Vasquez M, Vimont Y. Bounding the optimum for the problem of scheduling the photographs of an agile earth observing satellite [J]. Computational Optimization and Applications, 2010, 47(2): 307–333.

[210] Metropolis N, Rosenbluth A W, Rosenbluth M N. et al., Equation of state calculations by fast computing machines [J]. The Journal of Chemical Physics, 1953, 21: 1087–1091.

[211] Kirkpatrick S, Gelatt C D, Vecchi M P. Optimization by simulated annealing [J]. Science, 1983, 220: 611–680.

[212] Huang H, Zhang X Q. Study on imaging satellite task planning method based on graph theory model [J. Journal of Guilin University of Aerospace Technology, 2016, 21(2): 155–158. In Chinese.

[213] He R J, Gao P, Bai B C. et al., Models, algorithms and applications to the mission planning system of imaging satellites [J]. Systems Engineering Theory and Practice, 2011, 31(3): 411–422. In Chinese.

[214] Gao P, Li W, Yao F, et al. Simulated annealing algorithm for EOS scheduling problem with task merging [C]. Proceedings of the International Conference on Modelling, Identification and Control, Shanghai: IEEE, 2011: 547–522.

[215] Xu H, Zhu J H, Wang H L. Research on mission planning for electronic reconnaissance satellites based on simulated annealing [J]. Journal of The Academy of Equipment Command & Technology, 2010, 21(3): 62–66. In Chinese.

[216] Du Y H, Xing L N, Chen Y G. Integrated agile observation satellite scheduling problem considering different memory environments: A case study [J]. Journal of the Brazilian Society of Mechanical Sciences and Engineering, 2020, 42: 76.

[217] Huang S J, Xing L N, Guo B. Multi-satellites mission scheduling technique based on improved simulated annealing [J]. Science Technology and Engineering, 2012, 12(31): 8293–8298. In Chinese.

[218] Lin Z H Mission planning for electromagnetic environment monitors satellite based on simulated annealing algorithm [C]. Proceedings of the 28th Canadian Conference on Electrical and Computer Engineering, Halifax, Canada: IEEE, 2015: 530–535.

[219] Yao F, Li J F, Li W. et al., Study on dynamic capability assessment system of earth observation satellites [J]. Fire Control & Command Control, 2010, 35(12): 18–21. In Chinese.

[220] Moscato P. On evolution, search, optimization, genetic algorithms and martial arts: Towards memetic algorithms [R]. Caltech Con-Current Computation Program 158–79. Pasadena, USA: California Institute of Technology, 1989.

[221] Dawkins R. The Selfish Gene [M]. New York, USA: Oxford University Press, 1976.

[222] Li J, Li J, Jing N. et al., A satellite schedulability prediction algorithm for EO SPS [J]. Chinese Journal of Aeronautics, 2013, 26(3): 705–716.

[223] Liu S, Bai G Q, Chen Y W. Prediction method for imaging task schedulability of earth observation network [J. Journal of Astronautics, 2015, 36(5): 583–588. In Chinese.

[224] CPLEX Optimization Studio. CPLEX user's manual [M]. New York, USA: IBM Corporation, 2015.

[225] Xu Z L, Tan Y J. Area Target-oriented Mission Scheduling of Mapping Satellites [J]. Science Technology and Engineering, 2012, 12(28): 7303–7308. In Chinese.

[226] Wang P, Li J F, Tan Y J. Comparison of earth observation scheduling model for satellite formation [J]. Systems Engineering and Electronics, 2010, 32(8): 1689–1694. In Chinese.

[227] Orbit Logic. STK/Scheduler tutorial [M]. Maryland, USA: Orbit Logic, 2006.

[228] Orbit Logis. STK/Scheduler [EB/OL]. 2019 [2024-9-30]. http://www.orbitlogic.com/stk-scheduler.html.

[229] Li Y X, Liu Y, Fang Q. Realization of TDRSS mission scheduling based on STK/Schedule [J]. Modern Electronic Technology, 2012, 35(10): 122–125. In Chinese.

[230] Li Y F, Wu X Y. Application of STK/Scheduler in satellite data transmission scheduling [J]. Computer Simulation, 2008, 25(3): 70–74. In Chinese.

[231] Bai J P, Yan H, Gao Y M. et al., Application of space mission scheduling based on STK/Scheduler [J]. Journal of The Academy of Equipment Command & Technology, 2010, 21(3): 71–75. In Chinese.

[232] Li Y X, Qing F, Tan J B Application of relay satellite scheduling based on STK/X [C]. Proceedings of the 2011 IEEE CIE International Conference on Radar, Chengdu, China: IEEE, 2011: 288–291.

[233] Fisher W A, Herz E A flexible architecture for creating scheduling algorithms as used in STK Scheduler [C]. Proceedings of the 8th International Workshop on Planning and Scheduling for Space, California, USA: AAAI, 2013.

[234] Herz A F, Stoner F, Hall R, et al. SSA sensor tasking approach for improved orbit determination accuracies and more efficient use of ground assets [C]. Proceedings of the Advanced Maui Optical and Space Surveillance Technologies Conference, Hawaii, USA: The Maui Economic Development Board, 2013.

[235] NASA. EUROPA-2.6 [DB/OL]. 2011 [2024-9-30]. https://github.com/nasa/europa/tree/Releases/EUROPA-2.6.

[236] Muscettola N, Nayak P P, Pell B. et al., Remote agent: To boldly go where no AI system has gone before [J]. Artificial intelligence, 1998, 103(1–2): 5–47.

[237] Tran D, Chien S, Sherwood R, et al. The autonomous sciencecraft experiment onboard the EO-1 spacecraft [C]. Proceedings of the 4th International Joint Conference on Autonomous Agents and Multiagent Systems, Utrecht, Netherlands: ACM, 2004: 164–165.

[238] Chien S, Sherwood R, Rabideau G, et al. The Techsat-21 autonomous space science agent [C]. Proceedings of the 1st International Joint Conference on Autonomous Agents and Multiagent Systems, Bologna, Italy: ACM, 2002: 570–577.

[239] Frank J, Jonsson A, Morris R, et al. Planning and scheduling for fleets of earth observing satellites [C]. Proceedings of the 6th International Symposium on Artificial Intelligence, Robotics and Automation for Space, Montreal, Canada: Canadian Space Agency, 2001.

[240] Bedrax-Weiss T, Frank J, Jonsson A, et al. Europa 2: Plan database services for planning and scheduling applications [C]. Proceedings of the 14th International Conference on Automated Planning and Scheduling, California, USA: AAAI, 2004.

[241] Liu Y C, Zhong X Y, Fang Y S. et al., Dynamic simulation and modeling for AI planning based on europa [J]. Computer Engineering and Applications, 2012, 48(17): 211–214. In Chinese.

[242] NASA. Scheduling and planning interface for exploration [DB/OL]. 2015 [2024-9-30]. https://github.com/nasa/OpenSPIFe.

[243] Baike B. Engine [EB/OL] 2019 [2024-9-30]. https://baike.baidu.com/item/%E5%BC%95%E6%93%8E/2874935#viewPageContent.

[244] Burke E K, Bykov Y. The late acceptance hill-climbing heuristic [R]. Scotland, UK: University of Stirling, 2012.

[245] Burke E K, Bykov Y. The late acceptance hill-climbing heuristic [J]. European Journal of Operational Research, 2017, 258(1): 70–78.

[246] Golden B L, Levy L, Vohra R. The orienteering problem [J]. Naval Research Logistics, 1987, 34(3): 307–318.

[247] The orienteering problem: Test instances [DB/OL]. 2020 [2024-9-30]. https://www.mech.kuleuven.be/en/cib/op, 2020-07-24

[248] Chao I M, Golden B L, Wasil E A. A fast and effective heuristic for the orienteering problem [J]. European Journal of Operational Research, 1996, 88(3): 475–489.

[249] Kantor M G, Rosenwein M B. The orienteering problem with time windows [J]. Journal of the Operational Research Society, 1992, 43(6): 629–635.

[250] Verbeeck C, Vansteenwegen P, Aghezzaf E H. The time-dependent orienteering problem with time windows: a fast ant Colony System [J]. Annals of Operations Research, 2017, 254(1–2): 481–505.

[251] Fomin F V, Lingas A. Approximation algorithms for time-dependent orienteering [J]. Information Processing Letters, 2002, 83(2): 57–62.

[252] Yang P, Liu Y, Pei Y. Agent dynamic task allocation based on improved contract net protocol [J]. Fire Control & Command Control, 2011, 36(10): 77-80. (In Chinese).

[253] Li X L, Zhai J T, Dai Y W. a task allocation algorithm base on improved contract net protocol under the dynamic environment [J]. Science Technology and Engineering, 2013, 13(27): 8014–8019. In Chinese.

[254] Stanley K O, Miikkulainen R. Evolving neural networks through augmenting topologies [J]. Evolutionary Computation, 2002, 10(2): 99–127.

Appendix A: Explanation of mathematical symbols of general-purpose model for satellite task scheduling

Table A.1: Explanation of mathematical symbols of general-purpose model for satellite task scheduling.

Symbol	Description	Symbol	Description
A	The Euler angle of satellite attitude or station antenna	f_i^H	The ith constraint
$a(x_{ij})$	Whether satellite s_j in the ith timeslot is an anchor satellite	f_i^S	The ith soft constraint
$b(e_{ij})$	The beginning time of event e_{ij}	$f_i^H(X)$	The value of the scheme that violates constraint f_i^H (≤ 0)
$b(eo_{ij}^k)$	The beginning time of EO eo_{ij}^k	$f_i^S(X)$	The value of the scheme that violates soft constraint f_i^S (≤ 0)
$b(o_i)$	The beginning time of orbit o_i	G	The station set
$b(tw_{ij}^k)$	The beginning time of time window tw_{ij}^k	g_i	The ith station, while i is the station number
c_i^H	The constrained object of constraint f_i^H	$g(e_{ij})$	The station that executes event e_{ij}
D	The date set	$g(tw_{ij}^k)$	The station to which the time window tw_{ij}^k corresponds
d_i	The ith day of satellite task scheduling cycle	$l(e_{ij})$	The duration of event e_{ij}
$d(e_{ij})$	The day on which event e_{ij} is executed	$l_{Emax}(d_k, q_m)$	The maximum working time of payload q_m on day d_k
$d(o_i)$	The day to which orbit o_i corresponds	$l_{Emax}(o_k, q_m)$	The maximum working time of payload q_m in track o_k
E_i	The event set of task t_i	$l_{\Delta max}(d_k, s_m)$	The maximum maneuvering time of satellite s_m on day d_k
$E(o(e_{ij}))$	The electricity threshold of orbit $o(e_{ij})$ that executes event e_{ij}	$l_{\Delta max}(o_k)$	The maximum maneuvering time of the satellite in orbit o_k
EO	Event EO set	$l_{\Delta Cmax}(o_k)$	The maximum continuous maneuvering time of the satellite in orbit o_k
EO_{ij}	The EO set of event e_{ij}	$M(s(e_{ij}))$	The memory threshold of satellite $s(e_{ij})$ that executes event e_{ij}

https://doi.org/10.1515/9783111537191-009

Table A.1 (continued)

Symbol	Description	Symbol	Description
e_{ij}	The jth event of task t_i	$m(e_{ij})$	The amount of data generated by event e_{ij}
$e(e_{ij})$	The end time of event e_{ij}	$n^{\Delta}(e_{ij})$	The number of maneuvers required to execute event e_{ij}
$e(o_i)$	The end time of orbit o_i	$n_{Emax}(d_k, q_m)$	The maximum working times of payload q_m on day d_k
$e(tw_{ij}^k)$	The end time of time window tw_{ij}^k	$n_{Emax}(o_k, q_m)$	The maximum working times of payload q_m in orbit o_k
eo_{ij}^k	The kth EO of event e_{ij}	$n_{\Delta max}(d_k, s_m)$	The maximum number of maneuvers of satellite s_m on day d_k
F	Score set	$n_{\Delta max}(o_k)$	The maximum number of maneuvers of the satellite in orbit o_k
F^H	Constraint set	$n_{\Delta Cmax}(o_k)$	The maximum number of continuous maneuvers of the satellites in orbit o_k
$F^H(X)$	The total value of the scheme that violates the constraints (≤ 0)	$next(e_{ij})$	The next event that will be executed by the satellite executing event e_{ij}
F^S	Soft constraint set	O	The satellite orbit set
$F^S(X)$	The total value of the scheme that violates the soft constraints (≤ 0)	o_i	The ith satellite orbit, while i is the orbit number
f	Objective function	$o(e_{ij})$	The satellite orbit for executing event e_{ij}
f_i	The objective value of task t_i	$o(tw_{ij}^k)$	The orbit of the satellite to which time window tw_{ij}^k corresponds
P	Platform set	TW	The time window set
P_{Ci}	The property set (constant) of task t_i	TW_{ij}	The time window set of event e_{ij}
P_{Vi}	The variable set of task t_i	t_i	The ith task, while i is the task number
Q	Payload/equipment set	tw_{ij}^k	The kth time window of event e_{ij}
$Q(g_i)$	The payload/equipment set equipped by station g_i	$tw(eo_{ij}^k)$	The time window to which EO eo_{ij}^k corresponds
$Q(s_i)$	The payload/equipment set carried by satellite s_i	W	Window set
$q(e_{ij})$	The payload/equipment required to execute event e_{ij}	X	Decision matrix

Table A.1 (continued)

Symbol	Description	Symbol	Description
$q_j(g_i)$	The jth equipment equipped by station g_i	T	Task set
$q_j(s_i)$	The jth payload carried by satellite s_i	ω_i^H	The constraint relationship of constraint f_{iH}
R	Resource set	x_{ij}	The decision variable for the jth event of the ith task
S	The satellite set	y_i^H	The constraint threshold of constraint f_i^H
X_{Zij}	The jth subset of decision matrix X that the ith constraint may involve	$Z_i(X)$	The set of subsets of decision matrix X that the ith constraint may involve
STSP	Satellite task scheduling	$\theta(x_{i1})$	The satellite pitch angle when imaging event e_{i1} is executed
s_i	The ith satellite, while i is the satellite number	$\varepsilon(e_{ij})$	Power consumption of event e_{ij}
$s(e_{ij})$	The satellite that executes event e_{ij}	$\Delta(e_{ij}, e_{i'j'})$	The transition time allowed between events e_{ij} and $e_{i'j'}$
$s(o_i)$	The satellite in orbit o_i	$\delta(e_{ij}, e_{i'j'})$	The actual interval time between events e_{ij} and $e_{i'j'}$

Appendix B: Java program architecture of satellite task scheduling engine and the description

To facilitate scholars studying related fields to implement, improve, and further expand the satellite task scheduling engine in this book by programming means, the program's architecture of the Java-based satellite task scheduling engine is given below. This engine program mainly contains three modules such as .model, .normal-algorithm, and .urgentalgorithm, which are consistent with the design of the satellite task scheduling engine framework in this book decoupling "model-conventional algorithm-emergency algorithm." The descriptions of the main Java packages and classes in each module are shown in Tables B.1–B.3. The design concept of the program is consistent with the relevant contents in this book.

Table B.1: Description of the main Java packages and classes of .model module in the satellite task scheduling engine.

Package name	Subpackage name	Class	Description
.task(task package)	–	Task	Task class
		ImageEvent	Imaging event class
		DownlinkEvent	Downlink event class
		Property	Property class
		Variable	Variable class
.score (score package)	–	Score	Score class
		Constraint	Constraint class
		SoftConstraint	Soft constraint class
		Objective	Objective function class
.decision (decision package)	–	DecisionMatrix	Decision matrix class
		Scenario	Scenario class

https://doi.org/10.1515/9783111537191-010

Table B.1 (continued)

Package name	Subpackage name	Class	Description
.resource (resource package)	–	Resource	Resource class (parent class)
	.platform (platform package)	Platform	Platform class (parent class)
		Satellite	Satellite class
		Station	Ground station
	.payload (payload package)	Payload	Payload class (parent class)
		Antenna	Antenna class
		Battery	Battery class
		Camera	Camera class
		Memory	Memory class
	.window (window package)	Window	Window class (parent class)
		ImageWindow	Imaging window class
		DownlinkWindow	Downlink window class
		Orbit	Orbit class
		Day	Day class
	.opportunity (opportunity package)	ImageOpportunity	Imaging event EO class
		DownlinkOpportunity	Downlink event EO class

Table B.2: Description of the main Java packages and classes of .normalalgorithm module in the satellite task scheduling engine.

Package name	Subpackage name	Class	Description
.algorithm (conventional algorithm package)	–	Algorithm	Conventional algorithm class (parent class)
	.heuristic (heuristic package)	FirstInFirstService	First-in-first-service algorithm class
		Random	Random algorithm class
	.localsearch (local search package)	HillClimbing	Hill climbing algorithm class
		TabuSearch	Tabu search algorithm class
		SimulatedAnnealing	Simulated annealing algorithm class
		LateAcceptance	Late acceptance hill climbing algorithm class
		IteratedLocalSearch	Iterated local search algorithm class
	.evolution (EA package)	GeneticAlgorithm	Genetic algorithm class
		DifferentialEvolution	Differential evolution algorithm class
	.hybrid (hybrid algorithm package)	MemeticAlgorithm	Memetic algorithm class
		ParallelMA	Parallel memetic algorithm class
		AdaptiveParallelMA	Adaptive parallel memetic algorithm class
.operator (operator package)	–	Move	Move operator class (parent class)
		SwapMove	Swap operator class
		ReplaceMove	Replace operator class
		Selection	Selection operator class
		Crossover	Crossover operator class
		Mutation	Mutation operator class
		Repair	Repair operator class

Table B.3: Description of the main Java packages and classes of .urgentalgorithm module in the satellite task scheduling engine.

Package name	Subpackage name	Class	Statement
.assignment (assignment algorithm class)	–	TaskAssignment	Task assignment algorithm class
	.agent (agent package)	Agent	Agent class (parent class)
		CallerAgent	Caller agent class
		CompeteAgent	Compete agent class
	.message (message package)	Message	Message class (parent class)
		Book	Book class
.rescheduling (rescheduling algorithm class)	–	ReScheduling	Rescheduling algorithm
	.window (window package)	Window	Window class (parent class)
		LockedWindow	Locked window class
		RealTime-Window	Real-time window class
		FutureWindow	Future window class
.insert (insertion algorithm class)	–	Insert	Insertion algorithm class
		Prediction	Schedulability prediction model class
	.network (network package)	Network	Neural network class
		Node	Node class
		Connection	Connection class
	.operator (operator package)	Crossover	Crossover operator class
		Mutation	Mutation operator class
.deconflict (conflict-avoidance algorithm class)	–	Deconflict	Conflict-avoidance algorithm class
		ConflictDegree	Conflict degree class

Subject index

https://doi.org/10.1515/9783111537191-011

www.ingramcontent.com/pod-product-compliance
Lightning Source LLC
Chambersburg PA
CBHW061349210326
41598CB00035B/5924